Rachel Curiel

Fall 52

Chem 108

Basic Organic Chemistry

Basic Organic Chemistry

J. Rae Schwenck, Ch.E.

AND

Raymond M. Martin, M.S.

Sacramento Junior College
Sacramento, California

New York   THE BLAKISTON COMPANY   Toronto

Philadelphia

1951

PRINTED IN THE UNITED STATES OF AMERICA
BY THE MAPLE PRESS COMPANY, YORK, PA.

To Our Wives

We Affectionately Dedicate
This Book

"Basic Organic Chemistry" has been written to provide teachers and students with a textbook whose subject matter can be covered in one semester. Students will find that this textbook is not intended to serve as a reference work on Organic Chemistry and teachers will recognize that it is not a "boiled-down" edition of a text prepared earlier for chemistry majors. This book has been written by teachers of Organic Chemistry who have seriously studied the difficulties of students seeking a knowledge of this field as an adjunct and aid in the better understanding of their own major subjects in medicine, engineering, agriculture, home economics, and similar subjects.

Among the major difficulties experienced by students of Organic Chemistry, the following are notable: (a) A feeling of initial frustration arising from the overwhelming size of textbooks that cannot possibly be covered in one semester; (b) the bewildering newness of the field springing from a failure to correlate the principles of General Chemistry with those of Organic Chemistry; (c) a persistent perplexity due to the scrambled development of subject matter within the text, and (d) the Babel-like confusion of language arising from the indiscriminate use of different systems of nomenclature.

To minimize these major difficulties of the one-semester student, the size of this text has been limited to twenty short chapters, thereby allowing an average of about two lecture periods for each chapter. To correlate Organic Chemistry with the principles of General Chemistry, the electronic structure approach is used throughout the text. The subject matter is introduced by a study of the element carbon. To avoid any semblance of a scrambled development, each functional group is introduced and explained as far as possible in accord with a single master plan: naming, structure, preparation, reactions, and special compounds. To avoid the confusion of synonymous names for individual compounds, the I.U.C. system of nomenclature is used throughout the text. Only in those cases where students are familiar with common names (I.U.C. acceptable) have acetylene, acetone, formaldehyde, acetic acid, and a half-dozen others been used.

This text also incorporates such new features as (a) study questions arranged in the sequence of development of subject matter in the chapter, (b) the introduction of a separate chapter on systems of nomenclature other than that used in the text, (c) a unique, new presentation of subject matter on heterocyclics, (d) the use of classroom-proved charts as a summary of preparations and reactions at the end of many chapters, and (e) the chapter prologue. This latter innovation in nonpedagogic language is intended as a scientific *hors d'oeuvre* to add flavor to the meat of each chapter and to bring to receptive minds the high inspiration of an inspiring science.

The last three chapters have been designed to bring the text to a conclusion, not merely to an end. It is our hope that they who use this text will be enriched by a fuller understanding of the interrelationship of science with their own vocational objective, that they will recognize the ever-increasing importance of Organic Chemistry not merely in one particular field of industry or of biological advancement, but in their own personal living—that they may learn better the art of living well from the science that has added so much to worthwhile living.

The authors are especially grateful to Dr. Hal D. Draper, scholar, gentleman, and friend, who has critically evaluated the manuscript.

The authors also appreciate the courtesy of those industries and persons who have provided illustrations, tables, and similar materials. Specific acknowledgment for this assistance is given on the appropriate pages.

J. Rae Schwenck
Raymond M. Martin

Sacramento, California
Autumn, 1950

CONTENTS

CHAPTER PROLOGUE

Sometime in the eons preceding recorded history one of our ancestors emerged from his smoke-filled cave and gazed upon the charred sentinels of the forest that had sheltered wild game near his crude home. Awesome tongues of fire had driven him fearfully to the refuge of his cave and in their wake had left only the smoldering black image of the things he had loved.

In that moment man became acquainted with carbon in one of its forms. Later in the centuries man learned more about the different contrary and useful ways in which carbon exists. Only recently, as the ages go, has man become aware of the versatility of this remarkable substance. Less than two centuries have passed since man learned how to convert its latent energy into power and motion. Now many of the innermost secrets of its nature are known and that knowledge is being applied daily in wondrous new ways for the betterment of daily living.

This chapter will serve as an introduction to the field of Organic Chemistry, the Chemistry of Carbon. In the ever-brightening light of scientific progress we will look at the carbon atom and try to understand why carbon is so versatile and so useful.

Carbon is the most versatile chemical element known to man. As the free element it exists in more than one form. With most of the common elements carbon forms binary compounds. Many of these compounds are found abundantly in nature. The most unique property of carbon, however, is the ability of one carbon atom to unite directly with other carbon atoms in the formation of compounds. Because of this latter characteristic the number of chemical compounds containing carbon far exceeds the combined number of chemical compounds containing no carbon.

Carbon Chemistry. More than 400,000 compounds containing carbon are known. Because of this large number of compounds, organic chemistry has become a separate field of study. Early chemists thought that compounds containing carbon could be produced only by living organisms. The word "organic" is derived from this concept, called the Vital Force Theory. Woehler in 1828 disproved this theory by synthesizing urea from inorganic materials. But the name *Organic Chemistry* has persisted. The field of Organic Chemistry embraces the study of the compounds of carbon but the name *Carbon Chemistry* describes the field more adequately.

Kinds of Carbon. The element carbon is found in nature in at least three different allotropic forms. Two of these, diamond and graphite, have long been recognized as crystalline in structure. The remaining form, called *amorphous,* was for many years considered as having no definite crystalline form. Now, however, even amorphous carbon is known to have a crystalline structure different from both diamond and graphite.

As the uncombined element, carbon is a substance of strange contrasts. Diamond is the hardest substance known to man; graphite is one

1

of the softest solid materials. In the elemental form carbon resists chemical combination with most elements. It remains in the solid phase under conditions of high temperature. Most elements are capable of existing at fixed temperatures and pressures in three separate phases, as crystalline solids, as liquids, and as gases. Carbon has been momentarily melted at temperatures exceeding 3500°C but the exact temperature at which its vapor pressure becomes 760 mm. is not known.[1] Whether in the form of diamond or of graphite carbon apparently passes directly from the solid phase into the vapor phase.

Utility of Carbon. As an element, carbon in all its forms is extremely useful to man. As diamond it is one of man's best cutting tools; as graphite it is a lubricant of unusual qualities; in impure form as coke, charcoal, boneblack, and lampblack it serves mankind in dozens of different ways.

Since organic chemistry deals with the compounds of carbon, their properties and synthesis, our chief interest lies in the study of those compounds rather than of the element itself. Nevertheless our understanding of the compounds can best be developed by a consideration of carbon's position in the periodic table and of its atomic structure.

Carbon in the Periodic Table. The student will recall from his general chemistry that carbon is one of the simplest elements. When we compare its position in the periodic table with that of other simple atoms, some facts about the nature of atomic structure will serve as a bridge between general and organic chemistry.

Ten chemical elements are included in the first two groups of the periodic table. Hydrogen and helium are the only two members of the first group. The second group is made up of lithium, beryllium, boron, carbon, nitrogen, oxygen, fluorine, and neon. The first group differs from the second not only in nuclear content of protons but especially by having only a single shell (K-shell) of planetary electrons. Atoms of the second group have two shells of electrons, the K- and L-shells. Figure 1 shows the conventional representation of the first 10 elements of the periodic table as two dimensional drawings. The mass or weight of each atom is controlled chiefly by the number of protons and neutrons in each nucleus. Except in the case of hydrogen, the innermost shell (K-shell) always contains only two electrons. The L-shell may have from one to eight electrons. When an outermost shell of planetary electrons is completely filled, the atom is an inert gas. An atom of this kind is said to possess the "rare gas configuration."

Atomic inertness is thus based upon the complete filling of the outermost shell with electrons. Conversely, activity of an atom is due to an unfilled outer shell of electrons. The degree of an atom's activity is due to its proximity to the nearest rare gas configuration. Thus, lithium and

---

[1] The boiling point of a substance is the temperature at which the vapor pressure of the substance is equal to 760 mm. of mercury (standard pressure at sea level).

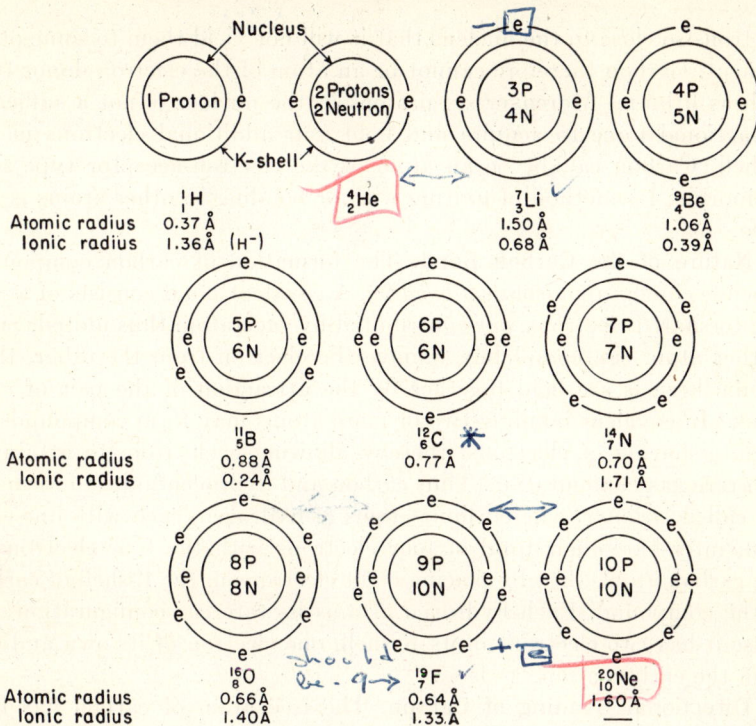

Fig. 1. The ten simplest atoms.[2]

fluorine, the most active elements in group two, are both close to the configurations of helium and neon respectively. When lithium loses its outermost or valence electron it becomes an ion. Only the kernel[3] of lithium atom remains. When fluorine accepts or gains an electron, it becomes an ion. Both lithium and fluorine ions are respectively like helium and neon except for nuclear content. Thus some atoms give electrons and some atoms accept electrons in attaining the rare gas configuration. They are ionic in their bonding with other atoms in the formation of most compounds.

Structure of the Carbon Atom. Carbon is unique among its neighbors of the periodic table because it lies exactly between helium and neon. It is equally far from two rare gas configurations. Carbon has six protons in its nucleus and six planetary electrons, two in the K-shell and four in the L-shell. The nuclear attraction of its six protons draws its four outer

[2] The following conventional method is used to designate the atomic weight and atomic number of a particular isotope of an element:

Atomic weight

Symbol, thus:— $^{12}_{6}C$

Atomic number

[3] The kernel of an atom is made up of all electrons, protons, and neutrons inside the unfilled outer shell of electrons.

electrons so close to the nucleus that it will not yield them to some other element. Carbon therefore cannot be an atom of the electron-donor type such as lithium. Moreover six protons in the nucleus is not a sufficient attractional force to acquire and hold four additional electrons in the L-shell. Carbon cannot be an atom of the electron-acceptor type such as fluorine. Its method of uniting with or bonding to other atoms is not ionic.

Nature of the Carbon Bond. The formation of carbon compounds depends chiefly upon *covalent bonding*. A covalent bond consists of a pair of intershared electrons, *one* contributed by each atom thus united. Since neither atom can completely remove the electron from the other, their atomic kernels are held together by the attraction of the pair of electrons.[4] In covalent bonding two or more atoms may form compounds by an intersharing of electrons, thereby allowing each atom to attain its own rare gas configuration. Thus carbon and hydrogen unite to form the gas called *methane*, wherein four atoms of hydrogen, each with one electron, intershare their total of four electrons with the four electrons of one carbon atom, thereby placing eight electrons in the L-shell of carbon in the compound. Each hydrogen attains its rare gas configuration[5] because it has two electrons in its K-shell, one electron of its own and one from the carbon atom.

Directional Bonding of Carbon. The free atom of carbon has four electrons in the L-shell. These electrons are mutually repellent, causing their equal distribution at a nuclear angle of 109°28'[6] from each other. In the formation of compounds each electron has a binding or attracting force for the electron to be intershared. When carbon unites with a univalent atom or radical, each bond retains the direction of the carbon electrons thereby giving rise to the common representation of carbon as a tetrahedron, as shown in Figure 3. Since an intershared pair of electrons constitutes a chemical bond, carbon in practically all cases has a valence of four.

Isotopes of Carbon. Carbon has an atomic weight of 12.01 and its atomic number is 6. The nucleus of the carbon atom has six protons and usually six neutrons. When the number of neutrons in the nucleus is greater or less than six, isotopes are formed having an atomic weight of higher or lower value than 12. Isotope 12 is most abundant in nature but

---

[4] The student may wonder why two electrons form a stable configuration when their like charges should repel each other. This phenomenon is due to the spin direction of the electrons. The spinning of electrons in opposite directions creates a magnetic field capable of holding the electrons in an equilibrium position, magnetic attraction balancing electrostatic repulsion.

[5] It will be recalled that helium, the simplest rare gas, requires only two electrons in the K-shell. Hydrogen attains this structure by intersharing one electron.

[6] In combination with other elements this angle may be altered somewhat but it remains close to 109°28'.

Radical = A group of atoms remaining unchanged during a series of reactions & which may be regarded as playing the part of a single atom.

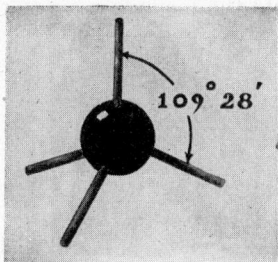

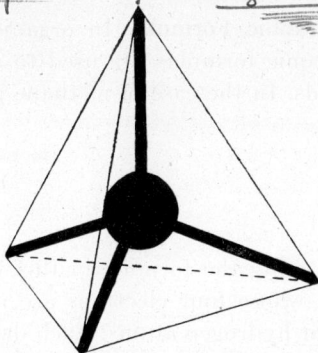

Fig. 2   Fig. 3

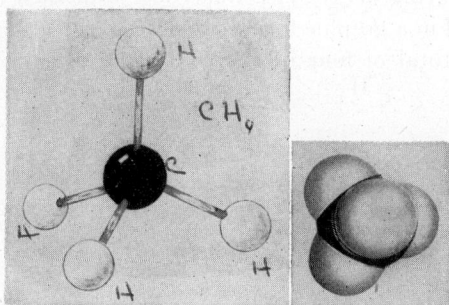

Fig. 4

Fig. 2. The carbon atom showing bond angle.
Fig. 3. Tetrahedral structure of carbon.
Fig. 4. Models of methane. (Left) Showing bonds. (Right)[7] Showing steric proportion.

isotope 13 has been isolated. Isotopes 11 and 14 have been prepared with the aid of the cyclotron.

Electronic Formula. Using the conventional system of representation wherein each dot indicates an electron and each symbol signifies the kernel of the atom, the electronic formula of methane gas is:

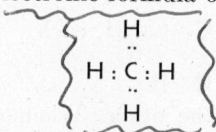

Such a representation is an abbreviated planar illustration of what actually occurs in three dimensions. Figure 4 is a photograph of models of methane. The photograph on the right shows relative sizes, respective locations, and space-filling properties of carbon and hydrogen atoms as they make up the molecule. This model is approximately 200,000,000 times larger than the actual molecule.

[7] This photograph was prepared from the Fisher-Hirschfelder-Taylor atom models (patented) and with other photographs of this type is reproduced by courtesy of the Fisher Scientific Company, Pittsburgh, Pa.

Graphic Formula. In organic chemistry graphic representations of electronic formulas are used to simplify the expression of various compounds. In the case of methane gas we write the formula

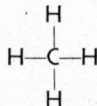

$$
\begin{array}{c}
\text{H} \\
| \\
\text{H}-\text{C}-\text{H} \\
| \\
\text{H}
\end{array}
$$

By writing the formula in this way we represent carbon as the central atom whose four electrons are mutually intershared with the electrons of four hydrogen atoms. Each dash or line between carbon and hydrogen indicates a single covalent bond between the respective atoms and represents the intersharing of two electrons, one from each atom. When two dashes are placed in a graphic formula between two atoms, a double bond is shown and a total of four intershared electrons is represented. Thus

ethene gas
$$
\begin{array}{cc}
\text{H} & \text{H} \\
\diagdown & \diagup \\
\text{C} & = & \text{C} \\
\diagup & \diagdown \\
\text{H} & \text{H}
\end{array}
$$
, is really a simplified writing of the electronic

formula
$$
\begin{array}{cc}
\text{H} & \text{H} \\
\ddot{\text{C}}::\ddot{\text{C}} \\
\text{H} & \text{H}
\end{array}
$$.

The use of graphic formulas with dashes in the place of dots also serves visually as an aid in understanding what occurs when two atoms are held together by a single bond, or by a double bond or by a triple bond. When two atoms are held together by a single bond there is a relative freedom of rotation of the carbon atoms of the molecule about the single bond. The single bond appears as an axis upon which the atom revolves. Thus when two carbon atoms are united by a single bond in the

formation of ethane gas,
$$
\begin{array}{cc}
\text{H} & \text{H} \\
| & | \\
\text{H}-\text{C}-\text{C}-\text{H} \\
| & | \\
\text{H} & \text{H}
\end{array}
$$, each $-\text{CH}_3$ group has a freedom

of rotation with respect to the other. When double or triple bonding

occurs as in the case of ethene
$$
\begin{array}{cc}
\text{H} & \text{H} \\
\diagdown & \diagup \\
\text{C} & = & \text{C} \\
\diagup & \diagdown \\
\text{H} & \text{H}
\end{array}
$$
or ethyne $\text{H}-\text{C}\equiv\text{C}-\text{H}$, the

position of each carbon atom with respect to the other becomes more rigid, further rotation of the carbon atom within the molecule becomes impossible, and each hydrogen atom acquires a fixed spacial relationship to all other hydrogen atoms within the molecule.

Graphic formulas are extremely useful because they show not only the kind of atoms within a molecule, but also the arrangement of the atoms and the manner of their bonding. Many compounds differ from other compounds having the same number and kind of atoms, only by the location of the atoms within the molecule. A molecule of ethanol (ethyl alcohol) has the same number and kinds of atoms as methoxymethane (dimethyl ether). Both have the same molecular formula, C_2H_6O. In ethanol the two carbon atoms are joined together directly by a single bond and the oxygen atom lies between one carbon atom and a hydrogen atom. In methoxymethane the oxygen lies between the two carbon atoms. The graphic formulas shown below illustrate in two dimensions the relative position of the atoms within the molecule.

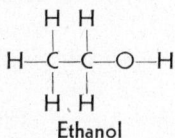

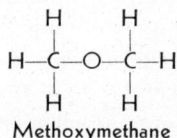

Ethanol   Methoxymethane

Skeletal Formula. Very often it is advantageous for simplicity and for speed of writing reactions not to write in the letter H for the hydrogen atoms *attached directly to carbon atoms.* In using such formulas it must be understood that hydrogen atoms occupy all spaces on the carbon atoms not otherwise designated. Examples of this skeletal formula in the case of ethanol and methoxymethane are   —C—C—O—H   (ethanol) and —C—O—C— (methoxymethane). It is important to note that the letter H is attached to the O in ethanol. By rule, only the H directly attached to the carbon atom may be omitted.

Structural Formula. Often it is advantageous to show atomic disposition within a molecule in a smaller space than is required by a graphic formula. Structural formulas show the relative position of atoms within the molecule and have the advantage of being written entirely on one horizontal line. Thus the structural formula of ethanol is CH_3CH_2OH and of methoxymethane CH_3OCH_3.

Molecular Formula. Molecular formulas show the exact number of each kind of atom within a molecule. Such molecular formulas show only the composition of the molecule but do not show the arrangement of the atoms within the molecules. The molecular formula of ethanol is identical with that of methoxymethane, C_2H_6O.

Empirical Formula. A simplified formula called the empirical formula may be derived by determining the percentage composition of a compound. Empirical formulas show only the proportional number of atoms of each element in a compound. They do not show the number of atoms nor the arrangement of the atoms within the molecule. Formaldehyde is a compound of carbon, hydrogen, and oxygen in the proportion represented by the formula CH_2O. One molecule of formaldehyde actually contains only one atom of carbon, two atoms of hydrogen, and one atom of oxygen. In this case the empirical formula is exactly the same as the molecular formula. Many other organic compounds such as glucose whose molecular formula is $C_6H_{12}O_6$, or $(CH_2O)_6$, have carbon, hydrogen, and oxygen present in the same proportions as formaldehyde.

Use of Formulas. Empirical formulas are useful primarily in the determining of molecular formulas. Molecular formulas also are not entirely adequate since they fail to show the arrangement of atoms within the molecule. Electronic formulas are used to clarify the structure and bonding within molecules. Later we shall see that knowledge of electron arrangement provides a basis for understanding and predicting reactions. Graphic formulas and structural formulas are used extensively by the student of organic chemistry. It is necessary, therefore, for the student to learn how to write the graphic formula of practically every compound used in this course. This recommendation can hardly be overemphasized and it applies not only to organic compounds but also to many of the inorganic compounds used in various reactions.

Calculation of Molecular Formulas. In the analysis of a carbon compound the percentage of each element may be determined. It follows that, if a compound contains 40.0 per cent carbon, 6.6 per cent hydrogen, and 53.3 per cent oxygen, the relative number of each kind of atom may be found by dividing the percentage by the atomic weight. In this example the relative number is obtained by dividing

$$\frac{40}{12} = 3.3 \text{ atoms of carbon}$$
$$\frac{6.6}{1} = 6.6 \text{ atoms of hydrogen}$$
$$\frac{53.3}{16} = 3.3 \text{ atoms of oxygen}$$

and by reducing

$$\frac{3.3}{3.3} = 1 \text{ atom of carbon}$$
$$\frac{6.6}{3.3} = 2 \text{ atoms of hydrogen}$$
$$\frac{3.3}{3.3} = 1 \text{ atom of oxygen}$$

The empirical formula therefore is CH_2O.

Since there are literally dozens of compounds having the empirical formula CH_2O, the true molecular formula $(CH_2O)_x$ must be obtained by some method for determining molecular weights. Any of the methods studied in general chemistry may be employed. For example, if the substance is soluble in water, its molecular weight may be determined by the lowering of the freezing point of a definite aqueous solution of the compound. It will be recalled that when one mole of a nonelectrolyte is dissolved in 1000 g. of water, the melting or freezing point of the solution becomes $-1.86°C$. Therefore if 5 g. of the substance analyzed above were dissolved in 200 g. of water and the melting point were found to be

$-0.31°C$, the weight of one mole of the substance would be calculated as follows:

[handwritten: 1 mole — 1 wt = -1.86°C] Lose
[handwritten: x = 200 = -.81°C]

$$5 \text{ g. of sample} \times \frac{1000}{200} \text{ g. of water} \times \frac{-1.86°}{-0.31°} = 150 \text{ g.} = \text{weight of 1 mole.}$$

[handwritten: 150/30 = 0]   *[handwritten: (?)x]*

The gram molecular weight of $(CH_2O)_1$ is the sum of the weights of atoms represented in the formula, or 30 g. Dividing the molecular weight of the compound by the weight of the empirical formula, we find $150/30 = 5$. Therefore this molecule is 5 times larger than represented by the empirical formula and may be written as $(CH_2O)_5$ or $C_5H_{10}O_5$. (Such a compound could be a simple sugar that is water soluble.)

Limitation of Graphic Formula. Graphic formulas are relatively true expressions of molecular structure within the limits of two dimensions available for writing formulas on paper or on a blackboard. Thus it is impossible to show the true direction of bonding. Actually molecules exist in three dimensions. Students must never forget that graphic formulas are shown in only two dimensions and that they do not show the correct relative proximity of one carbon atom to other carbon atoms within a molecule.

Direction in Bonding. In considering the types and the nature of bonds by which carbon atoms are held together, one of the most important concepts to fix in mind is based on the direction of bonds with respect to other bonds of the same carbon atom. Ordinarily graphic formulas of organic molecules are written in a straight line. One certain class of compounds is often called *straight-chain* compounds. Carbon compounds having three or more carbon atoms joined in a continuous chain are not straight-chain compounds in the sense that these carbon atoms lie in a straight line. Such a molecule is actually bent as shown by the diagrams in Figure 5. In the diagram two different directional types of carbon compounds are shown. Other directional types are also possible with four carbon atoms in the molecule. It is important to understand the effect of this directional bonding on the compounds formed. The

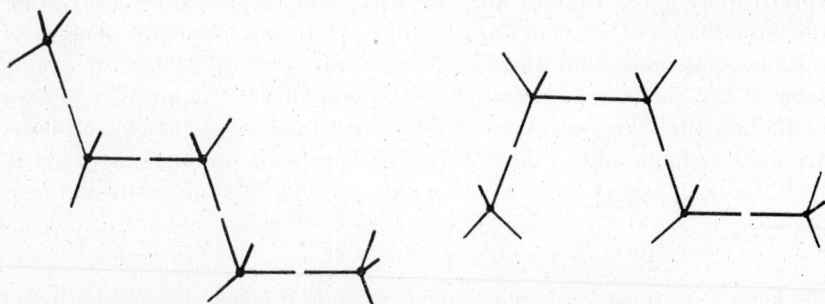

Fig. 5. Chains formed by carbon are not in a straight line.

length of the molecule is not a sum of the diameters of the carbon atoms as it would be if the carbon atoms actually lay in a straight line. Moreover it is possible for the end carbon atoms of a chain to be closer to each other than they are to carbon atoms only three atoms removed in the chain. This concept is especially necessary in understanding why cyclic compounds can be formed from open-chain compounds. In fact it will be the practice throughout this text *never* to refer to *straight-chain* compounds. The term *open chain* compounds is less misleading and will be used exclusively in referring to this type of compound.

Sources of Carbon Compounds. The sources of carbon compounds are both inorganic and organic. Carbonates of various kinds form an important part of the rocks that make up the surface of our earth. Our chief source of carbon compounds for industry, however, is from living things or the modified remains of things that have lived at some earlier age in the earth's existence.

The air surrounding our earth contains about 0.045 per cent carbon dioxide by weight (0.03 to 0.04 per cent by volume). Carbon dioxide is formed during the decay of animal and vegetable material by bacteria and mold. It is also formed and exhaled by animals having respiratory systems and it is one of the end products of all complete oxidation reactions of carbon compounds. It is slightly soluble in water and combines with water to form a weak acid called carbonic acid.

$$CO_2 + H_2O \rightarrow H_2CO_3 \rightarrow HCO_3^+ + H^+$$

Green plants also combine carbon dioxide and water in building plant tissues. When carbon dioxide and water are combined within plants under the catalytic influence of chlorophyll, energy is required to complete the reaction. Energy for this reaction is supplied by sunlight and the entire reaction is called "photosynthesis."

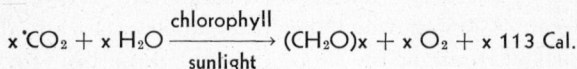

$$x\,CO_2 + x\,H_2O \xrightarrow[\text{sunlight}]{\text{chlorophyll}} (CH_2O)x + x\,O_2 + x\,113\ Cal.$$

For each mole of carbon dioxide and of water thus used in synthesis approximately 113 calories are captured from the radiant energy from the sun. Plants of all types are therefore not only abundant sources of carbon compounds, but they serve also as energy traps for preserving some of the vast energy radiated by the sun. Of vital importance to man is the fact that this energy is liberated in animal organisms when plants are used as food, and in our machines when plant products are used as fuel. In both instances a reaction opposite to that of photosynthesis occurs.

$$(CH_2O)x + x\,O_2 \rightarrow x\,CO_2 + x\,H_2O + x\,113\ Cal.$$

The entire process whereby carbon dioxide becomes a part of a living thing and eventually returns to carbon dioxide may be called the great

Table 1

DIVISION OF CARBON CHEMISTRY

		Elements Contained	Class	Subclass
Carbon compounds	Introduction	C only	Carbon	Carbon Atoms
		C and H only	Hydrocarbons	Alkanes Alkenes Alkynes
		C, H, and X	Halogen derivatives	Haloalkanes Polyhaloalkanes
	Acyclic (aliphatic) compounds	C, H, and O	First oxidation	Alcohols Polyalcohols Ethers
			Second oxidation	Aldehydes Ketones
			Third oxidation	Acids
		C, H, O, X, N	Acid derivatives	Salts Esters Anhydrides Acyl halides Amides
			Substituted acids	Halo- Hydroxy- Amino-
		C, H, N	Nitrogen derivatives	Nitriles Amines Quaternary
		C, H, O	Carbohydrates	Monosaccharides Disaccharides Polysaccharides
	Cyclic compounds	C, H with O, X, S, N on ring	Carbocyclic	Alicyclic Aromatic Aromatic with halogen Aromatic with sulfur Aromatic with nitrogen Phenols and alcohols Aromatic aldehydes Aromatic acids
		C, H with O, S, N in ring	Heterocyclic	With oxygen With sulfur With nitrogen Mixed

cycle of carbon in nature. During the ages this great cycle of carbon has been ever-present. At one time the earth supported an enormous growth of plants. Quantities of these plants and the living things which fed upon them eventually become deposits of coal, oil, and natural gas from which modern industry draws its energy. Thus the things that lived in another era have supplied us with a vast source of carbon compounds.

Divisions of Organic Chemistry. On the preceding page will be found a general outline of organic chemistry. From the very nature of the carbon atom and its manner of combining with other carbon atoms, it becomes apparent that many classes of compounds exist. Two broad divisions deal with those compounds formed with open-chains and those formed with a ring structure. The former are called *acyclic* or *aliphatic* and the latter *cyclic* or *alicyclic*. (ring) (chained)

Each of these classifications will be considered in relation to the various elements forming the compounds and in accord with the type of bonds found between the various carbon atoms. Thus compounds of carbon and hydrogen, called hydrocarbons, may be considered as open-chained (*acyclic*) or as ring (*cyclic*) compounds. The open-chain hydrocarbons are, according to the bonds between carbon atoms, single-bonded, double-bonded, or triple-bonded. Single-bonded hydrocarbons are saturated and are called *alkanes*. Double- and triple-bonded hydrocarbons are unsaturated and are called *alkenes* and *alkynes* respectively. Cyclic hydrocarbons may have single-bonded carbon atoms (*cycloalkanes*) or a specific pattern of bonding as exemplified by benzene and its related compounds (*aromatic compounds*). All cyclic hydrocarbons are said to be *carbocyclic*, carbon atoms only being the elements within the ring structure. When some other element than carbon is present in the ring structure, the compound is called *heterocyclic*.

In this book the nature and reactions of compounds containing carbon with other elements will be considered in the following order: hydrogen, halogens, oxygen, and nitrogen. Most compounds considered will also contain hydrogen. Other elements also combine with carbon and hydrogen and will be found in larger and less elementary books on organic chemistry. The classification of organic compounds given in Table 1 should be compared with the table of contents of this text.

Study and Review Questions

1. Why is Organic Chemistry a separate field of study?
2. Discuss briefly the suitability of the name Organic Chemistry as applied to this field of chemistry.
3. Name, and briefly list the differences between the three allotropic forms of carbon.
4. Make a two dimensional drawing of the $^{12}_{6}C$, showing the location of

all protons, neutrons, and electrons, and designating the nucleus, K-shell, and L-shell.

5. What is meant by the "rare gas configuration"? Give examples.

6. How does a comparison of carbon's position in the periodic table with that of other elements increase our knowledge of the nature of the carbon atom?

7. What is meant by "ionic bonding"?

8. What is meant by "covalent bonding"?

9. What is meant by "directional bonding" of carbon?

10. Make a drawing of $^{11}_6C$ and $^{14}_6C$ showing the location of all protons, neutrons, and electrons.

11. The molecular formula of ethane is C_2H_6. Write the electronic, graphic, skeletal, structural, and empirical formulas of ethane.

12. A compound containing 2.22 per cent hydrogen, 26.67 per cent carbon, and 71.11 per cent oxygen lowers the freezing point of a solution to $-0.62°C$ when 0.6 g. of the compound are dissolved in 20.0 g. of distilled water. Calculate the molecular formula of the compound.

13. Discuss briefly our chief sources of carbon compounds.

14. What is meant by "the great cycle of carbon"? Write the reactions occurring in plant and animal life showing why they are necessary for animal existence on earth.

15. Distinguish between the two main divisions of organic compounds.

Chapter 2

COMPOUNDS OF CARBON WITH HYDROGEN
I. SATURATED HYDROCARBONS: THE ALKANES

CHAPTER PROLOGUE

Even before our primitive ancestor first beheld carbon in an elemental form, many carbon atoms had lived a thousand lives and died a thousand deaths. The plant world for ages had witnessed the slow decay of giant trees and of frail ferns. In the decaying process many carbon atoms were liberated from complex molecular structures and became carbon dioxide—eventually to serve again as food for other growing plants. Many carbon atoms, however, did not secure even this fleeting freedom. Covered by the fallen foliage of a thousand summers, pressed flat and sodden by unnumbered spring rains, buried by countless tons of silt, the carbon atoms of great primeval forests became the future coal seams of industry. Upheavals in the crust structure of the earth subjected them to great pressure, squeezing from them oils that escaped into the porous rocks serving both as their oppressor and as their preserver. Less than a hundred years ago, man learned how to bring to the surface of the earth in quantity this treasure of another era. In the past half century man has learned how to use these treasures more wisely and more prodigally than any other natural resource. They have brought us light from their darkened tomb and power from their long period of inactivity. We are clothed by their woven molecules in gayly colored garments, cured of many ills by wondrous new drugs made from them, and we have been defended from the enemies of our way of life by their devastating might.

Civilization today is largely dependent upon the carbon compounds of a bygone age

Hydrocarbons. In presenting the divisions of organic chemistry (p. 11) it was pointed out that the simplest class is composed of carbon and hydrogen. These compounds are called *hydrocarbons*. When hydrocarbons form in such a way that the carbon atoms of the molecule unite to produce a ring structure, the name *cyclic* (Greek *kuklos*, circle) is used in referring to them, as, cyclohexane. If the hydrocarbons are open-chain compounds, they are called *acyclic* hydrocarbons (Greek *a-*, negative prefix, *kuklos*, circle—no circle, or ring). The name *acyclic* is not often used by organic chemists, however. Instead the synonymous word *aliphatic* applies to all hydrocarbons saturated and unsaturated, which have the open-chain structure.

Alkanes or Paraffins. When all the carbon atoms in a hydrocarbon are joined to other carbon atoms by single bonds forming an open-chain structure, the compound is described as a saturated aliphatic hydrocarbon. It may be called by either of the two synonymous names, an *alkane* or a *paraffin* hydrocarbon. The name paraffin (Latin *parum*, little, *affinis*, affinity) describes a quality or a lack of a quality in this type of compound, a lack of affinity for other substances. The generic name alkane is applied to this class of compounds in a systematic method of describing the structure of the compound. Specific simple compounds of this type are methane, ethane, and propane. In these substances the

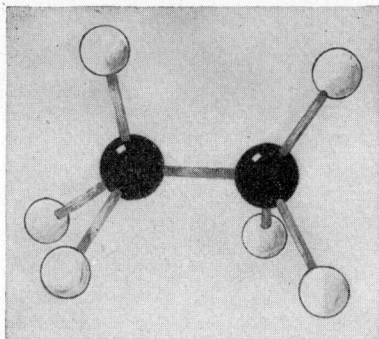

Fig. 6

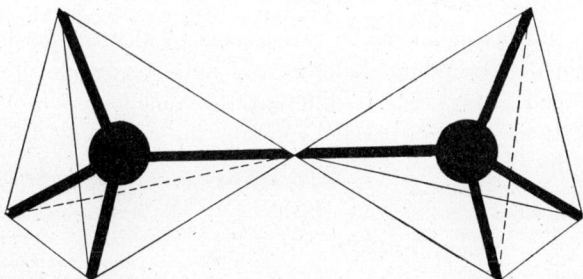

Fig. 7

Fig. 6. The ethane molecule.
Fig. 7. Tetrahedral structure of ethane showing single bond between carbon atoms.

suffix *-ane* denotes a saturated open-chain hydrocarbon. The word "alkane" is a positive scientific name given to those open-chain hydrocarbons in which the various carbon atoms are joined together by shared electron pair or covalent bonds. When derivatives are made from these simple hydrocarbons, the hydrocarbon portion of the molecule is called generically the *alkyl* part of the molecule.

Homologous Series. From the structure of alkanes a notable simplicity in organic compounds becomes apparent. The simplest possible saturated compound of carbon and hydrogen is one wherein one atom of carbon unites with all the atoms of hydrogen that it can hold, namely,

four. It is called methane, CH_4, and it has the graphic formula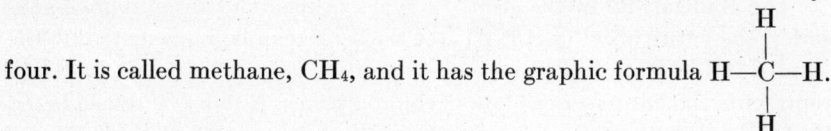

The next hydrocarbon is ethane, C_2H_6, and it has the graphic formula 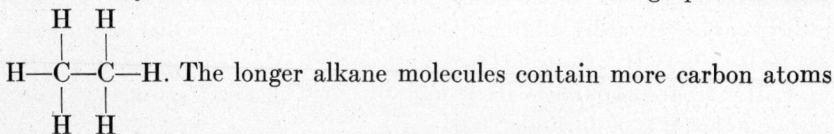 The longer alkane molecules contain more carbon atoms with hydrogen atoms joined to all the available bonds of the carbon atoms involved. When the physical properties of these various hydrocarbons are listed, small differences are noted that depend upon relative size of the molecules. The chemical properties are essentially the same for all alkanes. The structural variation and the only slight change in properties have led to a classification of hydrocarbons in an "homologous series." Open-chain hydrocarbons having only single bonds have the type formula C_nH_{2n+2}. Because of the very large number of such hydrocarbon compounds it would be impossible to study each substance individually. Fortunately it is not necessary to study compounds individually, except in a few special cases. Because of the parallel similarity of compounds in their structure, organic compounds in general are studied in families.

Students of organic chemistry must learn to identify the homologous series to which a compound belongs and must recognize the structural part (functional group) which differentiates one member of the series from others. If we write in parallel columns the members of two homologous series, the similarities and differences become quite clear.

Alkane series			Bromoalkane series		
CH_4	Methane	H H C H H	H H C Br H	Bromomethane	CH_3Br
C_2H_6	Ethane	H H H C C H H H	H H H C C Br H H	Bromoethane	C_2H_5Br
C_3H_8	Propane	H H H H C C C H H H H	H H H H C C C Br H H H	1-Bromopropane	C_3H_7Br
C_4H_{10}	Butane	H H H H H C C C C H H H H H	H H H H H C C C C Br H H H H	1-Bromobutane	C_4H_9Br

As each molecule increases in size, the difference in molecular weight of the next member of the series is 14, and the actual difference is called a methylene group, $-CH_2-$. The methylene group cannot exist by itself; however each member of an homologous series differs from the next by CH_2.

The Radical, R. In the study of types of reactions in an homologous series, the entire alkyl (p. 15) part of the series can be represented by the symbol "R." Thus RH represents the entire series of alkanes, and RCl represents the entire series of monochloroalkanes. It must be remembered that "R" does not refer to any specific alkyl radical. The "R" is always associated with some other atom or radical. The names of specific alkyl radicals are derived by adding the suffix "yl" to a particular root name, as: methyl ($-CH_3$) from methane; ethyl ($-C_2H_5$) from ethane; propyl ($-C_3H_7$) from propane (C_3H_8); and so on. The alkyl groups will correspond to the type formula (C_nH_{2n+1}).

Names of Alkanes. The alkanes derive their specific names from the number and arrangement of carbon atoms in each molecule. Some of these names were commonplace in the language of chemists and of non-chemists before any attempt was made to systematize their nomenclature. The names methane, ethane, propane, and butane, designating the four simplest alkanes having respectively one, two, three, and four carbon atoms in the hydrocarbon, have survived a century of chemical progress and will probably remain unchanged. In 1892 the International Union of Chemistry meeting at Geneva, Switzerland, adopted a new system of nomenclature for organic compounds. This system was modified at meetings in Liege, Belgium (1923), and again at Geneva (1930). Today we refer to it as the "I.U.C." or the "Geneva" system. The American Chemical Society has encouraged its more widespread adoption. With few exceptions the I.U.C. system of nomenclature will be used throughout this text. Other methods of naming compounds will be found in Chapter 12.

Naming the Alkanes. According to the I.U.C. system the names of hydrocarbons are derived from the longest continuous open-chain of carbon atoms in the compound. Greek prefixes are used to show the total number of such continuous carbon atoms and the suffix "-ane" is used to show that the compound is a saturated hydrocarbon. In the accompanying list all alkanes are named up to the twentieth carbon atom. After that each tenth compound is listed. All are "normal" compounds, that is, there are no "branch-chain" compounds included in this list.

CH_4	Methane	$C_{11}H_{24}$	Hendecane
C_2H_6	Ethane	$C_{12}H_{26}$	Dodecane
C_3H_8	Propane	$C_{13}H_{28}$	Tridecane
C_4H_{10}	Butane	$C_{14}H_{30}$	Tetradecane
C_5H_{12}	Pentane	$C_{15}H_{32}$	Pentadecane
C_6H_{14}	Hexane	$C_{16}H_{34}$	Hexadecane
C_7H_{16}	Heptane	$C_{17}H_{36}$	Heptadecane
C_8H_{18}	Octane	$C_{18}H_{38}$	Octadecane
C_9H_{20}	Nonane	$C_{19}H_{40}$	Nonadecane
$C_{10}H_{22}$	Decane	$C_{20}H_{42}$	Eicosane

$C_{30}H_{62}$	Triacontane
$C_{40}H_{82}$	Tetracontane
$C_{50}H_{102}$	Pentacontane
$C_{60}H_{122}$	Hexacontane
$C_{70}H_{142}$	Heptacontane

The longest continuous open-chain hydrocarbon reported in the literature thus far and having no branch alkyl groups is heptacontane. $C_{70}H_{142}$.

General Rules of Nomenclature. Hydrocarbons and hydrocarbon derivatives having branches of alkyl or other groups are named in accord with the following rules:

1. Write the skeletal formula (without H) of the compound, e.g.

[handwritten: 7C = Heptane]

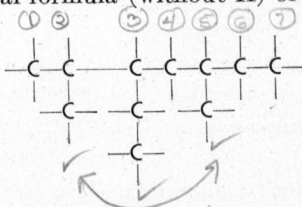

[handwritten: 2 methyl = dimethyl]

2. Count the total number of the longest continuous chain of carbon atoms and write down the correct alkane name. In the case shown there are seven carbon atoms in the longest chain: therefore, heptane.

3. Write down the number of the branched carbon atom closest to either end, making sure that the number is the lowest possible. In the case given the number would be 2. All further counting must start from the same end.

4. Write down the number of each carbon atom from which any branch starts, using only the previously chosen starting point, thus: 2,3,5.

5. After each number write the name of the alkyl group that forms each branch after its proper number, thus: 2-methyl-3-ethyl-5-methylheptane.

6. When the same kind of alkyl group occurs more than once, assemble the groups that are alike and show their total number and locations as follows: 2,5-dimethyl-3-ethylheptane. Write the name as a single word using commas to separate numbers and hyphens before and after numbers as shown in the example. The example given in this paragraph is the correct I.U.C. name. (Greek prefixes, *mono-*, *di-*, *tri-*, *tetra-*, *penta-*, etc., are used to denote how many of each specific group are present in the compound.)

7. When some atom or radical other than an alkyl group is branched from a carbon atom of the chain, the same procedure is followed using a root name for the element or radical and placing it immediately before the name of the hydrocarbon. A compound having the skeletal formula

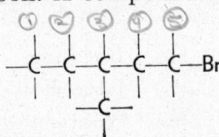

would be correctly named 1-bromo-3-methylpentane.

Isomerism. If the student will count the total number of carbon atoms in the alkane shown under Rule 1 for the naming of compounds, 2,5-dimethyl-3-ethylheptane, a total of eleven carbon atoms will be found. The total number of hydrogen atoms in this molecule will be twenty-four.

If all carbon atoms in the side or branch chains were so arranged that there would be one continuous chain of carbon atoms with no branches whatever, the molecule would still have eleven carbon atoms and twenty-four hydrogen atoms. The atomic content of the molecule and the molecular formula, $C_{11}H_{24}$, would be the same as in the case of the 2,5-dimethyl-3-ethylheptane. The unbranched molecule would be called "normal" hendecane and the name "normal" would be abbreviated to "n-", as n-hendecane. Thus 2,5-dimethyl-3-ethylheptane and n-hendecane are isomers. (Greek: *isos*, equal, *meros*, part, equal parts.)

Isomerism is common in organic chemistry. In the example given above it is called *chain* isomerism. The existence of the isomer depends upon the arrangement of atoms within the molecule. When the number of carbon atoms within the molecule permits only one arrangement, such a compound has no isomer. Among the alkanes this occurs in the case of methane, ethane, and propane. The atoms of hydrogen and carbon can be arranged in only one way.

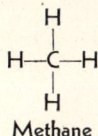

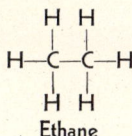

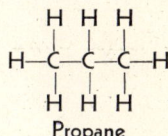

Methane   Ethane   Propane

The 4 carbon and 10 hydrogen atoms in butane, however, can be arranged in two different ways:

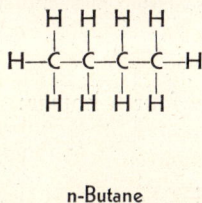

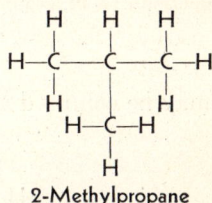

n-Butane   2-Methylpropane

Both of these compounds are gases and have the same molecular weight. A measurable difference, nearly 10°C, is found in their respective boiling points, the normal compound boiling at $-0.6°C$ and the other at $-10°C$.

As the number of carbon atoms increases in the alkanes, the number of possible arrangements, and therefore of isomers, increases very rapidly. Pentane has three isomers each with the same molecular formula, C_5H_{12}.

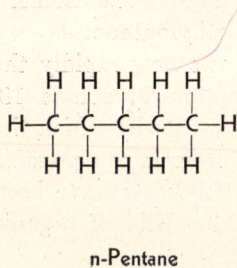

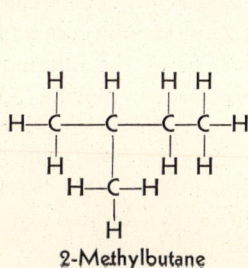

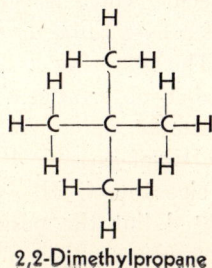

n-Pentane   2-Methylbutane   2,2-Dimethylpropane

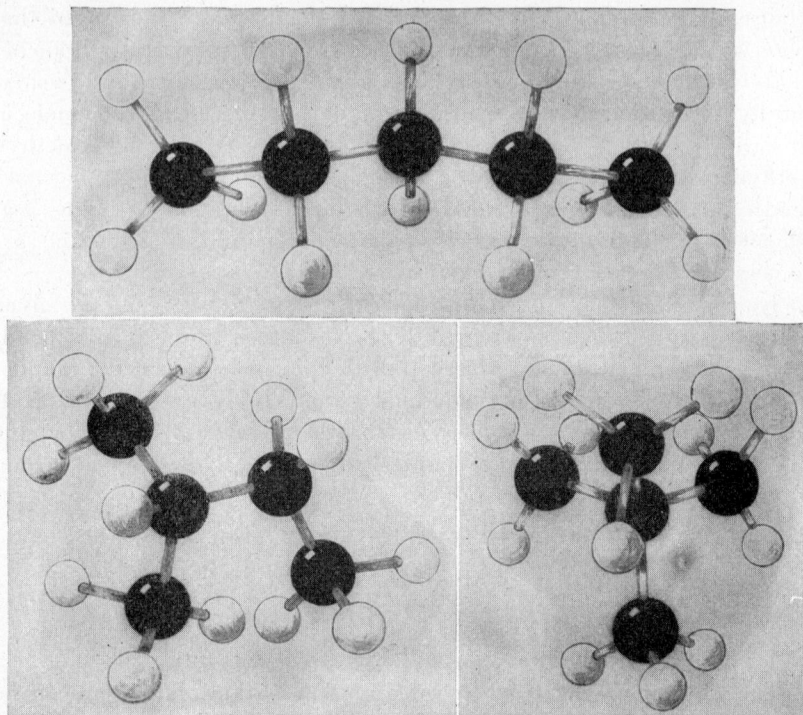

Fig. 8. The pentane isomers. (Top) Pentane. (Bottom, left) 2-Methylbutane. (Bottom, right) 2,2-Dimethylpropane.

Students may be confused at first and expect a fourth isomer:

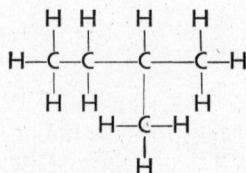

calling it 3-methylbutane. This difficulty will be overcome by recalling that the rules of nomenclature require counting from that end of the molecule which is closest to the branched chain. The "extra isomer" thus proves to be the same kind of molecule as 2-methylbutane.

The following list of isomers of the alkanes shows how rapidly the number of arrangements increases: hexane 5, heptane 9, octane 18, nonane 35, decane 75. According to the accepted formula by which the number of possible isomers is calculated, 366,319 isomers are possible for eicosane, $C_{20}H_{42}$. Obviously only a few of these isomers have been isolated but the possibility of future research in the field of organic

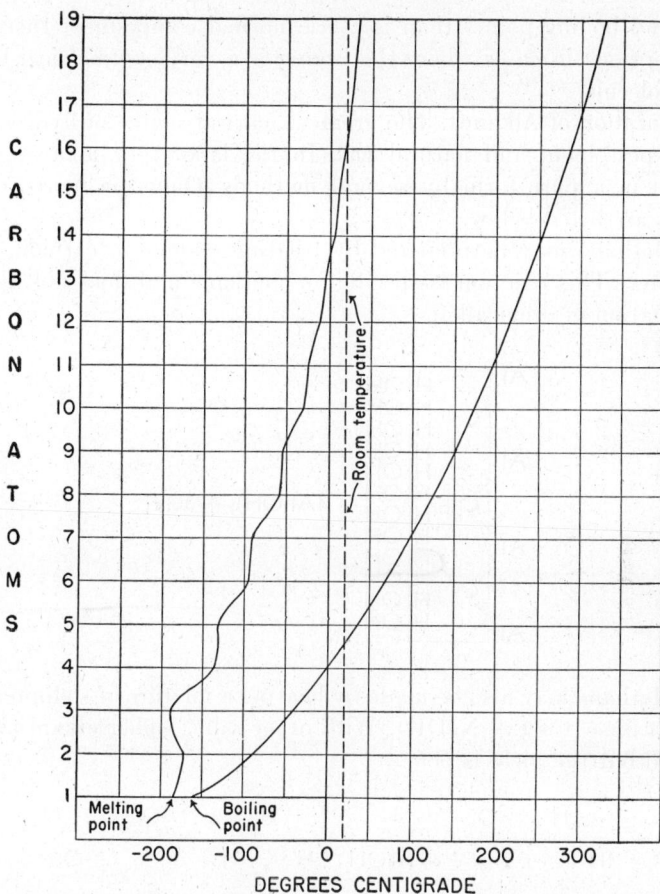

Fig. 9. Melting point and boiling point graph of n-alkanes.

chemistry is aptly illustrated by the enormous number of possible compounds in the alkane division alone.

At this point students will find it interesting to write out the skeletal formulas of all of the isomers of hexane and heptane and to name each substance. Students of petroleum chemistry will do well to write out all the isomers of octane also.

Physical Properties. The alkanes when pure are colorless and practically odorless. They are nearly insoluble in water but are soluble in most organic solvents. All alkanes have densities less than that of water. Methane is the only gaseous alkane less dense than air. As shown in Figure 9 of melting and boiling points, alkanes up to butane (C_1............C_4) are gases at room temperature (20°C), from pentane to hexadecane (C_5............C_{16}) are liquids, from heptadecane on (C_{17}............C_{70}) the alkanes are solids. As a rule branch-chain isomers have lower melting points and boiling points than isomeric normal compounds; there are a few exceptions in cases where the isomer is a very symmetrical branch-chain molecule.

Preparation of Alkanes. The greatest natural source of hydrocarbons is from petroleum and natural gas. In the laboratory alkanes can be prepared in a pure form by several methods. Only the more common methods are listed below.

1. Methane may be prepared by treating aluminum carbide (Al_4C_3) with water. This reaction is specific for methane and does not apply to the formation of other alkanes.

$$\rightarrow 4Al(OH)_3 + 3CH_4$$

2. Methane may also be made by heating a mixture of sodium acetate and soda lime (CaO + NaOH). With other salts, higher alkanes may be produced but the yield is poor.

$$H_3C{-}COONa + NaOH \rightarrow H{-}CH_3 + \begin{matrix} NaO \\ NaO \end{matrix}C{=}O$$

3. Alkanes may also be prepared by the Grignard[1] Reaction using an haloalkane, as bromoalkane, and magnesium metal in the medium of anhydrous ether to form alkylmagnesiumbromide which is later hydrolized to the alkane and basic magnesium bromide. By using the proper halo-alkane any desired alkane may be produced.

$$R{-}Br + Mg \xrightarrow{\text{Dry ether}} RMgBr$$

$$R{-}\boxed{MgBr + HO}H \rightarrow RH + Mg(OH)Br$$

4. For preparing any pure alkane having an even number of carbon atoms, the Wurtz[2] synthesis is used.

[1] Named after Victor Grignard (1872–1935).
[2] Named after Charles A. Wurtz (1817–1884).

$$\text{H}-\underset{\underset{\text{H}}{|}}{\overset{\overset{\text{H}}{|}}{\text{C}}}-\boxed{\text{X}+\overset{\text{Na}}{\underset{\text{Na}}{}}+\text{X}}-\underset{\underset{\text{H}}{|}}{\overset{\overset{\text{H}}{|}}{\text{C}}}-\text{H} \rightarrow \text{H}-\underset{\underset{\text{H}}{|}}{\overset{\overset{\text{H}}{|}}{\text{C}}}-\underset{\underset{\text{H}}{|}}{\overset{\overset{\text{H}}{|}}{\text{C}}}-\text{H} + 2\text{NaX}^3$$

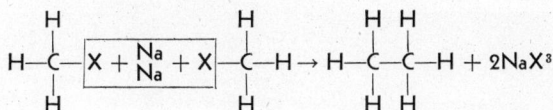

Alkanes having an odd number of carbon atoms can also be prepared by treating two kinds of haloalkanes with sodium using the Wurtz synthesis, but only a part of the product will have an odd number of carbon atoms. If CH_3X and C_2H_5X are treated with sodium metal, all three of the following reactions will take place:

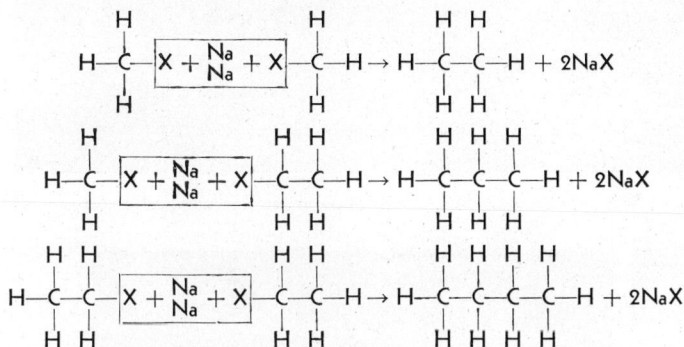

$$\text{H}-\overset{\text{H}}{\underset{\text{H}}{\text{C}}}-\boxed{\text{X}+\overset{\text{Na}}{\underset{\text{Na}}{}}+\text{X}}-\overset{\text{H}}{\underset{\text{H}}{\text{C}}}-\text{H} \rightarrow \text{H}-\overset{\text{H}}{\underset{\text{H}}{\text{C}}}-\overset{\text{H}}{\underset{\text{H}}{\text{C}}}-\text{H} + 2\text{NaX}$$

$$\text{H}-\overset{\text{H}}{\underset{\text{H}}{\text{C}}}-\boxed{\text{X}+\overset{\text{Na}}{\underset{\text{Na}}{}}+\text{X}}-\overset{\text{H}}{\underset{\text{H}}{\text{C}}}-\overset{\text{H}}{\underset{\text{H}}{\text{C}}}-\text{H} \rightarrow \text{H}-\overset{\text{H}}{\underset{\text{H}}{\text{C}}}-\overset{\text{H}}{\underset{\text{H}}{\text{C}}}-\overset{\text{H}}{\underset{\text{H}}{\text{C}}}-\text{H} + 2\text{NaX}$$

$$\text{H}-\overset{\text{H}}{\underset{\text{H}}{\text{C}}}-\overset{\text{H}}{\underset{\text{H}}{\text{C}}}-\boxed{\text{X}+\overset{\text{Na}}{\underset{\text{Na}}{}}+\text{X}}-\overset{\text{H}}{\underset{\text{H}}{\text{C}}}-\overset{\text{H}}{\underset{\text{H}}{\text{C}}}-\text{H} \rightarrow \text{H}-\overset{\text{H}}{\underset{\text{H}}{\text{C}}}-\overset{\text{H}}{\underset{\text{H}}{\text{C}}}-\overset{\text{H}}{\underset{\text{H}}{\text{C}}}-\overset{\text{H}}{\underset{\text{H}}{\text{C}}}-\text{H} + 2\text{NaX}$$

The extra work involved in purifying the final product makes this a possible, but not a desirable, method for preparing a pure alkane, having an odd number of carbon atoms.

5. A method for preparing pure alkanes is by the reduction of the proper haloalkane with hydrogen.

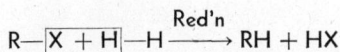

$$\text{R}-\boxed{\text{X}+\text{H}}-\text{H} \xrightarrow{\text{Red'n}} \text{RH} + \text{HX}$$

6. An iodoalkane may be treated with hydriodic acid to form an alkane and iodine. Other halogens than iodine will not enter into this type of reaction.

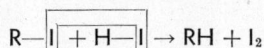

$$\text{R}-\boxed{\text{I}+\text{H}-\text{I}} \rightarrow \text{RH} + \text{I}_2$$

Alkanes from Natural Sources. Most of the hydrocarbons used today originate from petroleum, coal, or natural gas. Natural gas is chiefly methane, the simplest of the saturated hydrocarbons. By the hydrogenation of coal at high temperatures and pressures, many hydrocarbons can be synthesized. In Germany, before World War II, gasoline and other oil products were synthetically produced from coal. This method of producing desirable hydrocarbons from coal will probably be further de-

---

[3] The letter X will be used hereafter to represent a halogen.

(a) The chemist's dream comes true in the Research Laboratory.

(b) The engineer's proof is established in the Pilot Plant.

Fig. 10. Chemicals from petroleum. (Courtesy, Shell Development Company, Emeryville, Calif.)

(c) The commercial product flows from the vertical steel towers
of the Industrial Distillation Unit.
Fig. 10—(Continued). (Courtesy, Shell Development Company, Emeryville, Calif.)

veloped in the next decade even in countries where petroleum is at present found in fair abundance. Petroleum however remains our largest source of hydrocarbons.

Petroleum is not a chemical compound. It is a mixture of many compounds, mainly hydrocarbons. Although alkanes are abundant in petroleum there are also found in the mixture various unsaturated hydrocarbons and also a few ring-type hydrocarbons. Depending upon the geographical source of the petroleum, cyclic hydrocarbons may or may not be present in quantity. Petroleum from Pennsylvania and the eastern states is said to have a paraffin base, i.e. consist largely of alkanes and yield paraffin as an end product of distillation. Petroleum found in California, Texas, and Mexico is said to have an asphalt base; these oils yield a black tarlike residue upon distillation. Midcontinental oils have a mixture of asphalt and paraffin bases.

During the first 40 years after the discovery of oil by Drake in Pennsylvania (1859), kerosene, used largely as an illuminant, was the chief substance produced from petroleum. Since that time research in the field of petroleum chemistry has progressed simultaneously with the development of the internal combustion engine and especially with the widespread use of these engines in automobiles and airplanes. Yet, even today, chemists consider four chief products from petroleum in the order of their distillation from the crude oil. All substances which distill at temperatures up to 70°C are said to be in the *ligroin fraction*. The name "petroleum ether" is also applied to this mixture which is highly inflammable and is used chiefly as an organic solvent. The portion of the oil distilling between 85°C and 200°C is called the *gasoline fraction*. The so-called "straight-run gasolines" are made from this fraction. The distillate collected in the temperature range from 200°C to 300°C is called the *kerosene fraction*. The fraction boiling higher than 300°C is called the *fuel oil fraction*. Each fraction is a mixture of saturated and unsaturated hydrocarbons.

Cracking Process. In the past 30 years petroleum chemists have not only separated many of the individual hydrocarbons that naturally exist in petroleum; they have perfected methods whereby hydrocarbons having a long chain of carbon atoms may be broken down into simpler hydrocarbons. This procedure is called the "cracking process" because larger molecules are broken or cracked to form simpler compounds. To produce a gasoline from these higher boiling fractions by cracking, high temperatures and pressures are required. Two other names, "thermal decomposition" and "pyrolysis," are also applied to this process.

New Gasolines. The exacting fuel requirements of modern machines in World War II have led to further research. Not only are higher hydrocarbons cracked to make simpler compounds, many of the very simple products are then combined to produce motor fuels whose molecular structure permits power production at a greater efficiency per unit weight. "Tailor-made" gasoline of high octane rating is produced in this way. It has been found that "knocking" of motors could be reduced when a saturated, branch-chain hydrocarbon having eight carbon atoms is used as a fuel. The eight carbon atoms in this knockless gasoline are not arranged in the continuous chain of n-octane. An octane isomer, 2,2,4-trimethylpentane, has been found to give a result arbitrarily set as a standard of 100. The antiknock quality of n-heptane has been arbitrarily set as 0. Antiknock qualities of gasoline are measured from these standards. Some fuels are known that have octane ratings greater than 100 but most of our motor fuels have octane ratings somewhat less than 100. In 1922 it was found that the addition of lead tetraethyl, $Pb(C_2H_5)_4$, to straight-run gasolines reduced motor knock, thereby increasing octane rating. Such gasolines are called *ethyl gasolines*.

Reactions of Hydrocarbons. As indicated by the etymology of the word *paraffin*, there are very few reactions of alkanes. Six types of reactions will be listed here, (1) combustion, (2) halogenation, (3) nitration, (4) sulfonation, (5) decomposition, and (6) isomerization. Alkylation reactions will be considered in Chapter 3 on Alkenes.

1. COMBUSTION. The most important reaction of alkanes is oxidation. All alkanes react with oxygen, evolving much heat in the course of the reaction. The products formed by the reaction, when oxidation is complete, are carbon dioxide and water. When oxidation is not complete, water, carbon monoxide, carbon dioxide, and even free carbon may form. The following molecular reactions illustrate the molecular quantities of oxygen needed by one molecule of various alkanes undergoing complete combustion.

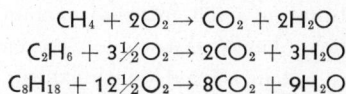

$$CH_4 + 2O_2 \rightarrow CO_2 + 2H_2O$$
$$C_2H_6 + 3\frac{1}{2}O_2 \rightarrow 2CO_2 + 3H_2O$$
$$C_8H_{18} + 12\frac{1}{2}O_2 \rightarrow 8CO_2 + 9H_2O$$

The energy evolved in these oxidation reactions of alkanes is important in engineering. The energy may be desired as heat, as is the case of burning fuel oil in a furnace; or it may be desired as mechanical energy, as is the case of the combustion of gasoline in an automobile motor.

2. HALOGENATION. The reaction of alkanes with the halogens introduces a new concept called *substitution*. When a halogen molecule, usually symbolized by X_2 or X—X, reacts with an alkane, one part of the halogen molecule reacts with one hydrogen atom from the alkane to form a hydrohalide and the other part of the halogen molecule substitutes in the place of the hydrogen, forming a new molecule called a haloalkane or alkyl halide. In writing the equation we picture the reacting substances as follows:

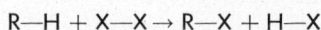

$$R—H + X—X \rightarrow R—X + H—X$$

One of the atoms of the halogen molecule substitutes for a hydrogen atom in the alkane molecule and the resulting molecule is accordingly a halogen derivative of the alkane.

In substituting the first halogen atom into the two simplest molecules, methane and ethane, it makes no difference which hydrogen atom is replaced. When their graphic formulas are written it is seen at once that each hydrogen atom has the same relationship to the molecule. All the

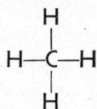

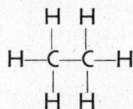

hydrogens in these two cases are alike. In the case of higher homologues beginning with propane, the hydrogen atoms are not all alike in their

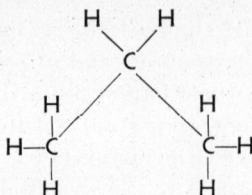

relationship to the molecule. The carbon atoms at each end of the chain have a freedom of rotation about the single bond as an axis. Carbon atoms inside the two end carbon atoms do not have the same freedom of rotation. Such carbon atoms are bonded at 109°28′ from each other. The two hydrogen atoms on each internal carbon atom are in a more fixed relationship to the molecule as a whole, whereas the hydrogen atoms on the end carbons are not in a fixed relationship.

Halogen substitution occurs more readily on the hydrogen atoms attached to the internal carbon atoms of the alkanes. A second substitution will usually occur in the place of a second hydrogen on the same carbon atom rather than in the place of hydrogen on a different carbon atom. A pure compound, however, cannot be made by direct halogenation. A mixture of many kinds of monohaloalkanes and polyhaloalkanes always results from direct halogenation. Methods of preparing pure haloalkanes will be presented in Chapter 4 on Halogen Compounds.

Thus far halogenation has been explained as a group reaction. Each halogen has its own peculiar condition of activity. Fluorine reacts violently with alkanes. Chlorine and bromine react directly with alkanes when heated, especially in the presence of a catalyst or of light. Iodine will not react directly with an alkane. It will be noted that the ease of reaction is related to the activity of the respective halogen.

3. NITRATION. In the gaseous phase hydrocarbons will react with nitric acid to form nitroalkanes. The gaseous phase is also used because the smaller homologues are gases at room temperatures. When higher liquid homologues are mixed with nitric acid in the presence of sulfuric acid as a catalyst, nitration also occurs.

In the case of methane and ethane one hydrogen can be replaced, and, since all the hydrogens are alike, only one nitroalkane is formed. In the case of methane and ethane, the carbon atoms of the molecules are not attached to more than one other carbon atom. When such a condition exists, these carbon atoms are called *primary* carbon atoms. In the case of propane, the carbon atoms at the end of each molecule are primary carbon atoms. The middle carbon atom of propane, however, is attached to two different carbon atoms and it is called a *secondary* carbon atom. If the carbon atom is attached to three different carbon atoms in a molecule it is called a *tertiary* carbon atom. Similarly, when a single carbon atom is attached to four different carbon atoms, the name *quaternary* carbon atom is applied. In the examples given below, the type of the middle carbon atom is written underneath the graphic formula. All other carbon atoms in these examples are primary carbon atoms.

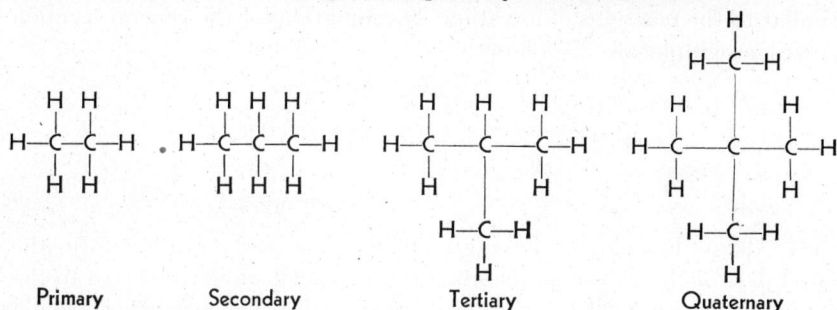

Primary   Secondary   Tertiary   Quaternary

In homologues larger than ethane, nitration will occur most readily in the place of the hydrogen atom attached to the tertiary carbon atom. If there is no tertiary carbon atom, the nitration will occur on the secondary carbon atom rather than on the primary carbon atom. Nitration on a primary carbon atom will take place, but the ease of nitration is notably less. In all cases only one nitro-group will be substituted directly. There is always the probability that the hydrocarbon molecule will "crack" and simpler nitrated products will be formed.

Typical reactions are shown below, illustrating that water is also one of the products of this reaction.

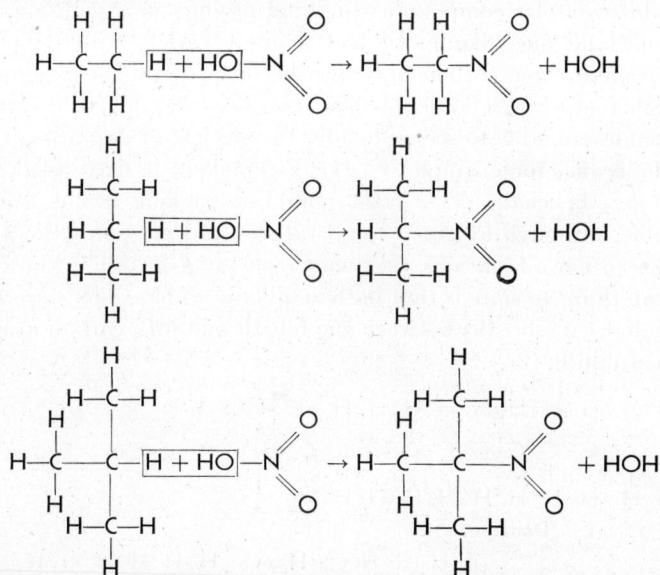

The names of the products of these reactions are respectively: nitro-ethane, 2-nitropropane, and 2-nitro-2-methylpropane.

4. SULFONATION. The reaction of alkanes with sulfuric acid has acquired a new importance in organic chemistry with the development of synthetic detergents. The reaction of an alkane with sulfuric acid is similar to the reactions of nitration. The equation for the general reaction is written graphically as follows:

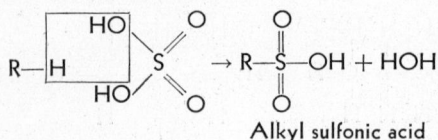

Alkyl sulfonic acid

If alkane homologues having more than ten carbon atoms form the alkyl part of the molecule, the alkyl sulfonic acid, when converted to the sodium salt with NaOH, has detergent properties. The molecular formula of such detergents is RSO_3Na. These salts are water soluble and the surface tension of the solution may be only one-third that of water. The solvent action of the alkyl part of the molecule and the decreased surface tension work effectively to remove dirt from surfaces. Hence the detergent action. (See p. 147.) These synthetic detergents are particularly more effective in hard water than soap, for soaps precipitate with the calcium and magnesium ions that cause water to be hard.

5. DECOMPOSITION. Thermal decomposition of alkanes has already been mentioned as a method for preparing alkanes from higher homologues in the gasoline cracking process. When larger alkane molecules are cracked into simpler compounds, the final product must contain at least one alkene and one alkane. In fact a great variety of products from hydrogen gas to free carbon may also be formed. Industrial procedures using high temperature and pressure have been developed by which the oil companies are able to secure a substantial yield of desirable products. When dodecane, for example, $C_{12}H_{26}$, is cracked, it may break at any point along the chain. The exact point of cracking is not accurately predictable, although at given temperatures and pressures the approximate percentage of various products from a given oil is known. The important thing to note is that both an alkane and an alkene are formed. Thus, if dodecane breaks between the fourth and fifth carbon atoms, the reaction would be:

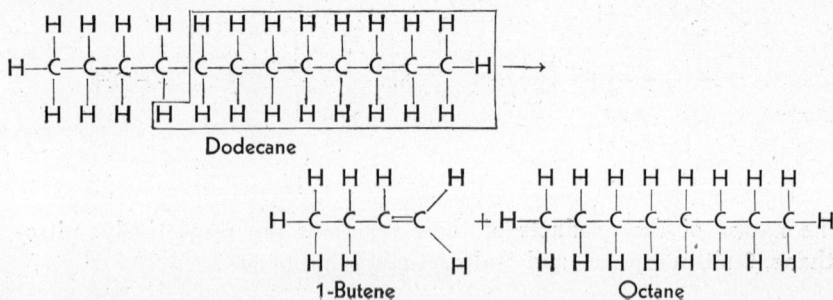

Dodecane

1-Butene   Octane

or

Dodecane

Butane + 1-Octene

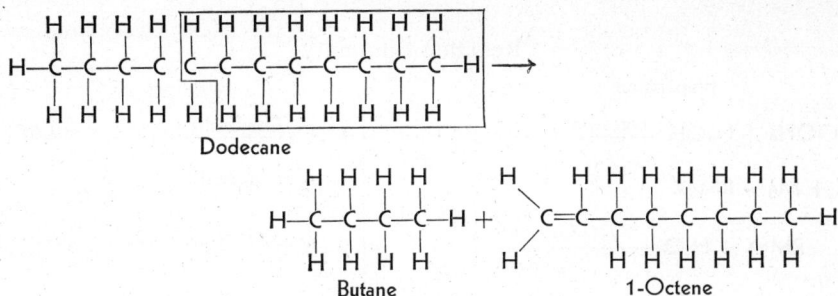

The compounds formed by the first cracking may undergo further cracking.

6. ISOMERIZATION. Some normal alkanes can be converted to an isomeric form by heating in the presence of a catalyst. One reaction of this type used by the gasoline industry has become very important in the manufacture of 2,2,4-trimethylpentane. This product can be made

by combining 2-methylpropane $CH_3\overset{\overset{\displaystyle CH_3}{|}}{C}HCH_3$ and 2-methylpropene

$CH_3\overset{\overset{\displaystyle CH_3}{|}}{C}{=}CH_2$ at room temperature in the presence of sulfuric acid. The two starting hydrocarbons, however, do not occur as abundantly in nature as normal butane. Moreover, n-butane and n-butene are products from cracked hydrocarbons, regular products of the cracking industry. It is possible to make 2-methylpropane (isobutane) from n-butane by heating the latter in the presence of aluminum chloride or aluminum bromide at a temperature between 100°C and 200°C. Not all the n-butane is converted to the isomer but the yield is sufficient to make this an important commercial reaction. To represent this reaction graphically the symbol Δ (Greek capital letter "delta") is written above the reaction arrow to indicate the addition of heat and the molecular formula of the catalyst is written either above or below the reaction arrow, as shown below:

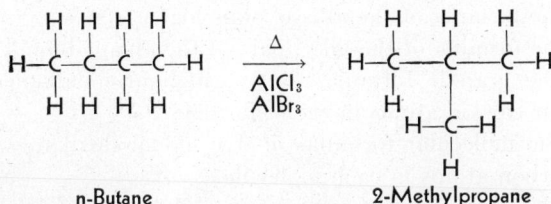

n-Butane   2-Methylpropane

Similarly, other normal alkanes will isomerize when heated with a catalyst but isomers formed by this reaction will revert in large part to the normal compound unless proper heat control is maintained.

Reaction Summary

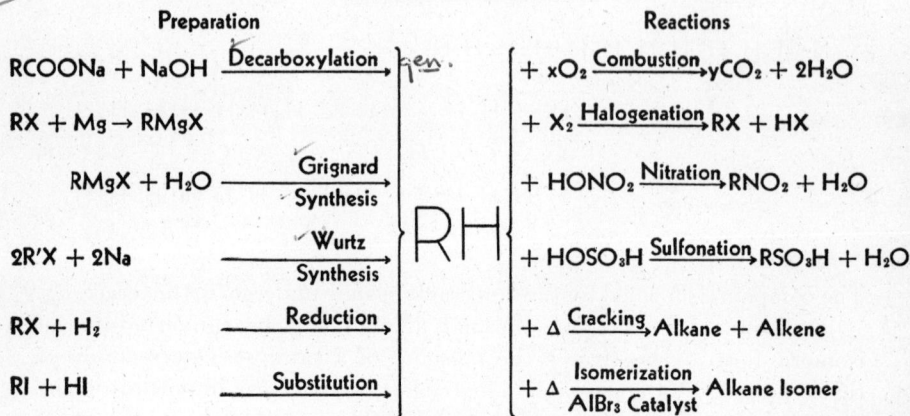

Preparation

$RCOONa + NaOH \xrightarrow{\text{Decarboxylation}}$ *gen.*

$RX + Mg \rightarrow RMgX$

$RMgX + H_2O \xrightarrow[\text{Synthesis}]{\text{Grignard}}$

$2R'X + 2Na \xrightarrow[\text{Synthesis}]{\text{Wurtz}}$

$RX + H_2 \xrightarrow{\text{Reduction}}$

$RI + HI \xrightarrow{\text{Substitution}}$

$\Big\}RH\Big\{$

Reactions

$+ xO_2 \xrightarrow{\text{Combustion}} yCO_2 + 2H_2O$

$+ X_2 \xrightarrow{\text{Halogenation}} RX + HX$

$+ HONO_2 \xrightarrow{\text{Nitration}} RNO_2 + H_2O$

$+ HOSO_3H \xrightarrow{\text{Sulfonation}} RSO_3H + H_2O$

$+ \Delta \xrightarrow{\text{Cracking}} \text{Alkane} + \text{Alkene}$

$+ \Delta \xrightarrow[\text{AlBr}_3 \text{ Catalyst}]{\text{Isomerization}} \text{Alkane Isomer}$

Specific Reactions

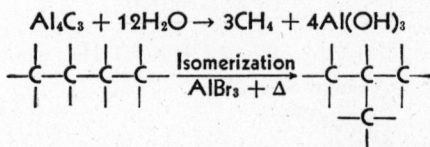

$$Al_4C_3 + 12H_2O \rightarrow 3CH_4 + 4Al(OH)_3$$

$$-\overset{|}{\underset{|}{C}}-\overset{|}{\underset{|}{C}}-\overset{|}{\underset{|}{C}}-\overset{|}{\underset{|}{C}}- \xrightarrow[\text{AlBr}_3 + \Delta]{\text{Isomerization}} -\overset{|}{\underset{|}{C}}-\overset{|}{\underset{|}{C}}-\overset{|}{\underset{|}{C}}-$$

Reaction summary charts are designed to provide the student with (1) a logical means of visualizing and remembering reactions; (2) a basis of study and review; (3) easy reference material on the preparation and reactions of the principal classes of compounds. The general reaction for the class of compound being studied is given first; important special reactions of specific compounds follow the general reactions. Only the principal reaction product is shown; the student should write out each reaction in complete form showing all products. It must be remembered that not all of the class reactions work efficiently. Constant cross-reference to text material is recommended during study and review.

Study and Review Questions

1. Express clearly in writing the meaning of the following terms: (a) Cyclic Hydrocarbons, (b) Aliphatic Hydrocarbons.
2. What is the etymological meaning of the word *paraffin?*
3. Write the formula of the alkane homologous series.
4. Write the formula of the alkyl part of the alkane homologous series.
5. Write the graphic formulas of five monobromoalkanes with more than four carbon atoms in each molecule.
6. Write the molecular formulas of five monoiodoalkanes with five or more carbon atoms in each molecule.
7. Write the correct I.U.C. name of each compound whose skeletal formulas are as follows:

Compounds & formulae

handwritten: Compounds & formulae

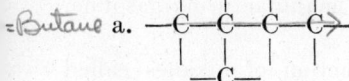

=Butane a. —C—C—C—C→

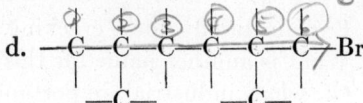
6 C = Hexane

d. —C—C—C—C—C—C→Br

2,5-dimethylhexane

methylbutane

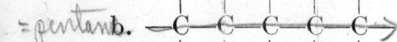

=pentane b. —C—C—C—C—C→

9 C = nonane

e.

4,5,7-Trimethylnonane

Tetra
2,4,4-Trimethylpentane

= Heptane c. —C—C—C—C—C—

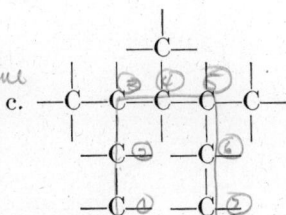

9 C - nonane

f. —C—C—C—C—C—C—C—

5 Trimethylheptane

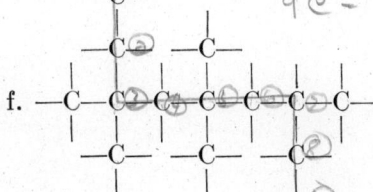

3,3,5,5,7 - Pentamethylnonane

8. Write the correct structural formula of each compound listed in question 7.

9. Write the skeletal formulas and give the correct I.U.C. names of all isomers of C_6H_{14}, C_7H_{16}, and C_8H_{18}.

10. Using a proper Grignard reagent, write a reaction for preparing n-pentane. (normal pentane).

11. Selecting your own reagents other than HI, write a reaction for preparing 2,2,5,5-tetramethylhexane.

12. With graphic formulas, write the reaction for preparing propane, using HI as one of the reagents.

13. Write the graphic and the structural formulas of the following compounds:

a. 2,2,4-trimethylhexane.
b. 2,3-dimethyl-3-ethylpentane.
c. 1-bromo-2,2,3,3-tetramethylbutane.
d. 2,2,3,4,4-pentamethylpentane.
e. 3,3-diethylpentane.

14. Make a list showing the approximate boiling point range of four chief fractions obtained from distilling petroleum. Using the graph on p. 21, classify all normal alkanes as high as nonadecane in one of these fractions.

15. Explain briefly the "cracking process" for manufacturing gasoline. What is another name for this process?

16. Of what industrial importance is the reaction of alkanes called "isomerization"?

17. List four halogens in the order of their readiness to react with alkanes.

18. In question 13, designate the type of each carbon atom correctly using a circle around all primary, a square around all secondary, and a triangle around all tertiary carbon atoms.

19. When the nitration of alkanes occurs, which type of carbon atom is more readily nitrated?

20. Explain briefly how detergents of the alkyl sulfonate type derive their cleansing quality.

21. Will halogenation readily occur by substitution of a halogen for any hydrogen atom in n-butane? Explain your answer briefly.

Chapter 3

COMPOUNDS OF CARBON WITH HYDROGEN
II. UNSATURATED HYDROCARBONS:
THE ALKENES AND ALKYNES

CHAPTER PROLOGUE

Thus far in our study of carbon compounds we have seen very little of the versatility of the carbon atom. Carbon atoms have been united only with other carbon atoms by a single bond and with hydrogen atoms. In this chapter carbon assumes a new role of far-reaching consequences. Like the versatile football player who can perform more than one task well on the field of contest, carbon is a triple-threat artist among the elements. When an insufficiency of electrons from other elements is present to form saturated compounds, carbon atoms may acquire the energy to unite with other carbon atoms by double bonding and even by triple bonding.

Triple-bonded acetylene is especially endowed with a marvelous versatility and vigor of combination. In its unique simplicity of structure this energy-packed molecule needs no other molecule to allure it into action. Merely the presence of a suitable companion catalyst will induce it to join hands with brother acetylene molecules often in the formation of a puzzling pattern of polymers. Yet in its activity acetylene is not only highly responsive; it is also highly selective in its associations. Like a dancer responding to the stimulus of various melodies, acetylene forms one type of polymer in the presence of cuprous ion and another in the presence of silver or of mercury.

In this chapter we shall study the configuration of these new types of molecules and try to understand the effect of these configurations upon the chemical activity of the molecules.

Alkenes

Names. The name *alkene* is used in referring to one class of unsaturated hydrocarbons in which one pair of carbon atoms is held together by a double bond. Late in the Eighteenth Century a group of Dutch artists used an oil made by chlorinating an unsaturated hydrocarbon. The unsaturated hydrocarbon was called an oil maker (Latin: *oleum*, oil, *fio*, make) or an olefiant. The name *olefin* is often used in referring to this type of compound, in contrast with the word *paraffin* used in referring to saturated hydrocarbons. Other names commonly used in referring to hydrocarbon homologues of this type are *ethene or ethylene series*. This name is taken from the simplest existing compound of the series, C_2H_4. For many decades the name ethylene has been used. In conformity with the request of the American Chemical Society the I.U.C. name, ethene, is now widely preferred. When larger hydrocarbon molecules have two pairs of carbon atoms, each pair united by a double bond, they are called *alkadienes*. When more than two pairs of carbon atoms are united each by double bonds, the compounds are called *alkapolyenes*. Diene and polyene compounds will be discussed briefly at the end of this section.

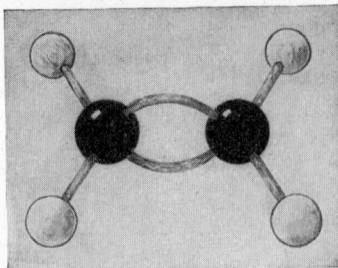

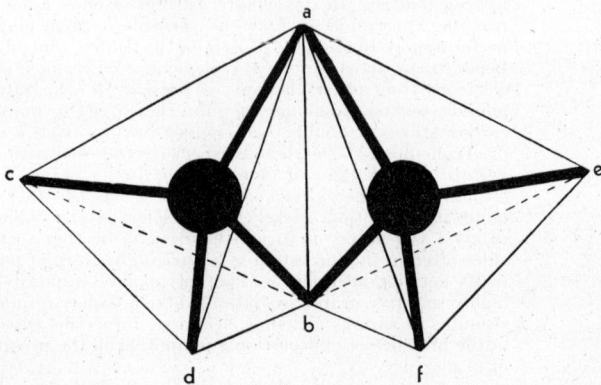

Fig. 11. (Top) The ethene molecule.
Fig. 12. (Bottom) Tetrahedral structure of ethene showing double bonds between carbon atoms.

Formula of Alkenes. Alkenes are represented by the type formula, C_nH_{2n}. As was pointed out on p. 5 in writing the electronic formula of alkenes, chemical symbols are used to represent the kernel of each atom and dots are used to symbolize electrons. Thus the electronic formula of

ethene is $\overset{\displaystyle H}{\underset{\displaystyle H}{\overset{..}{C}}} \overset{\displaystyle H}{\underset{\displaystyle H}{::\overset{..}{C}}}$. Graphically the formula of ethene has already

been shown (p. 6) written with dashes, each dash representing two mutually intershared electrons, $\overset{H}{\underset{}{C}} = \overset{H}{\underset{}{C}}$. The structural formula is CH_2CH_2

and the molecular formula is C_2H_4. Chemists often refer to the radical $=CH_2$ as methylene or methene, but it has never been isolated as a compound or as a free radical.

Structure and Bonding. In Chapter 1 on Carbon the geometrical structure of the carbon atom was given as a tetrahedron (p. 5). When saturated hydrocarbons are formed, each L-shell electron of the carbon atom is shared with a hydrogen atom or another carbon atom. When unsaturated hydrocarbons are formed one carbon atom may intershare two electrons with two electrons from its neighboring carbon atom. Diagrammatically the manner of sharing is represented in Figure 12.

Table 2

BOND ENERGIES IN KCAL/MOLE

Carbon to Carbon	Carbon and Nitrogen	Carbon and Oxygen	Carbon and Halogen	Hydrogen to:
C—C 58.6	C—N 49.6	C—O 70	C—I 45.5	H—I 71.4
C=C 100	C=N 94	C=O 149 (ald.)	C—Br 54.0	H—N 83.7
C≡C 123	C≡N 150	C=O 152 (ket.)	C—Cl 66.5	H—C 87.3
			C—F 107	H—Br 87.3
				H—Cl 102.7
	N—N 23.6	O—O 34.9	I—I 36.2	H—H 103.4
	N≡N 170	O=O 96	Br—Br 46.1	H—O 110.2
			Cl—Cl 57.8	H—F 147.5
			F—F 63.5	

Bond energies measure the amount of energy in kcal/mole required to dissociate the covalent bonding shown. These values are given by Pauling.

Although the scope of this text does not permit a comprehensive consideration of the physical-chemical significance of bond energies, their basic use in interpreting reactions will be clear to the student from the following example:

to dissociate a C—C linkage requires 58.6 kcal/mole.
to dissociate a C=C linkage requires 100 kcal/mole.

Thus the energy of the second bond $100 - 58.6 = 41.4$ kcal/mole as compared with 58.6 kcal/mole for the first bond. To break the third electron pair, only $123 - 100 = 23$ kcal/mole are required. Hence the order of activity among hydrocarbons is

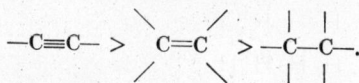

$$-C\equiv C- > \ \diagdown C=C\diagup \ > -\overset{|}{C}-\overset{|}{C}-.$$

When two adjoining carbon atoms are united at a and b, as shown in Figure 12, the nuclei of such atoms are brought closer together. In alkane compounds the nuclei of adjoining carbon atoms are 1.54 Å from each other. In alkenes the nuclei of adjoining double-bonded atoms are only 1.33 Å apart. By the very structure of an alkene molecule the stability of the compound is somewhat lessened and its reactive ability is increased. The two nuclei with their positive charges (mutual repelling forces) are brought closer together. Additional energy is required to hold the carbon nuclei in their closer mutual relationship, an energy at least equal to the repelling force of their nuclei. The bonding force of mutually inter-shared electrons plus this additional energy is sufficient to hold the molecule together. The compound formed is relatively stable. Examination of the data in Table 2 will show the extent to which the forces of nuclear repulsion have changed the bond energies of the intershared electrons. The result is that the alkene is far more active chemically than its alkane counterpart. Under favorable conditions the double bond is easily opened at one of its points of union, especially when one electron from each double-bonded carbon atom can be shared with an electron from two other atoms or radicals. The reactions on pp. 43 and 44 are of this type. It follows that kernel repulsion, mutual repelling force, of carbon atoms accounts for this greater reactivity.

Nomenclature. Ethene is the I.U.C. name for the simplest of the alkenes. All names of compounds in this series are derived by adding the suffix-*ene* to the root name of the corresponding alkane. The suffix-*ene*, according to the Geneva convention, designates double bonding once within the molecule. The names and graphic formulas of normal alkenes containing up to 10 carbon atoms are given in the following list:

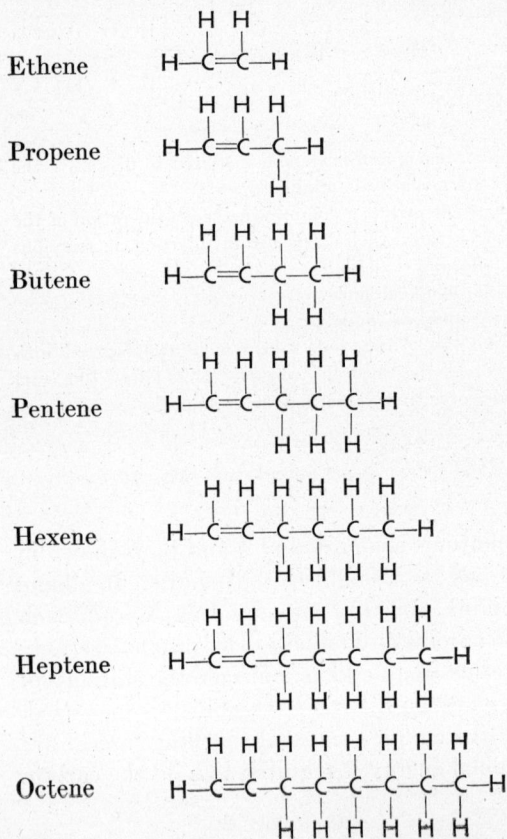

Ethene

Propene

Butene

Pentene

Hexene

Heptene

Octene

Nonene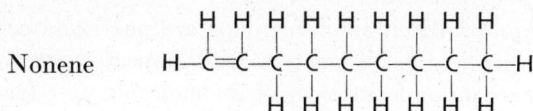

Decene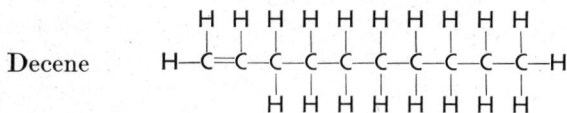

In the case of ethene and propene only one compound of each kind exists. By placing the double bond between different pairs of carbon atoms in butene, three butene isomers become apparent.

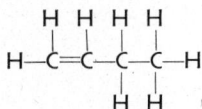

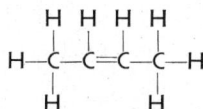

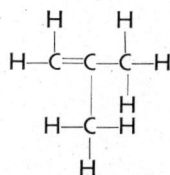

1-Butene   2-Butene   2-Methyl-1-propene

In the case of pentene there are five possible isomers[1] apparent.

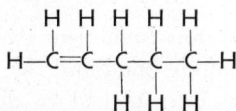

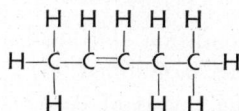

1-Pentene   2-Pentene

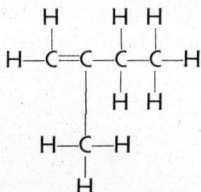

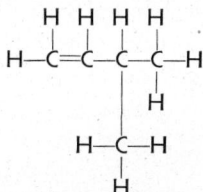

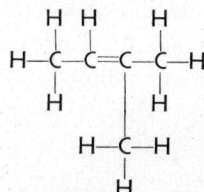

2-Methyl-butene   3-Methyl-1-butene   2-Methyl-2-butene

The manner of distinguishing one isomer from another is by using a number signifying the double-bonded carbon atom closest to either end of the chain. All other groups must be numbered from the chosen end of the molecule, and the number kept as low as possible. With the increase in the number of carbon atoms in a molecule, the number of isomers of alkenes increases much more rapidly than in the case of the alkanes. A comparative table showing possible isomers of alkanes and alkenes is shown in Table 3.

Cis-trans-, A Type of Geometric Isomerism. Alkene compounds show a further type of isomerism based upon the position of various groups in space with respect to the double bond. This kind of isomerism is called cis-trans-isomerism, depending upon which side of the molecule

[1] In addition 2-butene and 2-pentene each have two possible spacial structures accounted for by geometric isomerism also called cis-trans-isomerism.

carries the same kind of group. When similar groups are on the same side of the molecule, they are in the cis-position with respect to each other; when similar groups are on opposite sides of the molecule we refer

Table 3

COMPARISON OF ALKANE AND ALKENE
ISOMERS

Number of Carbon Atoms	Alkane Isomers	Alkene Nuclear Isomers
1	1	
2	1	1
3	1	1
4	2	3
5	3	5
6	5	13
7	9	27
8	18	66
9	35	153
10	75	377

to the condition as trans-isomerism. For cis-trans-isomerism two conditions must be fulfilled: (a) there must be a double bond and (b) the two remaining bonds of each carbon atom must be attached to dissimilar groups. In Figure 12 (p. 36) it will be noted that there is no freedom of rotation about the double bond. Hence the four groups, c, d, e, and f, retain fixed relationship to each other. The cis- and trans-isomers of 2-butene follow:

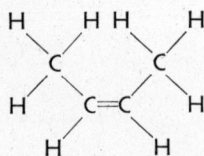

Cis-2-butene
(—CH₃ groups on same side
of the molecule)

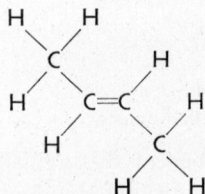

Trans-2-butene
(—CH₃ groups on opposite
sides of the molecule)

Physical Properties of Alkenes. The alkenes are colorless and practically odorless when pure. A comparison of the boiling and of the melting points of alkanes and alkenes shows, with few exceptions, only slight differences between corresponding compounds. All alkenes are practically insoluble in water and have a density less than that of water. Butene and lower alkenes are gases at room temperatures. From pentene through octadecene (C_5—C_{18}) the alkenes are liquids. Alkenes with more than 18 carbon atoms are solids.

Preparation. Alkenes are usually made from saturated alkane derivatives by removing one atom or group from each of two adjoining carbon atoms in the compound. Selected reagents and heat are generally required. Six of the more fundamental methods of preparing ethene or other alkenes are as follows: (1) dehydrogenation, (2) dealkylation or the "cracking" process, (3) dehydration of an alkanol, (4) removal of a hydrohalide, (5) dehalogenation, and (6) partial hydrogenation of an alkyne.

1. DEHYDROGENATION. When ethane gas is heated to 700°C one hydrogen atom from each of the two adjoining carbon atoms splits off and forms a molecule of hydrogen and a molecule of ethene.

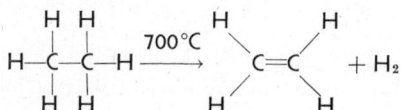

Other alkanes will likewise form alkenes under similar conditions but the point or points where dehydrogenation will take place cannot be predicted and a mixture of alkenes will result.

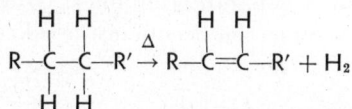

2. DEALKYLATION. When higher alkanes are heated under pressure, the larger molecules are cracked in the process, resulting in the formation of one alkane and one or more alkenes. This method and the uncertainty of producing specific alkenes has been discussed in Chapter 2 on Alkanes (p. 30).

3. DEHYDRATION OF AN ALKANOL. The commercial method of making specific alkenes is to pass a proper hot alcohol over heated alumina (Al_2O_3). For laboratory preparations of specific alkenes, H_2SO_4, H_3PO_4, and $AlPO_4$ are used as catalysts instead of Al_2O_3. When secondary alcohols are used, that is alcohols whose —OH group is on a secondary carbon atom, and when the group is at least three carbon atoms from the end of the chain, the trans-isomer is the predominant alkene formed.

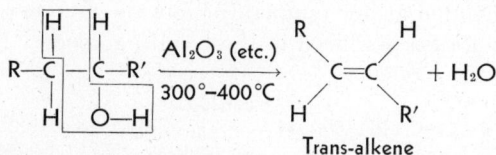

Trans-alkene

Control of the reacting temperature in the laboratory is essential for producing alkenes. When sulfuric acid is used as a catalyst in making ethene from ethanol (ethyl alcohol), an intermediate product (an inorganic ester) is formed at the lower temperatures. At temperatures greater than 160°C, the ester breaks down to form ethene and sulfuric acid.

4. REMOVAL OF HYDROHALIDE. Hydrohalides can be removed only when adjacent carbon atoms have respectively at least one hydrogen and one halogen atom. Strong alkali in alcoholic solution is used to insure the absence of water. Potassium hydroxide is favored because of its greater solubility in alcohol. The end product is the alkene together with potassium halide and water. Since the lower alkenes are gases, they are easily collected in a separate container. When proper secondary halides are used, trans-isomerism occurs in the alkene produced.

5. DEHALOGENATION. The halogens from certain dihaloalkanes, in which each halogen atom is attached to adjacent carbon atoms, can be removed by heating the compound with zinc dust. The cis-isomer is predominant in the final product.

6. PARTIAL HYDROGENATION OF AN ALKYNE. When an alkyne is partially hydrogenated at low temperatures in the presence of a catalyst, the cis-isomer is *always* produced in the resulting alkene.

Reactions of Alkenes. Five types of reactions will be considered for alkenes: (1) Reductive addition, (2) Oxidative addition, (3) Addition, (4) Alkylation, and (5) Polymerization.

1. Reductive Addition. The reduction reaction is really an addition reaction but it is placed here in contrast with the oxidation reaction which immediately follows. In the presence of suitable catalysts, hydrogen will add directly to alkenes to form alkanes. Platinum, palladium, and finely divided nickel are the catalysts ordinarily used. This reaction has attained commercial importance in the hydrogenation of certain oils in the manufacture of vegetable shortening and similar products.

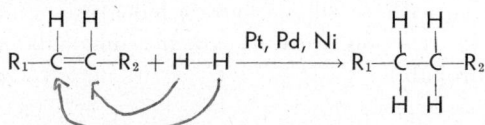

$$R_1-\overset{\overset{\displaystyle H}{|}}{C}=\overset{\overset{\displaystyle H}{|}}{C}-R_2 + H-H \xrightarrow{\text{Pt, Pd, Ni}} R_1-\overset{\overset{\displaystyle H}{|}}{\underset{\underset{\displaystyle H}{|}}{C}}-\overset{\overset{\displaystyle H}{|}}{\underset{\underset{\displaystyle H}{|}}{C}}-R_2$$

2. Oxidative Addition. Alkenes may be oxidized by potassium permanganate, $KMnO_4$ in basic solution. Because of the loss of the char-

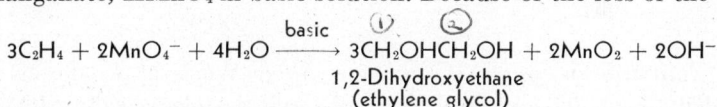

$$3C_2H_4 + 2MnO_4^- + 4H_2O \xrightarrow{\text{basic}} 3CH_2OHCH_2OH + 2MnO_2 + 2OH^-$$

1,2-Dihydroxyethane
(ethylene glycol)

acteristic purple permanganate color, this reaction is used as a general test for unsaturated hydrocarbons. It is called the *Bayer Test*.

3. Addition. In discussing the structure of alkenes earlier in this chapter, special emphasis was given to the ability of alkene to break the double bond and to add directly certain diatomic molecules. Since this may occur in the case of molecules made up of different kinds of atoms, the question arises: do alkenes show a selective ability in the placement by addition of certain atoms within the new molecule? Markownikoff has formulated a rule bearing his name for this type of addition, where one of the atoms of the added molecule is hydrogen. About a decade ago the original Markownikoff rule was modified when it was found that the presence of oxidizing agents reversed the results. The modified *Markownikoff Rule* may now be stated as: *In the absence of an oxidizing agent, when a substance capable of adding to an unsaturated hydrocarbon breaks the double bond, the more positive part of the added compound goes to the double-bonded carbon atom having the greater number of hydrogen atoms. When an oxidizing agent is present, the manner of combining is reversed.* This rule applies only to unsaturated hydrocarbons.

A. halogen acid or hydrohalide, hx. Markownikoff's rule applies as shown in the two reactions below. This reaction is especially advantageous in the exact placement of a halogen on a specific carbon atom.

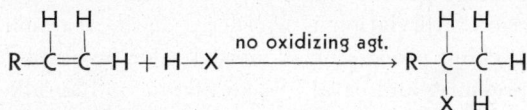

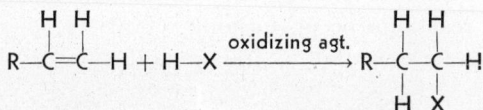

$$R-\overset{\overset{\displaystyle H}{|}}{C}=\overset{\overset{\displaystyle H}{|}}{C}-H + H-X \xrightarrow{\text{no oxidizing agt.}} R-\overset{\overset{\displaystyle H}{|}}{\underset{\underset{\displaystyle X}{|}}{C}}-\overset{\overset{\displaystyle H}{|}}{\underset{\underset{\displaystyle H}{|}}{C}}-H$$

$$R-\overset{\overset{\displaystyle H}{|}}{C}=\overset{\overset{\displaystyle H}{|}}{C}-H + H-X \xrightarrow{\text{oxidizing agt.}} R-\overset{\overset{\displaystyle H}{|}}{\underset{\underset{\displaystyle H}{|}}{C}}-\overset{\overset{\displaystyle H}{|}}{\underset{\underset{\displaystyle X}{|}}{C}}-H$$

B. OXIDATIVE ADDITION OF HALOGEN, X—X. In the addition of halogens directly to alkenes, chlorine is more active than bromine and bromine is more active than iodine. (See Bond Energies, p. 37.) The reaction takes place more rapidly in the presence of sunlight. The reaction of bromine dissolved in carbon tetrachloride is used as a test for unsaturated bonds. The brown color of bromine fades quickly when an alkene is tested with this solution. The reaction differs from the reaction of a halogen with an alkane. In reaction with an alkane, both a haloalkane and a hydrohalide are produced. In reaction with an alkene, dihaloalkane is produced without any hydrohalide.

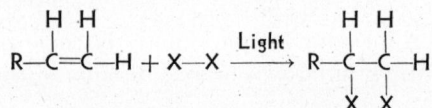

$$R-\overset{\overset{\displaystyle H}{|}}{C}=\overset{\overset{\displaystyle H}{|}}{C}-H + X-X \xrightarrow{\text{Light}} R-\overset{\overset{\displaystyle H}{|}}{\underset{\underset{\displaystyle X}{|}}{C}}-\overset{\overset{\displaystyle H}{|}}{\underset{\underset{\displaystyle X}{|}}{C}}-H$$

C. ADDITION OF HYPOHALOUS ACID, HOX. When hypochlorous acid is added to an alkene, Markownikoff's rule is followed. In HOX the halogen is the more positive part, HO^-X^+. With sodium bicarbonate,

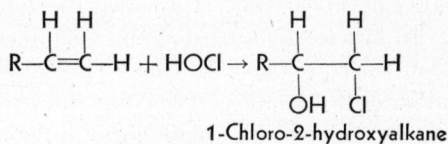

$$R-\overset{\overset{\displaystyle H}{|}}{C}=\overset{\overset{\displaystyle H}{|}}{C}-H + HOCl \rightarrow R-\overset{\overset{\displaystyle H}{|}}{\underset{\underset{\displaystyle OH}{|}}{C}}-\overset{\overset{\displaystyle H}{|}}{\underset{\underset{\displaystyle Cl}{|}}{C}}-H$$

1-Chloro-2-hydroxyalkane

1-chloro-2-hydroxyethane will produce 1,2-dihydroxyethane.

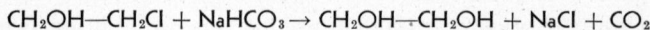

$$CH_2OH-CH_2Cl + NaHCO_3 \rightarrow CH_2OH-CH_2OH + NaCl + CO_2$$

D. ADDITION OF WATER, HOH. Water may be added to ethene indirectly by first forming an ester with sulfuric acid and then hydrolyzing to yield ethanol and sulfuric acid. This method is used in making alcohols from the unsaturated hydrocarbons obtained from petroleum. Some ethoxy-

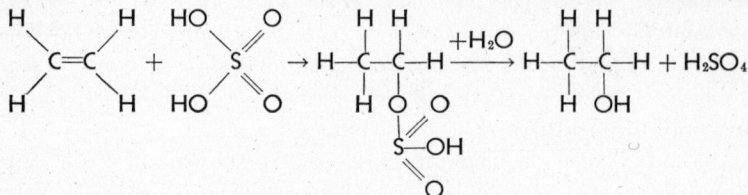

[reaction diagram: ethene + sulfuric acid → ethyl hydrogen sulfate $\xrightarrow{+H_2O}$ ethanol + H_2SO_4]

ethane (ether) is also produced by this reaction.

4. ALKYLATION. Alkylation is a reverse of the reaction which takes place in the cracking process of alkanes. Under favorable conditions of temperature, pressure, and catalysis, an alkane and an alkene will unite to form a higher alkane. A single final product is not formed by such reactions. Side reactions occur forming a variety of saturated compounds which can be separated by fractional distillation. It is interesting

to note that the alkylation process occurs at lower temperatures than the reverse process of cracking.

The following typical reaction is used by the gasoline industry for the manufacture of a high octane gasoline (p. 26).

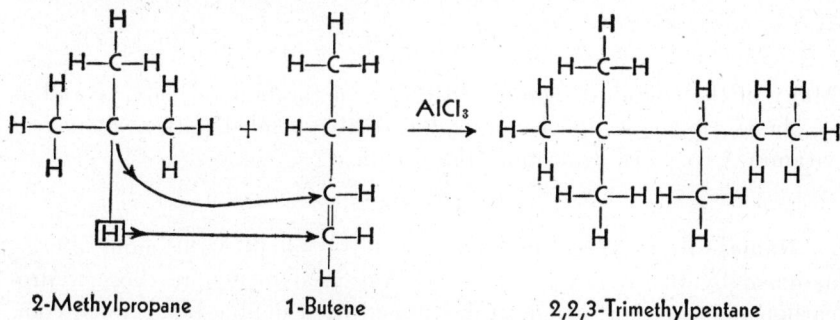

2-Methylpropane   1-Butene   2,2,3-Trimethylpentane

5. POLYMERIZATION. Under favorable conditions of temperature, pressure, and catalysis, alkene molecules will unite with one or several other alkene molecules of the same kind to form larger complex molecules. The role of polymerization in modern industry has attained a new importance with the manufacture of synthetic rubber, plastics, and similar synthetic compounds. Ethene does not polymerize easily; higher alkenes polymerize more readily.

DIENES

Alkene compounds are more active chemically than alkanes. So also dienes are more active than alkenes. Dienes are open-chain hydrocarbon compounds in which two pairs of carbon atoms in the molecule are united by double bonds. Two double bonds in a molecule are therefore characteristic of dienes. In naming dienes, numbers are used to designate the carbon atoms attached by double bonds, always keeping the numbers as low as possible. The compound

$$\begin{array}{cccc} H & H & H & H \\ C{=}C{-}C{=}C \\ H & & H & \end{array}$$

is named 1,3-butadiene.

The type formula of dienes is C_nH_{2n-2}. Dienes with double-bonded carbon atoms at the end of each molecule are especially useful in the polymerization of large molecules. Butadiene is the starting compound for certain processes in the synthetic rubber industry. Polymerization occurs according to the reaction:

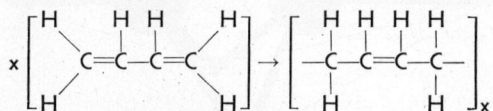

POLYENES

When a compound made up solely of hydrocarbons has more than two double-bonded pairs of carbon atoms, it is called a polyene. If the double bonds occur alternately with single bonds throughout the length of the molecule, the bonding is called *conjugate double bonding*. An example of conjugate double bonding is shown in the following skeletal formula:

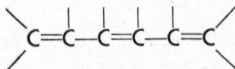

$$\diagdown C=C-C=C-C=C\diagdown$$

Many of the complex organic substances of the human body as well as of plants are made up of conjugated double-bonded polyenes, such as vitamin A (p. 175), β-carotene and others.

Alkynes

Names. In the preceding section of this chapter the name *alkene*, assigned by the Geneva convention was used to denote those hydrocarbons having two carbon atoms joined by a double bond. Such compounds are called unsaturated. When the degree of unsaturation is increased to a triple bond between two carbon atoms, the compounds are called *alkynes*. Acetylene is the simplest member of this homologous series and the name "acetylenes" is often used in referring to compounds having one or more triple bonds. In strict nomenclature, however, compounds having two triple linkages are known as *alkadiynes*.

Formula of Alkynes. The electronic formula of ethyne (acetylene) is $H:C::C:H$. Following the general principle of representing each pair of dots in the electronic formula by a line or dash, the graphic formula of ethyne is $H-C \equiv C-H$. The molecular formula is C_2H_2. The empirical formula $\equiv CH$ symbolizes methyne which like its alkene prototype methene ($=CH_2$) does not exist alone. All alkynes correspond to the

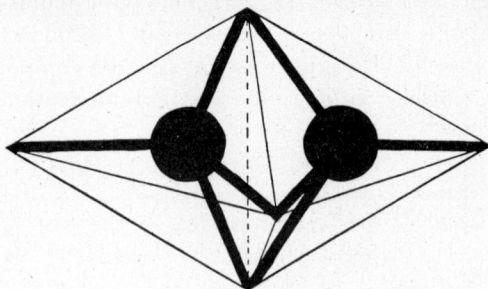

Fig. 13. (Top) The acetylene molecule. (Left) Showing bonds; (right) showing steric proportion.

Fig. 14. (Bottom) Tetrahedral structure of acetylene showing triple bonds between carbon atoms.

type formula C_nH_{2n-2}, but students must remember that alkadienes have this same type formula.

Structure and Bonding. In Chapter 2 on Alkanes, the geometric configuration of the carbon atom was shown to be tetrahedral. It also was shown in the section on Alkenes that two such tetrahedral carbon atoms could be united by the mutual intersharing of two electrons from each carbon atom. In support of this type of carbon union, the distance between the carbon nuclei in an alkene was found to be 1.33 Å, and the resulting configuration of the molecule appears as if one edge from each adjacent tetrahedron coincided. Moreover the nuclei of the carbon atoms and other attached atoms are found to exist in a single plane of space.

Similarly, when two adjacent carbon atoms are united by the mutual intersharing of three electrons from each carbon atom, the nuclei of the carbon atoms are brought even closer together than in the case of the alkenes. The distance between the carbon nuclei in an alkyne is only 1.205 Å. Geometrically speaking, the molecule is formed by the congruent superposition of one face of a carbon tetrahedron upon a face of the tetrahedron of its neighboring carbon atom. One result of such superposition is the arrangement of the carbon nuclei and the nuclei of the attached atoms in one straight line. Literally the molecule of ethyne is linear in shape, whereas the shape of the ethene molecule is planar, and that of ethane is three dimensional. Reference to Table 2, p. 37, will show that kernel repulsion has further changed the bond energies for this configuration. Hence the reactivity of acetylene is even greater than that of ethene.

When acetylene burns, its flame is hotter than that of ethene and therefore also than that of ethane. Ethyne has a higher energy content than either of its similar more saturated compounds. This excess energy is required to hold the molecule together. When neighboring nuclei are brought closer together the repellent force of one positively charged nucleus upon the other is increased. Energy is required to overcome this mutual repellent force and the energy must be equal to the repellent force if the molecule is to exist. In the case of ethyne, moreover, the effect of this increased proximity of the positively charged carbon nuclei causes the hydrogen atoms to be more labile, resulting in a noticeable ability of the hydrogen atoms to ionize. Thus the very structure of ethyne makes it an extremely active compound and the ionic nature of the molecule permits the formation of direct carbon linkages with metals.

Nomenclature. In naming alkynes all the previously given rules of nomenclature apply. The position of a triple bond is designated by a number only when some possible isomer exists. Thus ethyne and propyne require no numerical designation but the isomeric butynes, 1-butyne ($H-C\equiv C-CH_2-CH_3$) and 2-butyne ($CH_3-C\equiv C-CH_3$) are differentiated as shown. In compounds having both a double and a triple bond, the alkene name is usually given first with a number to locate the position of the double bond, followed by a number with the designation

"-yne" to mark the position of the triple bond, thus, vinyl-acetylene, $CH_2=CH-C\equiv C-H$, is named 1-butene-3-yne.

Isomerism and Resonance. Alkynes form isomers in the same way as the alkanes and alkenes by the rearrangement of groups within the molecule and by the change of position of the various kinds of bonds. The triple bond nature of alkynes permits another type of isomerism whereby a triple bond and a single bond become a pair of double bonds in an isomeric compound. Thus an alkyne is an isomer of an alkadiene. Both 1-butyne and 2-butyne are isomers of butadiene. The stability of many alkynes is attributable to this capability of existing in different electron structures and is explained by the same theory of resonance used in predicting the stability of ions in inorganic chemistry.

Physical Properties. Acetylene, propyne and 1-butyne are gases at room temperature ($20°C$); 2-butyne and higher homologues to octyne are liquids. Very little information is available on alkynes beyond octyne.

Acetylene has been known for more than a century. It is a colorless gas, very soluble in acetone. The solubility of acetylene in acetone is so very great (25 volumes to 1 at 1 atmosphere pressure and 300 volumes to 1 at 12 atmospheres pressure) that the gas for commercial purposes is sold in steel cylinders as an acetone solution. When the gas is dissolved under pressure in a cylinder packed with a porous substance saturated with acetone, it may be safely handled and transported. When prepared for commercial use, acetylene has a pungent odor somewhat like that of garlic but in the pure state it has a sweet ethereal odor and possesses anesthetic properties. In both the gaseous and liquid phases (B.P. $-85°C$) with oxidizing agents, acetylene is highly explosive.

Preparation of Alkynes. Until the accidental discovery of calcium carbide in 1892, acetylene and other alkynes were regarded as laboratory curiosities.

a. Early methods of preparation involved the removal of halogens from 1,1,2,2, tetrahaloalkanes with zinc dust in alcohol.

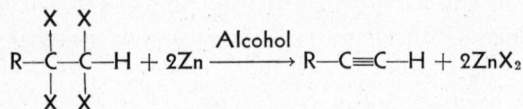

$$R-\underset{\underset{X}{|}}{\overset{\overset{X}{|}}{C}}-\underset{\underset{X}{|}}{\overset{\overset{X}{|}}{C}}-H + 2Zn \xrightarrow{\text{Alcohol}} R-C\equiv C-H + 2ZnX_2$$

b. Alkynes may be formed by the removal of two moles of hydrohalide from selected dihaloalkanes by an alcoholic solution of potassium hydroxide.

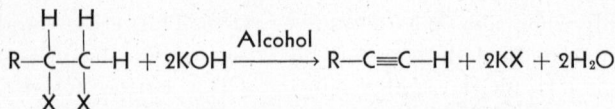

$$R-\underset{\underset{X}{|}}{\overset{\overset{H}{|}}{C}}-\underset{\underset{X}{|}}{\overset{\overset{H}{|}}{C}}-H + 2KOH \xrightarrow{\text{Alcohol}} R-C\equiv C-H + 2KX + 2H_2O$$

c. Two moles of hydrohalide may be removed from a proper dihaloalkane by sodamide in liquid ammonia.

$$\underset{\underset{X}{\overset{H}{|}}\ \underset{X}{\overset{H}{|}}}{R-C-C-H} + 2NaNH_2 \xrightarrow[\text{(liquid)}]{NH_3} R-C\equiv CH + 2NaX + 2NH_3$$

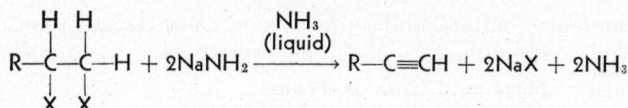

d. Since the discovery of calcium carbide, acetylene is readily prepared by hydrolysis of CaC_2. Higher alkynes may be made from acetylene

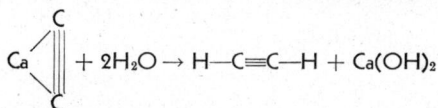

$$Ca\overset{C}{\underset{C}{\lVert}} + 2H_2O \rightarrow H-C\equiv C-H + Ca(OH)_2$$

by double decomposition of metallic acetylides with suitable haloalkanes.

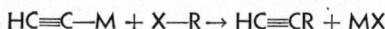

$$HC\equiv C-M + X-R \rightarrow HC\equiv CR + MX$$

e. Acetylene may also be prepared by the action of sodium metal upon trihaloalkanes. It will be noted that this reaction is a special case of the Wurtz synthesis.

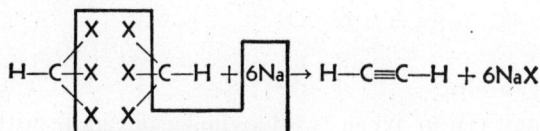

$$H-C\underset{X\ X}{\overset{X\ X}{|}}X\ \ X|C-H + 6Na \rightarrow H-C\equiv C-H + 6NaX$$

f. Another method of preparing acetylene is by the catalytic cracking of methane. Large reserves of methane are available from natural gas. When a commercial production method is perfected, a lower cost of acetylene may permit its more extensive use in many competitive organic synthetic processes.

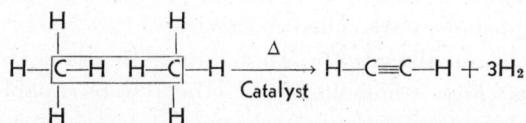

$$H-\underset{\underset{H}{|}}{\overset{\overset{H}{|}}{C}}-H\ \ H-\underset{\underset{H}{|}}{\overset{\overset{H}{|}}{C}}-H \xrightarrow[\text{Catalyst}]{\Delta} H-C\equiv C-H + 3H_2$$

Reactions of Alkynes. The reactions of alkynes most often listed in textbooks are those of acetylene. Higher homologues are usually made by synthesis from acetylene. Three types of acetylene reactions are (1) addition, (2) polymerization, and (3) substitution.

1. ADDITION. Hydrogen, halogens, and hydrohalides are directly added by the acetylene molecule. Further additions to the alkene compounds will occur in accord with the rules for adding to alkenes.

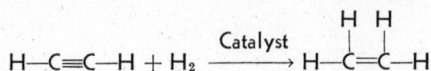

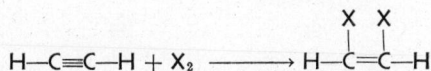

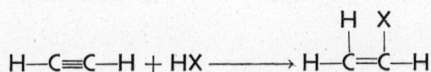

$$H-C\equiv C-H + H_2 \xrightarrow{\text{Catalyst}} H-\underset{\underset{}{\overset{H}{|}}}{C}=\underset{\underset{}{\overset{H}{|}}}{C}-H$$

$$H-C\equiv C-H + X_2 \longrightarrow H-\underset{\underset{}{\overset{X}{|}}}{C}=\underset{\underset{}{\overset{X}{|}}}{C}-H$$

$$H-C\equiv C-H + HX \longrightarrow H-\underset{\underset{}{\overset{H}{|}}}{C}=\underset{\underset{}{\overset{X}{|}}}{C}-H$$

With mercuric sulfate and sulfuric acid as catalysts, acetylene will add water directly to form ethanal (acetaldehyde), an intermediate in manufacturing acetic acid from acetylene.

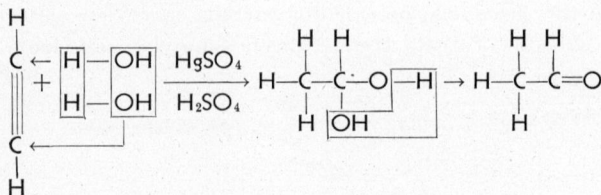

In the presence of aluminum chloride as a catalyst, acetylene and arsenous chloride form Lewisite, one of the most feared agents in chemical warfare.

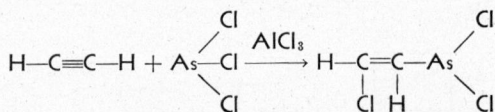

2. Polymerization. When two like molecules unite with each other to form a compound, the resulting compound is called a *dimer.* Similarly, three like molecules may form a *trimer.* When many like molecules form a compound, the resulting compound is a *polymer.* The word *polymerization* has acquired even a broader meaning. In the manufacture of plastics a new molecule may be formed by a merging reaction between two different kinds of molecules and the resulting product will polymerize. The two original substances are called *copolymers.*

In the section on alkenes it was noted that ethene does not polymerize easily, but that higher homologues of ethene were capable of polymerization. Acetylene with the aid of selected catalysts polymerizes readily.

With iron as a catalyst acetylene will polymerize at higher temperatures (600–700°C) to form benzene. This reaction serves as a link between open-chain and cyclic hydrocarbons.

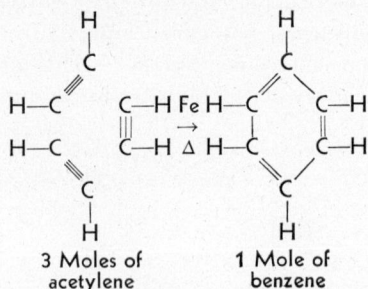

3 Moles of acetylene   1 Mole of benzene

Some catalysts permit controlled polymerization, notably cuprous chloride in ammoniacal solution. With it acetylene forms the dimer, 1-butene-3-yne (vinyl acetylene) and the reaction can be controlled without the formation of much 1,5-hexadiene-3-yne (divinyl acetylene). Vinyl acetylene is an important commercial compound capable of adding hydrogen chloride to form 2-chloro-1,3-butadiene (chloroprene). Chloroprene molecules when exposed to air polymerize into neoprene, a substance having many properties like natural rubber. Substances such as neoprene are often called synthetic rubbers but are more properly known as *elastomers*. Neoprene is a particularly valuable synthetic rubber because of its resistance to decomposition by hydrocarbons.

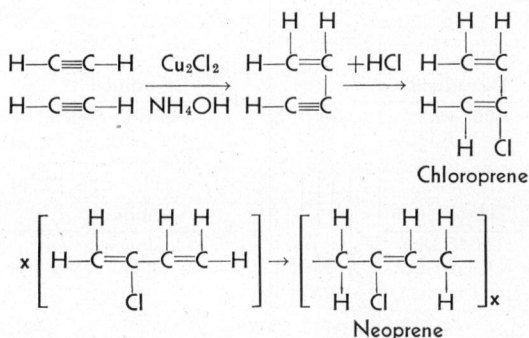

Chloroprene

Neoprene

3. SUBSTITUTION. Acetylene is peculiar among the hydrocarbons in its ability to ionize slightly (p. 47). Many metallic salts react with acetylene to form acetylides. Sodamide in liquid ammonia forms monosodium and disodium acetylide. Cuprous and silver salts in ammoniacal solution react with acetylene to form respectively cuprous and silver acetylide precipitates, both dangerous explosives when dry. Reactions with these two salts are definite tests for acetylene, the cuprous acetylide being reddish brown and the silver acetylide being white when first formed. When heated slightly these acetylides explode. This test is valid for all triple-bonded hydrocarbons *only when* at least one of the triple-bonded carbon atoms has a hydrogen attached to it, i.e. when the compound is a 1-yne. Salts of mercury, antimony, phosphorus, and magnesium also react with acetylene. Several acetylides seem to serve as intermediate compounds in the formation of higher hydrocarbons, as has been exemplified under polymerization.

Reaction Summaries

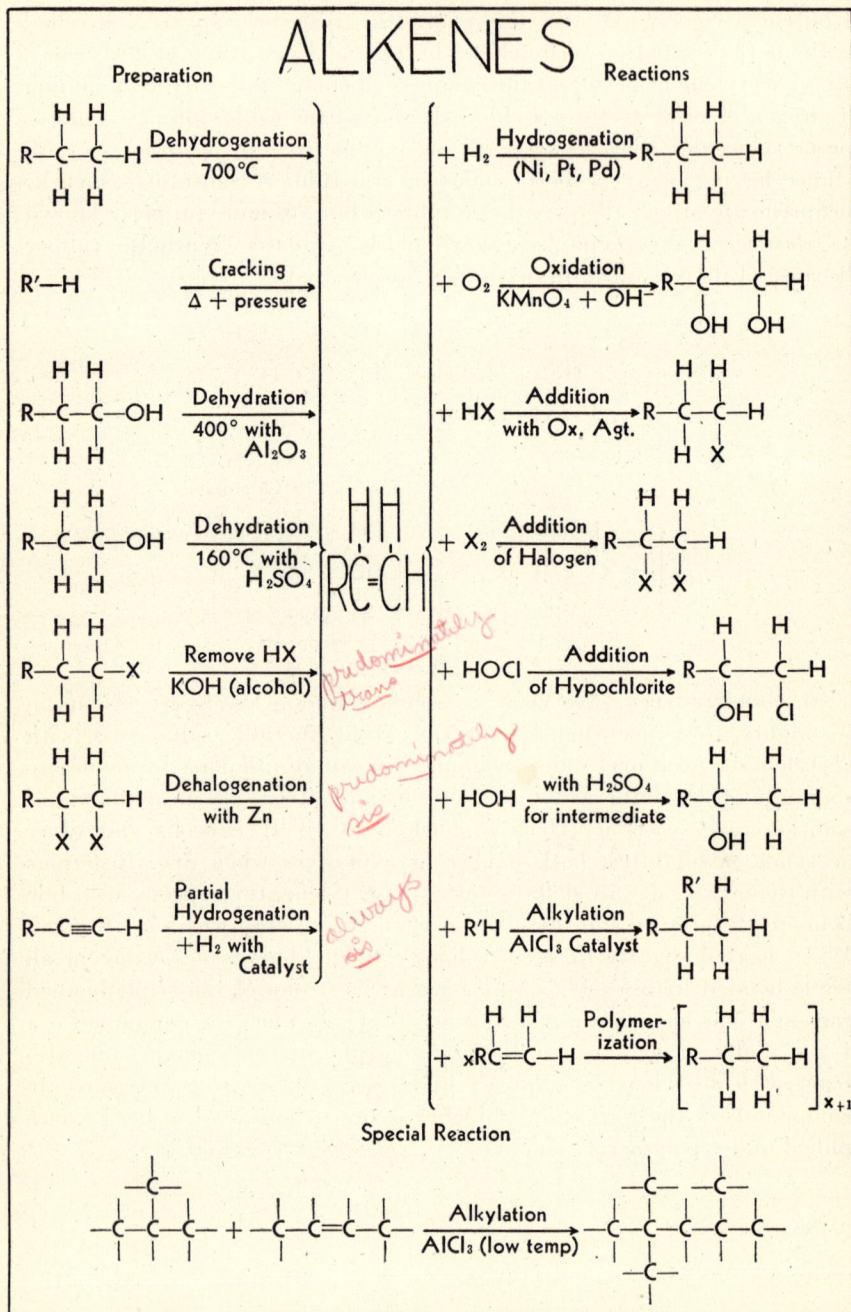

ALKENES

Preparation

$$R-\underset{\underset{H}{|}}{\overset{\overset{H}{|}}{C}}-\underset{\underset{H}{|}}{\overset{\overset{H}{|}}{C}}-H \xrightarrow[\text{700°C}]{\text{Dehydrogenation}}$$

$$R'-H \xrightarrow[\Delta + \text{pressure}]{\text{Cracking}}$$

$$R-\underset{\underset{H}{|}}{\overset{\overset{H}{|}}{C}}-\underset{\underset{H}{|}}{\overset{\overset{H}{|}}{C}}-OH \xrightarrow[\substack{400° \text{ with} \\ Al_2O_3}]{\text{Dehydration}}$$

$$R-\underset{\underset{H}{|}}{\overset{\overset{H}{|}}{C}}-\underset{\underset{H}{|}}{\overset{\overset{H}{|}}{C}}-OH \xrightarrow[\substack{160°C \text{ with} \\ H_2SO_4}]{\text{Dehydration}}$$

$$R-\underset{\underset{H}{|}}{\overset{\overset{H}{|}}{C}}-\underset{\underset{H}{|}}{\overset{\overset{H}{|}}{C}}-X \xrightarrow[\text{KOH (alcohol)}]{\text{Remove HX}}$$

$$R-\underset{\underset{X}{|}}{\overset{\overset{H}{|}}{C}}-\underset{\underset{X}{|}}{\overset{\overset{H}{|}}{C}}-H \xrightarrow[\text{with Zn}]{\text{Dehalogenation}}$$

$$R-C\equiv C-H \xrightarrow[\substack{+H_2 \text{ with} \\ \text{Catalyst}}]{\substack{\text{Partial} \\ \text{Hydrogenation}}}$$

Center

$$\underset{R\overset{}{C}=\overset{}{C}H}{\overset{H\;H}{\underset{|\;\;|}{}}}$$

Reactions

$$+ H_2 \xrightarrow[\text{(Ni, Pt, Pd)}]{\text{Hydrogenation}} R-\underset{\underset{H}{|}}{\overset{\overset{H}{|}}{C}}-\underset{\underset{H}{|}}{\overset{\overset{H}{|}}{C}}-H$$

$$+ O_2 \xrightarrow[\text{KMnO}_4 + OH^-]{\text{Oxidation}} R-\underset{\underset{OH}{|}}{\overset{\overset{H}{|}}{C}}-\underset{\underset{OH}{|}}{\overset{\overset{H}{|}}{C}}-H$$

$$+ HX \xrightarrow[\text{with Ox. Agt.}]{\text{Addition}} R-\underset{\underset{H}{|}}{\overset{\overset{H}{|}}{C}}-\underset{\underset{X}{|}}{\overset{\overset{H}{|}}{C}}-H$$

$$+ X_2 \xrightarrow[\text{of Halogen}]{\text{Addition}} R-\underset{\underset{X}{|}}{\overset{\overset{H}{|}}{C}}-\underset{\underset{X}{|}}{\overset{\overset{H}{|}}{C}}-H$$

$$+ HOCl \xrightarrow[\text{of Hypochlorite}]{\text{Addition}} R-\underset{\underset{OH}{|}}{\overset{\overset{H}{|}}{C}}-\underset{\underset{Cl}{|}}{\overset{\overset{H}{|}}{C}}-H$$

$$+ HOH \xrightarrow[\text{for intermediate}]{\text{with } H_2SO_4} R-\underset{\underset{OH}{|}}{\overset{\overset{H}{|}}{C}}-\underset{\underset{H}{|}}{\overset{\overset{H}{|}}{C}}-H$$

$$+ R'H \xrightarrow[\text{AlCl}_3 \text{ Catalyst}]{\text{Alkylation}} R-\underset{\underset{H}{|}}{\overset{\overset{R'}{|}}{C}}-\underset{\underset{H}{|}}{\overset{\overset{H}{|}}{C}}-H$$

$$+ xR\overset{\overset{H}{|}}{C}=\overset{\overset{H}{|}}{C}-H \xrightarrow[\text{ization}]{\text{Polymer-}} \left[R-\underset{\underset{H}{|}}{\overset{\overset{H}{|}}{C}}-\underset{\underset{H}{|}}{\overset{\overset{H}{|}}{C}}-H \right]_{x+1}$$

predominately trans (handwritten)

predominately cis (handwritten)

always cis (handwritten)

Special Reaction

$$-\overset{|}{\underset{|}{C}}-\;-\overset{|}{\underset{|}{C}}-\overset{|}{\underset{|}{C}}-\overset{|}{\underset{|}{C}}- + -\overset{|}{\underset{|}{C}}-C=C-\overset{|}{\underset{|}{C}}- \xrightarrow[\text{AlCl}_3 \text{ (low temp)}]{\text{Alkylation}} $$

ACETYLENE

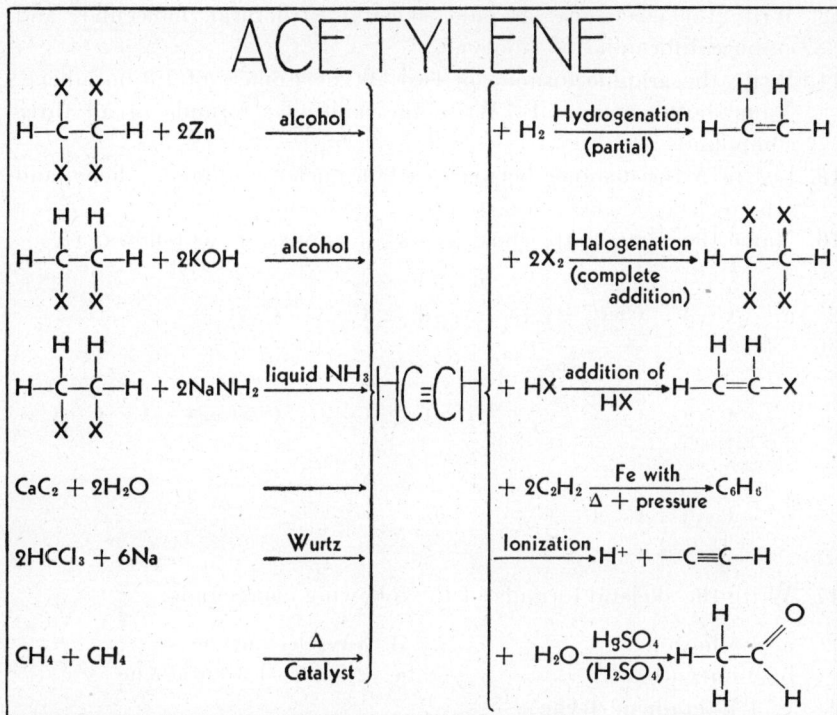

$$H-\overset{\displaystyle X}{\underset{\displaystyle X}{C}}-\overset{\displaystyle X}{\underset{\displaystyle X}{C}}-H + 2Zn \xrightarrow{\text{alcohol}}$$

$$H-\overset{\displaystyle H}{\underset{\displaystyle X}{C}}-\overset{\displaystyle H}{\underset{\displaystyle X}{C}}-H + 2KOH \xrightarrow{\text{alcohol}}$$

$$H-\overset{\displaystyle H}{\underset{\displaystyle X}{C}}-\overset{\displaystyle H}{\underset{\displaystyle X}{C}}-H + 2NaNH_2 \xrightarrow{\text{liquid } NH_3}$$

$$CaC_2 + 2H_2O \longrightarrow$$

$$2HCCl_3 + 6Na \xrightarrow{\text{Wurtz}}$$

$$CH_4 + CH_4 \xrightarrow[\text{Catalyst}]{\Delta}$$

$$HC{\equiv}CH$$

$$+ H_2 \xrightarrow[\text{(partial)}]{\text{Hydrogenation}} H-\overset{\displaystyle H}{C}{=}\overset{\displaystyle H}{C}-H$$

$$+ 2X_2 \xrightarrow[\text{addition)}]{\substack{\text{Halogenation}\\ \text{(complete}}} H-\overset{\displaystyle X}{\underset{\displaystyle X}{C}}-\overset{\displaystyle X}{\underset{\displaystyle X}{C}}-H$$

$$+ HX \xrightarrow[\text{HX}]{\text{addition of}} H-\overset{\displaystyle H}{C}{=}\overset{\displaystyle H}{C}-X$$

$$+ 2C_2H_2 \xrightarrow[\Delta + \text{pressure}]{\text{Fe with}} C_6H_6$$

$$\xrightarrow{\text{Ionization}} H^+ + {-}C{\equiv}C-H$$

$$+ H_2O \xrightarrow[(H_2SO_4)]{HgSO_4} H-\overset{\displaystyle H}{\underset{\displaystyle H}{C}}-C\overset{\displaystyle O}{\underset{\displaystyle H}{\big<}}$$

Study and Review Questions

1. Explain the name *olefin*. Give two synonymous expressions for *olefins*.
2. Write the type formula of alkenes.
3. Write the electronic, graphic, skeletal, structural, molecular, and empirical formulas of 2-butene.
4. Write the skeletal formulas of 13 hexenes.
5. List two conditions to be fulfilled for cis-trans-isomerism.
6. Make a comparative table showing the phase (solid, liquid, gas) at room temperature (20°C) of normal alkanes and normal alkenes having from 2 to 19 carbon atoms.
7. Why cannot pure 1-butene be made by heating butane to 700°C?
8. List two methods for preparing cis-alkenes.
9. List two methods for preparing trans-alkenes.
10. State the modified Markownikoff rules on addition of hydrohalides to alkenes.
11. Show by equations how water can be added to an alkene to form an alkane derivative.
12. Write the skeletal formula of an octapolyene with conjugated double bonds. Name the compound correctly.

13. Write the electronic, graphic, skeletal, structural, molecular, and empirical formulas of 2-butyne.

14. Write the graphic formula of two alkyne isomers of 1,3-butadiene. Name both compounds. Write the molecular formula of all three compounds.

15. List in Å the distance between carbon nuclei in ethane, ethene, and ethyne.

16. Name the compounds whose skeletal formulas are as follows:

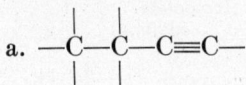

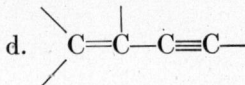

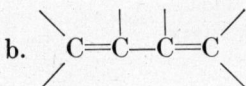

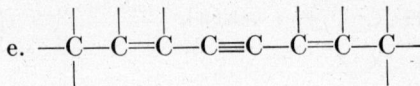

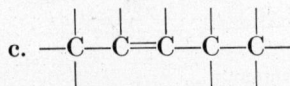

a. $-\overset{|}{C}-\overset{|}{C}-C\equiv C-$

d. $\overset{\diagdown}{\underset{\diagup}{C}}=\overset{|}{C}-C\equiv C-$

b. $\overset{\diagdown}{\underset{\diagup}{C}}=\overset{|}{C}-\overset{|}{C}=\overset{\diagup}{\underset{\diagdown}{C}}$

e. $-\overset{|}{C}-\overset{|}{C}=\overset{|}{C}-C\equiv C-\overset{|}{C}=\overset{|}{C}-\overset{|}{C}-$

c. $-\overset{|}{C}-\overset{|}{C}=\overset{|}{C}-\overset{|}{C}-\overset{|}{C}-$

17. Write the skeletal formula of the following compounds:

    a. 1,5-hexadiene   d. 3-methylbutyne

    b. 2-butyne   e. 2-chloro-1,3-butadiene

    c. 1,7-octadiene-4-yne

Chapter 4

COMPOUNDS OF CARBON WITH HYDROGEN
AND HALOGENS
THE HALOALKANES

CHAPTER PROLOGUE

An outstanding fact about haloalkanes is that they do not exist in nature. They are man-made chemical compounds. Their apparent unimportance to nature, however, is in striking contrast to the useful role which they play in the field of Organic Chemistry. For more than a century chemists have used haloalkanes and polyhalo-alkanes as important building blocks in the synthesis of other organic compounds. Haloalkanes are used especially in the laboratory preparation of large molecules of definite structure. The ability of the halogen part of the molecule to react with sodium, potassium, silver, zinc, and magnesium has led to readily controlled processes for making unique products not found in nature and for synthesizing pure products that are found in nature.

Polyhaloalkanes are versatile compounds. The kind of halogen may be changed and different kinds can exist in the same molecule at the same time. Without these compounds chemists would have made much slower progress in their study of the nature of organic compounds.

In this chapter we shall investigate the nature and action of some of these man-made tools of synthesis.

Haloalkanes

Names. A haloalkane is usually considered as a product resulting from the substitution of one hydrogen atom from an alkane by one atom of a halogen. Another name for this type of compound is *alkyl halide*. Both names suggest that each halo-alkane is composed of two parts, an alkyl radical and a halogen atom. As such it may be looked upon as the double decomposition product of an alcohol (ROH) with an acid (as HCl) yielding a halide salt of the aliphatic alcohol. Since the symbol R refers to any alkyl radical and the symbol X refers to any halogen atom, the type formula of haloalkanes is RX.

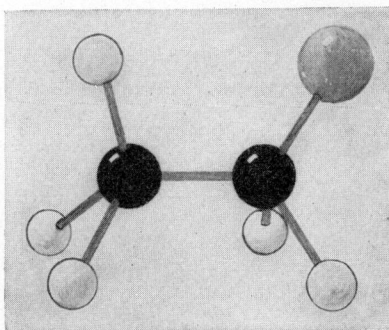

Fig. 15. The bromoethane molecule.

Kinds of Haloalkanes and Their Activity. In considering haloalkanes we must study both parts of the molecule. Since R may represent any alkyl radical from the simple methyl and ethyl radicals to complex branch-chained radicals, it is necessary to know which hydrogen from an alkane is more apt to be substituted by a halogen. In Chapter 2 on Alkanes under the reaction for nitration (p. 28) we saw that the capacity for reaction of different hydrogen atoms varied with the kind of carbon atom to which the hydrogen was bonded. Similarly the halogen atoms replace hydrogen atoms more readily from a tertiary carbon atom, next from a secondary carbon atom, and finally from a primary carbon atom.

From the point of view of the halogen involved, there are varying degrees of activity in the direct formation of a haloalkane by direct halogenation, and conversely in the degree of chemical activity of the resulting molecule. As a general rule the halogen which is more reactive in producing a haloalkane forms a compound which is less reactive than homologues of other slower acting halogens. Thus chlorine reacts more vigorously with alkanes than does bromine. Conversely, bromoalkanes give up their bromine and react more readily than chloroalkanes. The activity order of the halogens is fluorine, chlorine, bromine, and iodine.

$$F_2 > Cl_2 > Br_2 > I_2$$

The chemical activity order of haloalkanes of the same alkyl content is iodoalkane, bromoalkane, chloroalkane, and fluoroalkane. In general halomethanes are considerably more active than haloethanes. Higher

$$RI > RBr > RCl > RF$$

homologues are not appreciably more or less active than the haloethanes.

Structure and Bonding. In atomic structure halogens are all alike in one detail. They have seven electrons present in their outer shells. Thus in covalent structures they are capable of mutually intersharing one electron with another element to form a compound. In technical language the halogens have a covalence of one. Halogens differ from each other by their atomic weight and by the number of shells of electrons in the atom, or in other words by their atomic weight and by their atomic radii through which attractional forces of their respective nuclei must act. The covalent radius and atomic weight of various halogen atoms are as follows:

Name of Atom	Atomic Radius	Atomic Weight
Fluorine	0.64 Å	19.00
Chlorine	0.97 Å	35.46
Bromine	1.13 Å	79.92
Iodine	1.35 Å	126.92

The difference in reactive ability of halogens and in the stability of the compounds formed by them may be explained in part by the structure of the halogen atom involved. Molecules formed by a carbon-chlorine linkage are formed more rapidly and are more stable than bromo- and iodoalkanes because the attractional force of the chlorine nucleus is exerted through a shorter distance. Since attractional forces vary inversely as the square of the distance through which they operate, the atomic radius of the respective halogen atoms is one key to their order of activity and to the relative stability of homologous compounds. A second key lies in the mass or atomic weight of each halogen. In the case of iodine compounds, attractional forces not only act through greater distances but must also support a greater mass than in the case of other halogen compounds. Retardation of reaction because of particle size or spacial arrangement is called *steric hindrance*.

From the foregoing discussion it would seem that fluoroalkanes should be extremely stable compounds. The monofluoroalkanes are very unstable. Higher homologues readily decompose forming an alkene and hydrogen fluoride (H_2F_2). This occurs because of the even greater stability of hydrogen fluoride. Throughout the remainder of this chapter the name haloalkane, except where specifically noted, will refer only to chloro-, bromo-, and iodoalkanes. For most reactions the bromoalkane is more convenient to use than any other.

Table 4

FORMULAS OF HALOALKANE ISOMERS

Formula	1-Halobutane (attached to primary C)	2-Halobutane (attached to secondary C)	2-Methyl-2-halopropane (attached to tertiary C)
Electronic........	H H H H H : C : C : C : C : X : H H H H	H H H H H : C : C : C : C : H H H : X : H	H X H H:C : C : C:H H:C:H H
Graphic..........	H H H H H—C—C—C—C—X H H H H	H H H H H—C—C—C—C—H H H X H	H X H H—C—C—C—H H H H—C—H H
Skeletal..........	—C—C—C—C—X	—C—C—C—C— X	X —C—C—C— —C—
Structural........	$CH_3CH_2CH_2CH_2X$	$CH_3CH_2CHXCH_3$	$CH_3C(CH_3)XCH_3$
Molecular........	C_4H_9X	C_4H_9X	C_4H_9X
Type............	RCH_2X	R CHX R′	R R′—CX R″

Formulas of Haloalkanes. In Table 4 are shown the electronic, graphic, skeletal, structural, molecular, and type formulas of three halobutanes.

Nomenclature. To name a specific haloalkane, the root name of the proper halogen is prefixed to the name of the proper alkane. The entire name is written as a single word. Root names of the halogens are fluoro-, chloro-, bromo-, and iodo-. Thus when a chlorine atom is substituted for a hydrogen atom of methane or ethane, the compound is called chloromethane or chloroethane. In the case of higher alkane substitutions, the location of the substituted atom must be given in a manner similar to the naming of branch-chain alkanes. Thus if a substitution of bromine for a hydrogen on the second carbon atom of propane occurs, the compound is called 2-bromopropane.

Physical Properties. Chloromethane, chloroethane and bromomethane are gases at room temperature (20°C). Other homologues of monohaloalkanes are liquids well beyond halodecane. In the pure state haloalkanes are colorless and have a pleasant odor. Like their alkane homologues, haloalkanes are practically insoluble in water but are soluble in benzene, ether, and other organic solvents. The monohaloalkanes will burn in air. Fluoroalkanes and chloroalkanes are lighter than water. Bromo- and iodoalkanes have a specific gravity greater than that of water, the densities become less as the proportion of the alkyl part increases. A notable regularity of change of boiling point exists between homologues of haloalkanes, indicating that a regular change of boiling temperatures is caused by a change in either functional group, the alkyl part or the halogen part of the molecule.

Table 5

COMPARATIVE BOILING POINTS OF HALOALKANES IN °C

Name	Formula	—Cl	—Br	—I
Halomethane	CH_3X	−23.7	4.6	42.3
Haloethane	CH_3CH_2X	13.1	38.4	72.3
1-Halopropane	$CH_3CH_2CH_2X$	46.4	71.0	102.0
2-Halopropane	$(CH_3)_2CHX$	36.6	59.5	89.4
1-Halobutane	$CH_3CH_2CH_2CH_2X$	78.1	101.6	127.0
1-Halo-2-methylpropane	$(CH_3)_2CHCH_2X$	68.9	91.5	120.4
2-Halobutane	$CH_3CH_2CHXCH_3$	68.0	91.3	117.3
2-Halo-2-methylpropane	$(CH_3)_3CX$	51.0	73.3	100.0

Preparation. Four of the more common methods of preparing haloalkanes are as follows: (1) by addition of a hydrohalide to an alkene, (2) by substitution in direct halogenation of an alkane, (3) by reaction of alcohols with certain hydrohalides, and (4) by the reaction of alcohols with phosphorus trihalides.

1. Haloalkanes may be prepared by the addition of a hydrohalide to an alkene. The position taken by the halogen is in accord with Markownikoff's rule (p. 43).

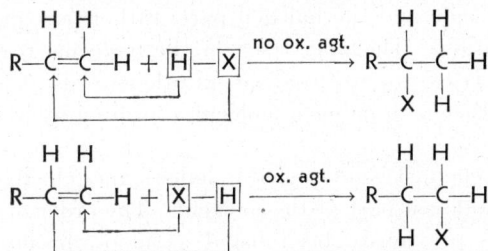

$$
\underset{\begin{array}{cc} \text{H} & \text{H} \\ | & | \\ \text{R—C=C—H} \end{array}}{} + \boxed{\text{H}}—\boxed{\text{X}} \xrightarrow{\text{no ox. agt.}} \underset{\begin{array}{cc} \text{H} & \text{H} \\ | & | \\ \text{R—C—C—H} \\ | & | \\ \text{X} & \text{H} \end{array}}{}
$$

$$
\underset{\begin{array}{cc} \text{H} & \text{H} \\ | & | \\ \text{R—C=C—H} \end{array}}{} + \boxed{\text{X}}—\boxed{\text{H}} \xrightarrow{\text{ox. agt.}} \underset{\begin{array}{cc} \text{H} & \text{H} \\ | & | \\ \text{R—C—C—H} \\ | & | \\ \text{H} & \text{X} \end{array}}{}
$$

2. Haloalkanes may be prepared by the direct halogenation of an alkane. Hydrogen halide is always produced by this reaction. Iodine will not react in this manner to form iodoalkane. One of the difficulties

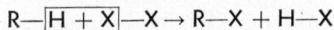

$$ \text{R—}\boxed{\text{H + X}}\text{—X} \rightarrow \text{R—X + H—X} $$

of halogenation lies in controlling the reaction. Neither the amount of halogenation nor the position halogenated can be controlled, especially in preparing the lower homologues. When various types of carbon atoms are present in an alkane, substitution will occur in greater part on the tertiary carbon atom, followed by the secondary and finally by the primary carbon atoms. This type of halogenation is usually accompanied by the formation of mixtures of polyhaloalkanes. Substitutions occur on each of the three types of carbon atoms, though in different molecules, resulting in a mixture of haloalkanes whose separation by distillation into pure compounds is fairly long and difficult.

3. Specific chloroalkanes and bromoalkanes may be prepared from proper alcohols by the use of gaseous hydrogen chloride with anhydrous zinc chloride or sulfuric acid as a dehydrating agent, or of concentrated hydrogen bromide in aqueous solution with sulfuric acid as a catalyst. Hydrogen bromide in the latter case is usually made in the reacting vessel by using sodium bromide and sulfuric acid.

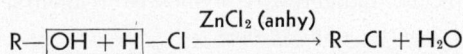

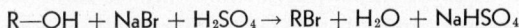

$$ \text{R—}\boxed{\text{OH + H}}\text{—Cl} \xrightarrow{\text{ZnCl}_2 \text{ (anhy)}} \text{R—Cl + H}_2\text{O} $$

$$ \text{R—OH + NaBr + H}_2\text{SO}_4 \rightarrow \text{RBr + H}_2\text{O + NaHSO}_4 $$

4. Specific haloalkanes may be made from proper alcohols with phosphorus trihalide. Phosphorous acid is also produced by this reaction. A specific iodoalkane may be made from a proper alcohol in which

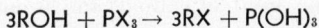

$$ 3\text{ROH + PX}_3 \rightarrow 3\text{RX + P(OH)}_3 $$

red phosphorus is in suspension simply by adding iodine crystals.

Phosphorus pentachloride and thionyl chloride also react with proper alcohols to form specific chloroalkanes.

$$ROH + PCl_5 \rightarrow RCl + HCl + POCl_3$$
$$ROH + SOCl_2 \rightarrow RCl + HCl + SO_2$$

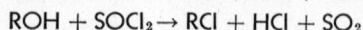

Reactions of Haloalkanes. Haloalkanes have two parts, the alkyl part of the molecule and the halogen part. Either part may react under favorable conditions. The alkyl part of the molecule reacts very much like an alkane. It can be oxidized, or, in other words, it will burn in air. It can react further with halogen molecules forming hydrogen halide and polyhaloalkanes.

The organic chemist is interested in haloalkanes chiefly because of the activity of the halogen part of the molecule. The order of activity of the haloalkanes has previously been listed as iodo-, bromo-, and chloro-alkanes. Further generalizations are: (1) the shorter the alkyl portion of the molecule the more rapid and easy the reaction, and (2) tertiary compounds react more rapidly than secondary while both are more rapid than primary, due to the instability of tertiary derivatives and the greater ease of forming alkenes. The more important types of reactions of haloalkanes may be listed as reactions, (1) by addition, (2) by condensation, (3) by removal of hydrogen halide, and (4) by various double decomposition reactions.

1. BY ADDITION. Many substances react with haloalkanes to form complex molecules by direct addition. Of greatest importance is the direct addition of magnesium by a haloalkane to form a type of compound called the "Grignard Reagent," RMgX. The reaction takes place in a solution of absolute ether. Reactions of Grignard reagents have been

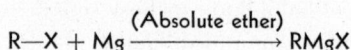

$$R{-}X + Mg \xrightarrow{\text{(Absolute ether)}} RMgX$$

of widespread importance in the synthesis of organic compounds. We have already referred to the use of Grignard reagents in preparing alkanes (p. 22), also their importance will be noted individually in later chapters under the preparation and reactions of various organic compounds. Use of the Grignard reagent has replaced the older reactions of synthesis involving zinc alkyls, dangerously explosive reagents, formed in the direct addition of zinc by iodoalkanes.

Some basic organic compounds such as amines, to be seen later under nitrogen compounds, may be added to form alkylammonium halides (p. 145).

2. BY CONDENSATION. Haloalkanes react with sodium to form alkanes and sodium halide. This reaction is known as the "Wurtz Reaction" (p. 22). When only one kind of haloalkane is present an alkane twice the size of the alkyl part of the molecule is formed. The reaction may be summarized here as follows:

$$2RX + 2Na \rightarrow RR + 2NaX$$

Haloalkanes will also react with Grignard reagent to form an alkane and magnesium halide.

$$R'—X + R''—MgX \rightarrow R'—R'' + MgX_2$$

3. BY REMOVAL OF HX. One of the methods listed for preparing alkenes was the reaction of a haloalkane with potassium or sodium hydroxide in alcoholic solution (p. 42).

4. BY DOUBLE DECOMPOSITION. Haloalkanes react with certain metallic hydroxides in aqueous solution and with certain salts to form a number of interesting products. Usually the metallic hydroxides and salts are those (a) forming precipitates or (b) highly soluble and ionic salts with the halogen part of the haloalkane. Silver, sodium, and potassium salts are cited as examples.

a. An alcohol may be formed by the reaction of a haloalkane with aqueous solution of strong bases. Such reactions are ordinarily referred to as hydrolysis in basic solution.

$$R—\boxed{X + Na}—OH \xrightarrow{\text{Aqueous}} ROH + NaX$$

b. An ether is formed when a haloalkane reacts with sodium alkoxide (the sodium salt of an alcohol).

$$R—\boxed{X + Na}—OR' \xrightarrow{\text{Alcohol}} ROR' + NaX$$

c. When a haloalkane reacts with the sodium salt of an organic acid, an ester is formed.

d. A nitrile is formed by the reaction of a haloalkane with potassium cyanide. This reaction is of importance in certain syntheses for adding a

$$R—\boxed{X + K}—CN \rightarrow RCN + KX$$

single carbon atom to a molecule.

Of commercial importance is the reaction of chloroethane with an alloy of lead and sodium. Tetraethyl lead is manufactured by this reaction. The use of tetraethyl lead as an antiknock agent in gasoline

$$4C_2H_5Cl + 4PbNa \rightarrow (C_2H_5)_4Pb + 4NaCl + 3Pb$$

motor fuels has increased the octane rating of gasoline and the efficiency of automobile and aviation motors. Tetraethyl lead is highly toxic. Gasolines containing it should not be used in gasoline stoves or lights nor should they be used as cleaning solvents.

Polyhaloalkanes

Polyhaloalkanes are compounds in which more than one hydrogen atom of an alkane have been replaced by halogen atoms. After the first halogen has replaced a hydrogen atom, further halogenation on the same carbon atom occurs more readily than on other carbon atoms, until halogenation of that carbon atom is completed. This phenomenon is due to the fact that the electronegativity of the halogen induces more positive character in the carbon atom to which it is attached. Hence these atomic hydrogens, being also positive, are more readily released for substitution than others in the molecule.

There are many polyhaloalkanes. We shall discuss two general types: (1) polyhalomethanes and (2) polyhaloethanes.

POLYHALOMETHANES

1. Dihalomethanes (methylene halides) CH_2X_2, are methane derivatives in which two atoms of halogen have replaced two atoms of hydrogen in the molecule. All three dihalomethanes, dichloro-, dibromo-, and diiodomethane, may be prepared by the reduction of the proper trihalomethane, using sodium arsenite in alkaline solution.

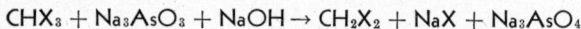

$$CHX_3 + Na_3AsO_3 + NaOH \rightarrow CH_2X_2 + NaX + Na_3AsO_4$$

A. DICHLOROMETHANE (methylene chloride), CH_2Cl_2, is usually prepared by the reduction of tetrachloromethane (carbon tetrachloride). It is used extensively as a solvent. It boils at 40.1°C and has a density of 1.34.

B. DIBROMOMETHANE (methylene bromide), CH_2Br_2, boils at 97.8°C and has a density of 2.46.

C. DIIODOMETHANE (methylene iodide), CH_2I_2, decomposes at 180°C. Its density, 3.33, is the highest of all organic liquids. In fact mercury is the only liquid substance at room temperature that is heavier than diiodomethane. The principal use of diiodomethane is in the separation of minerals containing substances of low and high densities, such as slate and coal.

2. Trihalomethanes. The three trihalomethanes, trichloro-, tribromo-, and triiodomethane, are commonly called chloroform, bromoform, and iodoform. As a group they are also called "haloforms." In their respective preparations one type of reaction, called "the haloform reaction," may be used by varying the kind of sodium hypohalite used with acetone. The haloform reaction also occurs with a few other com-

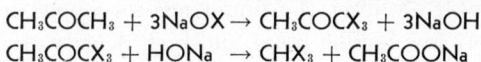

$$CH_3COCH_3 + 3NaOX \rightarrow CH_3COCX_3 + 3NaOH$$
$$CH_3COCX_3 + HONa \rightarrow CHX_3 + CH_3COONa$$

pounds as will be seen in later chapters.

A. TRICHLOROMETHANE (chloroform), $CHCl_3$, is prepared commercially by the reduction of CCl_4 using iron and water as the source of hydrogen. Trichloromethane boils at 61°C and its density is 1.489.

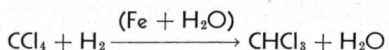

$$CCl_4 + H_2 \xrightarrow{\text{(Fe + H}_2\text{O)}} CHCl_3 + H_2O$$

It is considered to be a noninflammable liquid, though under special conditions its vapors will burn. More than a century ago, in 1848, trichloromethane was first used as a general anesthetic. It should always be kept in brown bottles completely filled, for on standing some dichloroxymethane (phosgene), $COCl_2$, one of the war gases, may be formed. To help prevent the formation of this poison a small amount of ethanol (1 per cent) is added to commercial preparations of chloroform.

Chloroform is an excellent solvent for fats and rubber. A test for trichloromethane is the addition of aniline and sodium hydroxide. Phenyl isocyanide, a toxic compound, whose odor is extremely nauseating is formed by this reaction.

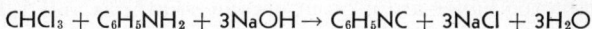

$$CHCl_3 + C_6H_5NH_2 + 3NaOH \rightarrow C_6H_5NC + 3NaCl + 3H_2O$$

B. TRIBROMOMETHANE boils at 149.6°C and its density is 2.865. Because of its high density it finds some use in the flotation of certain minerals.

C. TRIIODOMETHANE is a yellow solid of crystalline structure, practically insoluble in water. It melts at 119°C and its density is 4.1. It has a characteristic odor. For many years it was used as an antiseptic powder.

3. Tetrahalomethanes. Of the tetrahalomethanes tetrachloromethane (carbon tetrachloride), CCl_4, has attained notable commercial importance. It boils at 76.8°C and its density is 1.575. Its melting point is −23°C. Its vapors are toxic, having a bad effect on the heart.

Tetrachloromethane is prepared commercially from coke, sulfur, and chlorine. Coke and sulfur are combined in the electric furnace to form carbon disulfide. Carbon disulfide in which a few crystals of iodine are dissolved reacts with chlorine to form tetrachloromethane and sulfur monochloride. Other catalysts are antimony pentachloride, aluminum chloride, and ferric chloride.

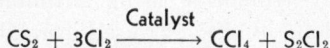

$$CS_2 + 3Cl_2 \xrightarrow{\text{Catalyst}} CCl_4 + S_2Cl_2$$

Tetrachloromethane is used extensively in the cleaning industry. It is an excellent solvent for many resins. In the laboratory it is used in the separation from aqueous solution of organic substances more soluble in it than in water. Neither the liquid nor its vapors will burn in air. Its comparatively low boiling point and high density assist in making it an excellent fire extinguisher (pyrene). The liquid, when sprayed on a fire, forms dense fumes which extinguish the fire by excluding oxygen. When tetrachloromethane is sprayed on hot metal, in the presence of air or oxygen, highly poisonous phosgene ($COCl_2$) is often formed.

Polyhaloethanes

Polyhaloethanes are substances in which two or more halogen atoms have replaced two or more hydrogen atoms of ethane. Since the halogens may replace hydrogen atoms on the same carbon atom or on both carbon atoms, it is necessary to designate the location of the halogens by prefixing numbers to the name of the compound. Thus 1,1-dihaloethane, CH_3CHX_2, is a different compound from 1,2-dihaloethane, CH_2XCH_2X. The latter is prepared by the reaction of halogen with ethene. The former is pre-

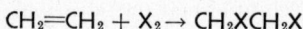

$$CH_2{=}CH_2 + X_2 \rightarrow CH_2XCH_2X$$

pared by the reaction of ethanal with phosphorus pentahalide.

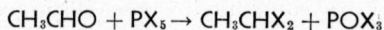

$$CH_3CHO + PX_5 \rightarrow CH_3CHX_2 + POX_3$$

Both dichloroethanes are used in organic synthesis and 1,2-dibromoethane is used in ethyl gasoline to remove the lead part of tetraethyl lead as a vaporous lead bromide in motor exhaust fumes. Tetrahaloethane is formed by the halogenation of acetylene. Hexahaloethane is the product of complete halogenation of ethane. Hexachloroethane was used during the recent war as one of the constituents for white smoke screens. During the war the polyhaloethanes, particularly 1,2-dichloroethane, have been used in tremendous tonnages for the production of solutions for metal cleaning. This use now is carried over into peace time industrial activity.

Polyhaloalkanes may contain more than one kind of halogen.

Fluoroalkanes

Fluoroalkanes are compounds in which fluorine replaces the hydrogen of alkanes. Among the fluoroalkanes two have already achieved commercial importance. Dichlorodifluoromethane (Freon 12), CCl_2F_2, is used as a refrigerant. It is an odorless, noninflammable, practically nontoxic, noncorrosive gas which boils at $-30°C$. It is prepared from tetrachloromethane and antimony trifluoride using antimony pentafluoride as a catalyst. Hexafluoroethane, C_2F_6, ; a used in relatively large quantities

$$3CCl_4 + 2SbF_3 \xrightarrow{\text{SbF}_5} 3CCl_2F_2 + 2SbCl_3$$

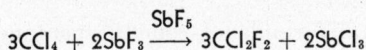

by Manhattan Project during World War II.

Extensive research in the field of fluoro-organic compounds has resulted in the development of alkane derivatives in which —CF_3 groups have replaced methyl groups in branch chain compounds. Application to the field of lubrication has resulted in the manufacture of longer lasting oils and greases.

Reaction Summary

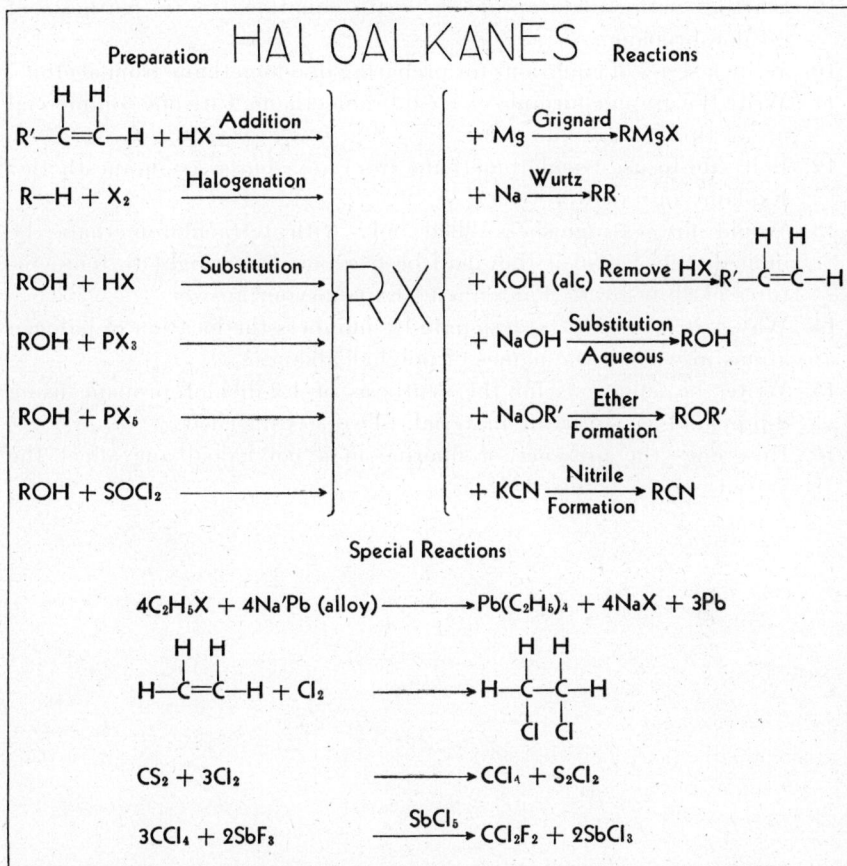

HALOALKANES

Preparation

$$R'—\overset{\overset{\displaystyle H}{|}}{C}=\overset{\overset{\displaystyle H}{|}}{C}—H + HX \xrightarrow{\text{Addition}}$$

$$R—H + X_2 \xrightarrow{\text{Halogenation}}$$

$$ROH + HX \xrightarrow{\text{Substitution}}$$

$$ROH + PX_3 \longrightarrow$$

$$ROH + PX_5 \longrightarrow$$

$$ROH + SOCl_2 \longrightarrow$$

RX

Reactions

$$+ Mg \xrightarrow{\text{Grignard}} RMgX$$

$$+ Na \xrightarrow{\text{Wurtz}} RR$$

$$+ KOH \text{ (alc)} \xrightarrow{\text{Remove HX}} R'—\overset{\overset{\displaystyle H}{|}}{C}=\overset{\overset{\displaystyle H}{|}}{C}—H$$

$$+ NaOH \xrightarrow[\text{Aqueous}]{\text{Substitution}} ROH$$

$$+ NaOR' \xrightarrow[\text{Formation}]{\text{Ether}} ROR'$$

$$+ KCN \xrightarrow[\text{Formation}]{\text{Nitrile}} RCN$$

Special Reactions

$$4C_2H_5X + 4Na'Pb \text{ (alloy)} \longrightarrow Pb(C_2H_5)_4 + 4NaX + 3Pb$$

$$H—\overset{\overset{\displaystyle H}{|}}{C}=\overset{\overset{\displaystyle H}{|}}{C}—H + Cl_2 \longrightarrow H—\overset{\overset{\displaystyle H}{|}}{\underset{\underset{\displaystyle Cl}{|}}{C}}—\overset{\overset{\displaystyle H}{|}}{\underset{\underset{\displaystyle Cl}{|}}{C}}—H$$

$$CS_2 + 3Cl_2 \longrightarrow CCl_4 + S_2Cl_2$$

$$3CCl_4 + 2SbF_3 \xrightarrow{\text{SbCl}_5} CCl_2F_2 + 2SbCl_3$$

Study and Review Questions

1. Define haloalkane. Define polyhaloalkane.
2. Name and write the graphic formula of one polyhaloalkane having two kinds of halogen atoms in the molecule.
3. List the halogens in the order of their reactivity with alkanes.
4. Make a list of halomethanes in the order of their activity.

5. Write the skeletal formula of 3-methylpentane and then write in the place of the hydrogen atoms the numerals 1, 2, or 3 to show the order of halogenation.

6. Write the graphic formula of 1,2-dichlorobutane. Write the equation for a method of preparing it from 1-butanol.

7. Write the graphic formula of 2-chlorobutane and write the equations for two different methods of preparing it.

8. Explain the type of reaction for which Victor Grignard received the Nobel Prize Award in 1912.

9. Starting with 1-chloropropane, write equations for a synthesis of 2-chloropropane.

10. Write a series of equations for preparing diiodomethane from acetone.

11. Write the graphic formula of 1,2-dibromoethane. Cite one commercial use of this compound.

12. Is it safe to use trichloromethane from any source as an anesthetic? Explain your answer.

13. Would fire extinguishers filled only with tetrachloromethane be immediately useful if they had been exposed overnight to temperatures of 20 below 0° Fahrenheit? Explain your answer.

14. Why is it necessary to designate by numbers the location of halogen atoms in writing the names of polyhaloalkanes?

15. Write the equations for the synthesis of 1,2-dichloropropane using 2-propanol as a starting material. (Two steps needed.)

16. How does the presence of fluorine in a polyhaloalkane affect the activity of the compound?

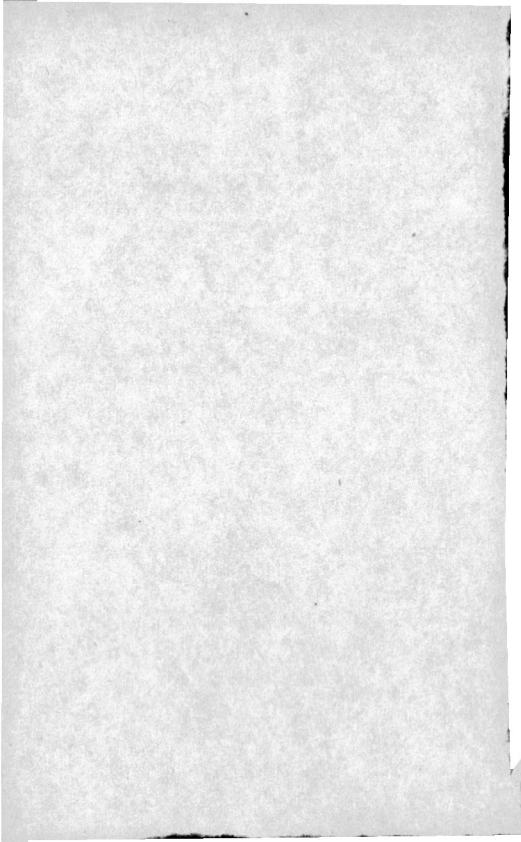

Chapter 5

COMPOUNDS OF CARBON
WITH OXYGEN AND HYDROGEN
I. FIRST OXIDATION: ALCOHOLS; ETHERS

CHAPTER PROLOGUE

Primitive man made alcoholic beverages in the earliest days of recorded history; modern man made ether in the later years of his development. Nature provided the ingredients and started the reaction to produce the first alcoholic beverage; to produce ether man had first to discover other strong chemicals needed for the reaction. During the ages man has often brought pain and suffering to his associates by abusing this gift of nature in riotous living; during the past century man has used his own product to lessen pain and suffering in his fellow men.

Only in very recent years with the development of organic synthesis has man learned how to use various alcohols and especially his old social friend and enemy, ethyl alcohol, for the preparation of many substances beneficial to his fellow men. Significantly the production of industrial alcohol in the United States ranks among the leaders in organic raw materials.

One distressing association always accompanies the word alcohol, the connotation that only one important alcohol exists. In this chapter the student will learn of other alcohols and see something of their importance in modern synthesis.

That alkanes can be oxidized is evident from the end products of burning in air such gases as methane. Water and carbon dioxide are produced. Not immediately evident is the fact that such oxidation may be considered as occurring in four different stages. This is true because the intermediate products of oxidation are not obtained by direct burning in air. Only after examining the molecular configuration of substances produced by less violent and slower methods of oxidation are we able to know that such substances as alcohols, aldehydes, and acids are intermediate stages in the complete oxidation of alkanes. In the course of oxidation, if each hydrogen atom in methane is oxidized separately, the following types of molecules are theoretically formed:

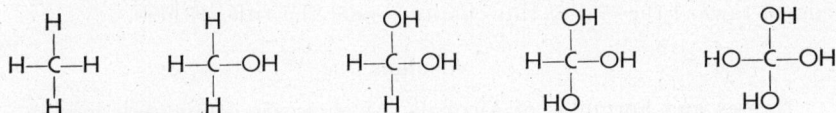

$$
\begin{array}{ccccc}
\text{H} & \text{H} & \text{OH} & \text{OH} & \text{OH} \\
| & | & | & | & | \\
\text{H--C--H} & \text{H--C--OH} & \text{H--C--OH} & \text{H--C--OH} & \text{HO--C--OH} \\
| & | & | & | & | \\
\text{H} & \text{H} & \text{H} & \text{HO} & \text{HO}
\end{array}
$$

One peculiarity of the carbon atom is that, with very few exceptions and for reasons explained in those individual cases, two or more hydroxyl groups (—OH) cannot exist as such on the same carbon atom. Their proximity to each other induces the splitting out of water and the leaving of a double-bonded oxygen for each pair of hydroxyls, thus:

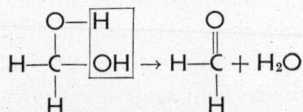

$$
\begin{array}{ccc}
\text{O--H} & & \text{O} \\
| & & \| \\
\text{H--C--OH} & \rightarrow & \text{H--C} + H_2O \\
| & & | \\
\text{H} & & \text{H}
\end{array}
$$

$$\text{H}-\underset{\overset{|}{\text{HO}}}{\overset{\overset{\text{O}-\boxed{\text{H}}}{|}}{\text{C}}}-\boxed{\text{OH}} \rightarrow \text{H}-\underset{\overset{|}{\text{HO}}}{\overset{\text{O}}{\text{C}}} + \text{H}_2\text{O}$$

$$\boxed{\text{HO}}-\underset{\overset{|}{\boxed{\text{H}}-\text{O}}}{\overset{\overset{\text{O}-\boxed{\text{H}}}{|}}{\text{C}}}-\boxed{\text{OH}} \rightarrow \underset{\overset{||}{\text{O}}}{\overset{\overset{\text{O}}{||}}{\text{C}}} + 2\text{H}_2\text{O}$$

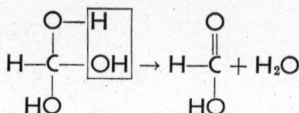

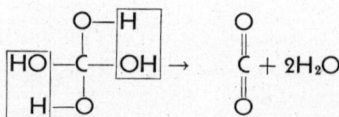

Products of these four stages of oxidation are known by the characteristic group of each molecule.

	First Oxidation	Second Oxidation	Third Oxidation	Complete Oxidation
Graphic Formula	$\text{H}-\underset{\overset{|}{\text{H}}}{\overset{\overset{\text{H}}{|}}{\text{C}}}-\text{OH}$	$\text{H}-\underset{\text{H}}{\overset{\text{O}}{\text{C}}}$	$\text{H}-\text{C}\overset{\text{O}}{\underset{\text{OH}}{}}$	$\text{C}\overset{\text{O}}{\underset{\text{O}}{}}$
Characteristic Group	$\overset{\diagdown}{\underset{\diagup}{}}\text{C}-\text{OH}$	$\overset{\diagdown}{\underset{\diagup}{}}\text{C}=\text{O}$	$-\text{C}\overset{\overset{\text{O}}{||}}{\underset{\text{OH}}{}}$	CO_2
Name of Group	Carbinol	Carbonyl	Carboxyl	
Name of Compounds	Alkanol (Alcohol)	Alkanal or Alkanone	Alkanoic Acid	

Each stage of oxidation short of completion will be considered in this and the following chapters.

In the first oxidation of a carbon atom, one covalent bond exists between the oxygen atom and the carbon atom. Furthermore the oxygen atom may lie between a carbon and a hydrogen to form a carbinol group or it may lie between two carbon atoms. The latter condition will be discussed toward the end of this chapter under the title: Ethers.

Alcohols

Names and Formulas of Alcohols. Alcohols are compounds having a carbinol group. They are called alkanols in the I.U.C. system of nomenclature, and alcohols or carbinols in other systems of naming. (See Chapter 12.) If only one carbinol group exists in a molecule, the compound is called a monohydric alcohol. The name polyhydric alcohols applies to compounds having two or more carbinol groups.

The name primary, secondary, or tertiary alcohol is applied to a compound in accord with the location of the hydroxyl group on a primary, secondary, or tertiary carbon atom (p. 28). The graphic and structural formulas of three isomeric pentanols are shown on next page.

Hydrocarbons

Prep. of.

1. decarboxylation

2. Grignard Synthesis

3. Wurtz Synthesis

4. Reduction

5. Substitution

Reactions of Hydrocarbons

1. Combustion

2. Halogenation

3. Nitration

4. Sulfonation

(over)

5) Cracking process

6) Isomerization

Primary	Secondary	Tertiary

```
    H   H   H  OH              H   H  OH  H              H  OH  H  H
    |   |   |   |              |   |   |   |             |   |   |  |
H—C—— C—— C— C—H          H—C—— C—— C—— C—H          H—C—— C—— C—C—H
    |   |   |   |              |   |   |   |             |   |   |  |
    H   |   H   H              H   |   H   H             H   |   H  H
        |                          |                        |
      H—C—H                      H—C—H                    H—C—H
        |                          |                        |
        H                          H                        H
```

CH₃CH(CH₃)CH₂CH₂OH   CH₃CH(CH₃)CH(OH)CH₃   CH₃C(CH₃)(OH)CH₂CH₃

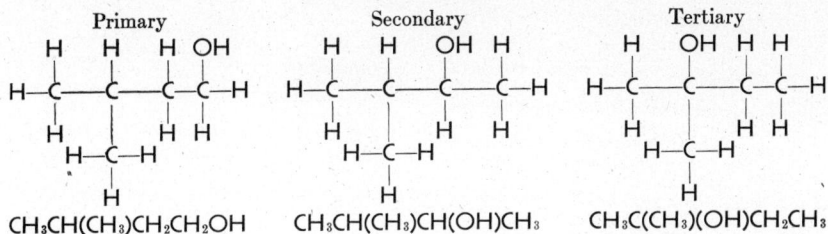

The general type formula of alcohols is ROH, in which R- may be any alkyl group. When alcohols are distinguished among themselves, the general type formula of primary alcohols is ROH, of secondary alcohols R_2CHOH, and of tertiary R_3COH.

Structure and Bonding. Carbon, hydrogen, and oxygen atoms of alcohols are united by covalent bonding as shown in the conventional electronic formula. In the vapor phase alcohols have this particular

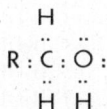

$$
\begin{array}{c}
\text{H} \quad \ \ \cdot\cdot \\
\text{R} : \text{C} : \text{O} : \\
\cdot\cdot \quad \cdot\cdot \\
\text{H} \quad \text{H}
\end{array}
$$

structure similar to the electronic formula of water in the vapor phase,

H : O : . Molecules in the vapor phase are individuals or units. They are

H

called "monomers" since they exist alone and separately from other molecules. In the liquid or solid phase, molecules are so very close to one another that new attracting and binding forces cause the association of several molecules to form polymerlike structures whose unit formula is $(ROH)_x$.

In this respect alcohols are similar to water. In the vapor phase both are monomers but in the liquid and solid phases both are made up of

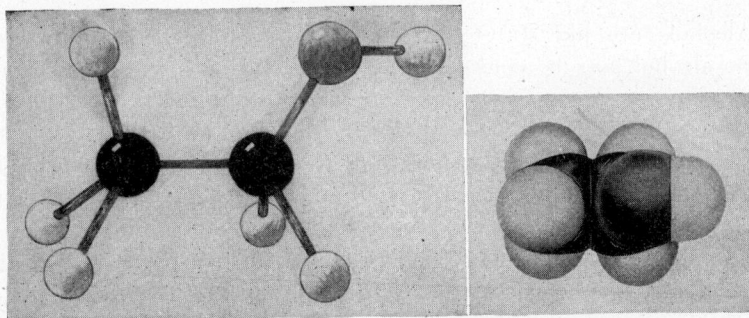

Fig. 16. The ethanol molecule.

Table 6

IMPORTANT ALCOHOLS

Name	Molecular Formula	Structural Formula	Boiling Point*
NORMAL ALCOHOLS			°C
Methanol....................	CH_3OH	CH_3OH	64.7
Ethanol....................	C_2H_5OH	CH_3CH_2OH	78.3
Propanol....................	C_3H_7OH	$CH_3CH_2CH_2OH$	97.2
Butanol....................	C_4H_9OH	$CH_3(CH_2)_2CH_2OH$	117.7
Pentanol....................	$C_5H_{11}OH$	$CH_3(CH_2)_3CH_2OH$	138
Hexanol....................	$C_6H_{13}OH$	$CH_3(CH_2)_4CH_2OH$	155
Octanol....................	$C_8H_{17}OH$	$CH_3(CH_2)_6CH_2OH$	194
Decanol....................	$C_{10}H_{21}OH$	$CH_3(CH_2)_8CH_2OH$	233
ISOMERIC ALCOHOLS (SECONDARY)			
2-Propanol....................	C_3H_7OH	$CH_3CH(OH)CH_3$	82.3
2-Butanol (Branched chain).............	C_4H_9OH	$CH_3CH(OH)CH_2CH_3$	95.5
2-Methyl-2-propanol. (Branched)...	C_4H_9OH	$(CH_3)_3COH$	
2-Methyl-1-propanol....................	C_4H_9OH	$(CH_3)_2CHCH_2OH$	107.9

* Note that the boiling points given show an almost constant increment of variation for the normal alcohols. Note also that the secondary and branched chain alcohols have boiling points lower than the isomeric normal alcohol.

associated molecular units, the binding force being due primarily to the association of a hydrogen (proton portion) from one molecule with an unshared pair of electrons of the oxygen atom. This is termed "hydrogen bonding."

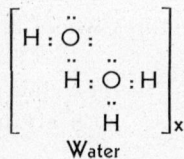

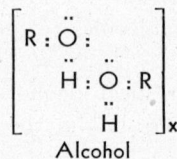

$$\left[\begin{array}{c} H : \ddot{O} : \\ H : \ddot{O} : H \\ H \end{array} \right]_x \qquad \left[\begin{array}{c} R : \ddot{O} : \\ H : \ddot{O} : R \\ H \end{array} \right]_x$$

Water   Alcohol

Alcohols are like water also in structure, in that the alkyl group of the alcohol may be considered as a substitution for one hydrogen in water. This similarity of structure is further exhibited in some physical properties of water and alcohols. Both water and alcohols have melting and boiling points much higher than would normally be expected of compounds of similar molecular size and weight. The abnormally high boiling point of water is explained by hydrogen bonding within the liquid, additional heat energy being required to free the molecules from this bonding. Alcohols are held together by a hydrogen bond in both the liquid and solid phases. Since, however, alcohols have only one hydrogen (the one in the —OH group) that exerts this bonding quality, the effect on boiling point and freezing point is not so pronounced as for water. The extra energy in the form of heat required to break this bonding is reflected in higher boiling points of alcohols as compared with their ether isomers. In the case of ethyl ether there is no hydrogen attached

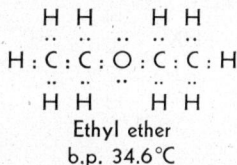

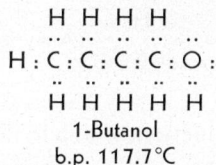

```
  H  H     H  H                    H  H  H  H
  ··  ··  ··  ··  ··               ··  ··  ··  ··  ··
H : C : C : O : C : C : H        H : C : C : C : C : O :
  ··  ··  ··  ··  ··               ··  ··  ··  ··  ··
  H  H     H  H                    H  H  H  H  H
  Ethyl ether                      1-Butanol
  b.p. 34.6°C                      b.p. 117.7°C
```

to an oxygen which could associate itself with the oxygen of other ether molecules. In the case of 1-butanol, hydrogen of the hydroxyl group can associate itself with the oxygen of other butanol molecules, since oxygen has unshared pairs of electrons, thereby establishing the intermolecular hydrogen bonding. Thus an alcohol exhibits normal covalent intrastructural bonding within its monomer and additional hydrogen interstructural bonding in its liquid and solid phases.

Nomenclature. In the I.U.C. system alcohols are called alkanols. The letter *e* in an alkane is dropped and the suffix -*ol* is added to the root name, as methanol, ethanol, etc. The specific carbon atom to which an —OH is attached is designated by the smallest number of carbon atoms counting

$$
\begin{array}{ccccc}
 & & & & OH \\
 | & | & | & | & | \\
\text{from either end of the molecule, thus, 2-pentanol is} & -C{-}C{-}C{-}C{-}C{-}. \\
 | & | & | & | & | \\
\end{array}
$$

When more than one —OH group is in a molecule, the suffix is -diol, -triol, -polyol, with numbers to designate location of the groups.

Physical Properties. Alkanols are liquids from C 1 through C 12. Above C 12 they are solids at room temperatures. The lower alcohols are all soluble in water decreasing to about two parts per 100 with pentanols. Usually the addition of —OH to an organic molecule contributes a degree of sweetness to the product, and increases its water solubility.

Preparation. Some few alcohols require specific methods of preparation, especially when commercial quantities are produced. In general, however, the preparation of alcohols depends upon the kind of alcohol required.

1. Primary alcohols are prepared by:

a. Alkaline hydrolysis of a primary haloalkane. This method was

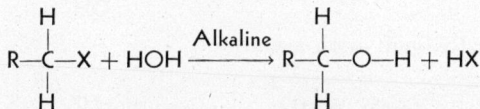

$$
\begin{array}{ccc}
H & & H \\
| & \xrightarrow{\text{Alkaline}} & | \\
R{-}C{-}X + HOH & \longrightarrow & R{-}C{-}O{-}H + HX \\
| & & | \\
H & & H \\
\end{array}
$$

given as a reaction of primary haloalkanes (p. 61). Secondary and tertiary alcohols may be produced by the hydrolysis of corresponding halides. Tertiary formation is easiest but is also more apt to produce alkenes.

b. Grignard reagent with formaldehyde.

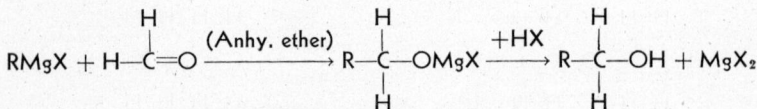

$$RMgX + H-\overset{\overset{\displaystyle H}{|}}{\underset{}{C}}=O \xrightarrow{\text{(Anhy. ether)}} R-\overset{\overset{\displaystyle H}{|}}{\underset{\underset{\displaystyle H}{|}}{C}}-OMgX \xrightarrow{+HX} R-\overset{\overset{\displaystyle H}{|}}{\underset{\underset{\displaystyle H}{|}}{C}}-OH + MgX_2$$

c. Reduction of aldehydes.

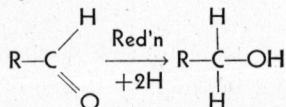

$$R-\overset{\overset{\displaystyle H}{\diagup}}{\underset{\underset{\displaystyle O}{\diagdown}}{C}} \xrightarrow[+2H]{\text{Red'n}} R-\overset{\overset{\displaystyle H}{|}}{\underset{\underset{\displaystyle H}{|}}{C}}-OH$$

d. Reduction of esters.

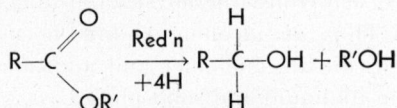

$$R-\overset{\overset{\displaystyle O}{\diagup}}{\underset{\underset{\displaystyle OR'}{\diagdown}}{C}} \xrightarrow[+4H]{\text{Red'n}} R-\overset{\overset{\displaystyle H}{|}}{\underset{\underset{\displaystyle H}{|}}{C}}-OH + R'OH$$

2. Secondary alcohols are prepared by:
a. Hydration of a proper alkene (p. 44).

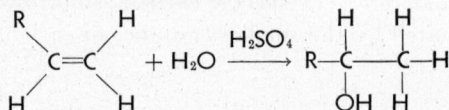

$$\overset{\overset{\displaystyle R}{\diagup}}{\underset{\underset{\displaystyle H}{\diagdown}}{C}}=\overset{\overset{\displaystyle H}{\diagup}}{\underset{\underset{\displaystyle H}{\diagdown}}{C} } + H_2O \xrightarrow{H_2SO_4} R-\overset{\overset{\displaystyle H}{|}}{\underset{\underset{\displaystyle OH}{|}}{C}}-\overset{\overset{\displaystyle H}{|}}{\underset{\underset{\displaystyle H}{|}}{C}}-H$$

b. Grignard Reagent with an aldehyde other than formaldehyde (p. 87).

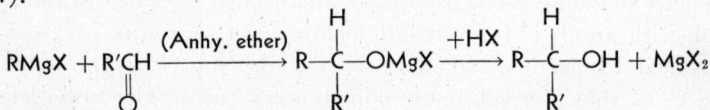

$$RMgX + R'CH \underset{\underset{\displaystyle O}{\|}}{} \xrightarrow{\text{(Anhy. ether)}} R-\overset{\overset{\displaystyle H}{|}}{\underset{\underset{\displaystyle R'}{|}}{C}}-OMgX \xrightarrow{+HX} R-\overset{\overset{\displaystyle H}{|}}{\underset{\underset{\displaystyle R'}{|}}{C}}-OH + MgX_2$$

c. Reduction of a ketone (p. 90).

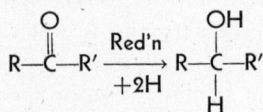

$$R-\overset{\overset{\displaystyle O}{\|}}{C}-R' \xrightarrow[+2H]{\text{Red'n}} R-\overset{\overset{\displaystyle OH}{|}}{\underset{\underset{\displaystyle H}{|}}{C}}-R'$$

3. Tertiary alcohols are prepared by:
a. Hydration of a suitable alkene (p. 44).

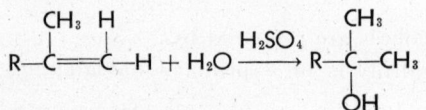

$$R-\overset{\overset{\displaystyle CH_3}{|}}{C}=\overset{\overset{\displaystyle H}{|}}{C}-H + H_2O \xrightarrow{H_2SO_4} R-\overset{\overset{\displaystyle CH_3}{|}}{\underset{\underset{\displaystyle OH}{|}}{C}}-CH_3$$

b. Grignard Reagent with a ketone.

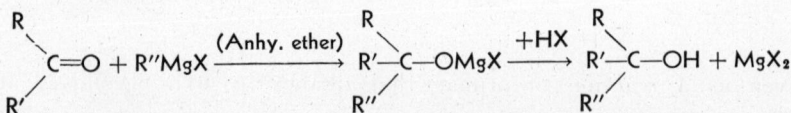

$$\overset{\overset{\displaystyle R}{\diagdown}}{\underset{\underset{\displaystyle R'}{\diagup}}{C}}=O + R''MgX \xrightarrow{\text{(Anhy. ether)}} \overset{\overset{\displaystyle R}{\diagdown}}{\underset{\underset{\displaystyle R''}{\diagup}}{C}}-OMgX \xrightarrow{+HX} \overset{\overset{\displaystyle R}{\diagdown}}{\underset{\underset{\displaystyle R''}{\diagup}}{C}}-OH + MgX_2$$

4. Methods of Preparing Specific Alcohols: a. Methanol. Methanol is prepared in two ways. The first method is by the destructive distillation of wood in the absence of air, followed by a separation of the distillate products. Many different products are formed in the watery mixture called "pyroligneous acid," such as methanol, acetic acid, and acetone. This source of methanol led to its common name "Wood Alcohol" because commercial quantities were formerly made chiefly by this process. The yield of methanol by this method is small.

Since great quantities of methanol are required for industry, a synthetic method of manufacture has now been developed. Methanol is now made synthetically by the Patart Process. In this method water gas, consisting mainly of carbon monoxide and hydrogen is produced by spraying hot coke with steam. The gases formed and some added hydrogen are forced at high pressure (200 atmospheres) over a zinc chromite catalyst at temperatures of 350–400°C. Ninety per cent of

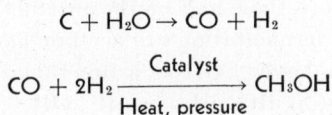

$$C + H_2O \rightarrow CO + H_2$$

$$CO + 2H_2 \xrightarrow[\text{Heat, pressure}]{\text{Catalyst}} CH_3OH$$

our methanol is made by this process.

b. Ethanol. The oldest method of preparing ethanol is by fermentation. It is also the cheapest method since it often utilizes waste products and by-products of other industries as starting materials. When starch is used as a starting material, certain enzymes (p. 277), such as enzyme of malt found in sprouted barley, convert the starch to sugar. Another

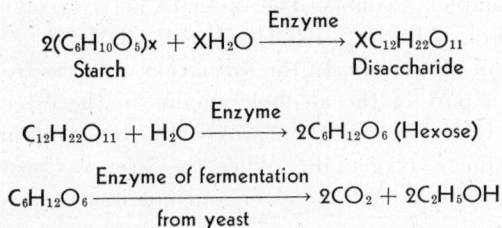

$$\underset{\text{Starch}}{2(C_6H_{10}O_5)x} + XH_2O \xrightarrow{\text{Enzyme}} \underset{\text{Disaccharide}}{XC_{12}H_{22}O_{11}}$$

$$C_{12}H_{22}O_{11} + H_2O \xrightarrow{\text{Enzyme}} 2C_6H_{12}O_6 \text{ (Hexose)}$$

$$C_6H_{12}O_6 \xrightarrow[\text{from yeast}]{\text{Enzyme of fermentation}} 2CO_2 + 2C_2H_5OH$$

enzyme produced by growing yeast plants ferments the simple sugar, forming ethanol and carbon dioxide.

Modifications of this method of producing ethanol have been used since the dawn of history in the production of alcoholic beverages. The alcoholic content of beverages resulting from direct fermentation seldom exceeds 10 per cent, but the product can be concentrated to 95 per cent alcohol by fractional distillation. Thus: wines and beers that are produced by direct fermentation contain from 4 to 11 per cent alcohol; fortified sweet wines, to which distilled liquors are added, contain about 20 per cent; brandies and whiskies are distilled to about 50 per cent

(100 proof) alcohol content; the ethanol of commerce is concentrated to 95 per cent alcohol. When the concentration of ethanol reaches 95 per cent with 5 per cent water, a constant boiling mixture (azeotropic mixture) results, and further concentration by fractional distillation is impossible. Absolute ethanol (water free) for use in synthesis and research is prepared by chemical dehydration of the constant boiling mixture.

Synthetic ethanol prepared by the hydration of ethene (p. 44) is more expensive than ethanol made by fermentation, but it is expected that the synthetic production will become progressively more important.

c. 1-Butanol. 1-Butanol is also prepared in large quantities by fermentation. Starch from corn, rice, rye, wheat, and other similar sources may be used. To produce this specific alcohol a special bacterial culture, *Clostridium acetobutylicum* Weizmann, is placed in a sterile starch paste made from finely ground grain and water. The germ of the grain is removed before the paste is made. Carbon dioxide is released as fermentation progresses and more than half of the product formed is 1-butanol. Other products of this fermentation are acetone and ethanol.

Reactions. Alcohols have —OH as a functional group. The molecule may react in two ways, as RO—H or as R—OH.

1. REACTIONS AS RO—H: A. WITH METALS. Sodium metal reacts with alcohols to liberate hydrogen and form a sodium alkoxide, RONa.

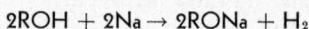

$$2ROH + 2Na \rightarrow 2RONa + H_2$$

Primary alcohols react more readily than secondary and these in turn react more readily than tertiary alcohols in the forming of sodium alkoxide (alcoholate). Likewise simple, short-chain alcohols react more readily than complex alcohols. The order of activity of the alcohols in their —H reaction is $ROH > R_2CHOH > R_3COH$.

B. FORMATION OF ESTERS. In the formation of esters from alcohol and acid the oxygen part of the alcohol remains in the ester and the acid provides the —OH. This has been proved by preparing an alcohol with radioactive isotope "Oxygen 18." When the ester was formed, the radioactive oxygen was found in the ester and not in the water formed during the reaction.

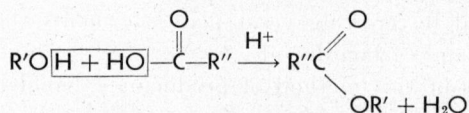

$$R'O\boxed{H + HO}\overset{\displaystyle O}{-\!\!\overset{\|}{C}\!-}R'' \xrightarrow{H^+} R''C\overset{\displaystyle O}{\diagup}\underset{\diagdown}{} \!\! OR' + H_2O$$

2. REACTIONS AS R—OH: A. REMOVAL OF —OH. The hydroxyl group may be partially removed by a dehydrating agent such as concentrated sulfuric acid with the formation of an ether, provided temperatures are carefully controlled. The dehydration is called partial, because one molecule of water is removed from two molecules of alcohol. The reaction occurs in two steps, the first being the formation of an inorganic ester.

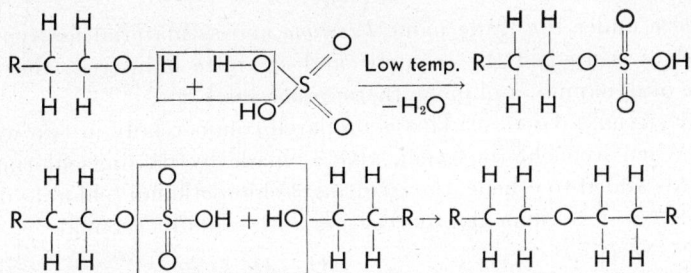

The —OH may be completely removed from the same starting material together with a hydrogen atom from the adjacent carbon atom, resulting in the formation of an alkene if the temperature is increased. The higher temperature supplies the required energy for the formation and stability of the new unsaturated bond. In the formation of alkenes from secondary or tertiary alcohols, the alkene will be formed predomi-

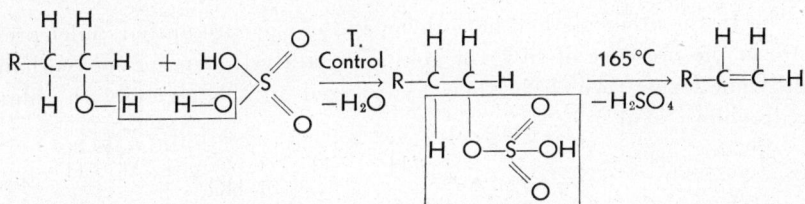

nantly with the carbon atom having the smaller number of hydrogen atoms. The order of activity of alcohols in the formation of alkenes is $R_3COH > R_2CHOH > ROH$.

B. REPLACEMENT OF —OH. Haloalkanes may be made from alcohols with the replacement of —OH by a halogen. Review the reactions of HBr, HCl, and PX_3 with alcohols as given in Chapter 4, pp. 59 and 60. The activity order of alcohols in promoting the replacement of an —OH group is $R_3COH > R_2CHOH > ROH$.

3. OXIDATION. Primary and secondary alcohols can be oxidized to aldehydes and ketones respectively.

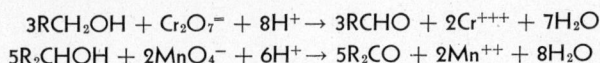

$$3RCH_2OH + Cr_2O_7^- + 8H^+ \rightarrow 3RCHO + 2Cr^{+++} + 7H_2O$$
$$5R_2CHOH + 2MnO_4^- + 6H^+ \rightarrow 5R_2CO + 2Mn^{++} + 8H_2O$$

Later (Chapter 6) we shall see that aldehydes formed by the oxidation of the primary alcohol may be further oxidized by strong reagents. When ketones are oxidized the molecule is split and two aldehydes or acids are formed. These reactions provide very interesting laboratory reactions for the student of organic chemistry.

Polyhydric Alcohols. Two polyhydric alcohols have gained such commercial importance that they must be included in this chapter. They are the dihydric alcohol 1,2-ethanediol (ethylene glycol) and the trihydric alcohol 1,2,3-propanetriol (glycerol). The former is used extensively as an antifreeze under the trade name *Prestone* and as material for synthesis. The latter is used in the cosmetic and cigarette industries, the manufacture of dynamite, and in synthetic processes.

1. ETHYLENE GLYCOL. This is prepared commercially in two ways.

a. When hypochlorous acid, HOCl (made by passing chlorine into water), is added to ethene, the resulting 2-chloroethanol (chlorohydrin) is converted to 1,2-ethanediol by reacting with sodium carbonate.

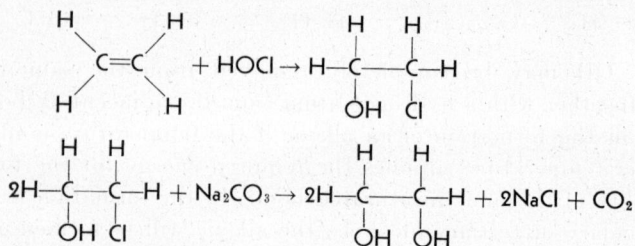

$$
\underset{\substack{\text{H} \\ \diagup \\ \text{H}}}{\overset{\substack{\text{H} \\ \\ }}{\text{C}}}=\underset{\substack{\text{H} \\ }}{\overset{\substack{\text{H} \\ }}{\text{C}}} + \text{HOCl} \rightarrow \underset{\text{OH}}{\overset{\text{H}}{\text{H}-\text{C}}}\underset{\text{Cl}}{\overset{\text{H}}{-\text{C}-\text{H}}}
$$

$$
2\text{H}-\underset{\text{OH}}{\overset{\text{H}}{\text{C}}}-\underset{\text{Cl}}{\overset{\text{H}}{\text{C}}}-\text{H} + \text{Na}_2\text{CO}_3 \rightarrow 2\text{H}-\underset{\text{OH}}{\overset{\text{H}}{\text{C}}}-\underset{\text{OH}}{\overset{\text{H}}{\text{C}}}-\text{H} + 2\text{NaCl} + \text{CO}_2
$$

b. In the other method hot ethene and oxygen are united under pressure in the presence of silver or gold as a catalyst, forming epoxyethane (ethylene oxide) which is then hydrolyzed to the glycol with dilute hydrochloric acid.

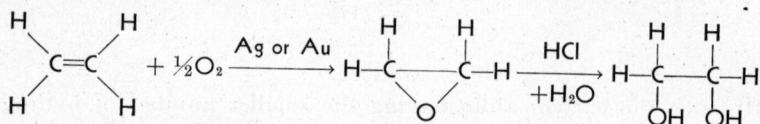

$$
\underset{\substack{\text{H} \\ \diagup \\ \text{H}}}{\overset{\substack{\text{H} \\ }}{\text{C}}}=\underset{\text{H}}{\overset{\text{H}}{\text{C}}} + \tfrac{1}{2}\text{O}_2 \xrightarrow{\text{Ag or Au}} \text{H}-\underset{O}{\overset{\text{H}}{\text{C}}}\underset{}{\overset{\text{H}}{\text{C}}}-\text{H} \xrightarrow[+\text{H}_2\text{O}]{\text{HCl}} \text{H}-\underset{\text{OH}}{\overset{\text{H}}{\text{C}}}-\underset{\text{OH}}{\overset{\text{H}}{\text{C}}}-\text{H}
$$

2. GLYCEROL. Glycerol is prepared commercially (a) as a product of the soap industry. Animal fats and certain vegetable oils are esters of fatty acids with glycerol. Hydrolysis of fat with sodium hydroxide produces glycerol and sodium salts of fatty acids (soaps). This reaction is called saponification.

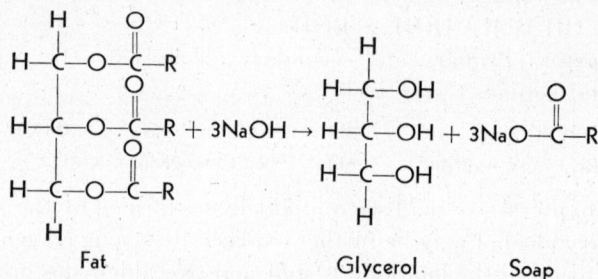

$$
\begin{array}{c}
\text{H} \quad \text{O} \\
\text{H}-\text{C}-\text{O}-\text{C}-\text{R} \\
\quad\quad\quad\quad \text{O} \\
\text{H}-\text{C}-\text{O}-\text{C}-\text{R} + 3\text{NaOH} \rightarrow \\
\quad\quad\quad\quad \text{O} \\
\text{H}-\text{C}-\text{O}-\text{C}-\text{R} \\
\text{H}
\end{array}
\quad
\begin{array}{c}
\text{H} \\
\text{H}-\text{C}-\text{OH} \\
\text{H}-\text{C}-\text{OH} + 3\text{NaO}-\overset{\text{O}}{\text{C}}-\text{R} \\
\text{H}-\text{C}-\text{OH} \\
\text{H}
\end{array}
$$

Fat   Glycerol   Soap

(b) Glycerol is also produced synthetically from propene, a product of the petroleum cracking industry, because of a peculiar reaction of propene with chlorine. At temperatures exceeding 300°C chlorine does not add to the double bond but replaces one of the hydrogens of the methyl group in propene, forming 3-chloro-1-propene (allyl chloride). When hypochlorous acid is added to the double bond, 1,3-dichloro-2-

propanol is formed. With soda lime this is converted to 3-hydroxyepoxy-propane which is then hydrolyzed with dilute hydrochloric acid to glycerol.

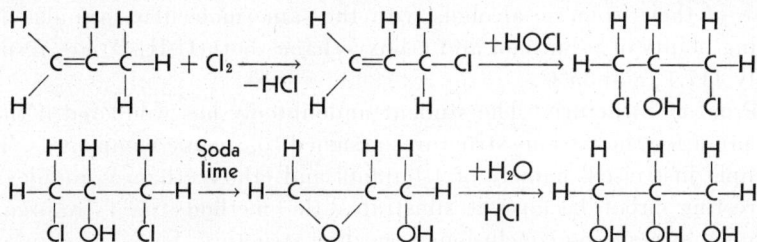

The nitric acid ester of glycerol, trinitroglycerol, absorbed in porous material to form dynamite, has served mankind for nearly a century as a comparatively safe explosive.

Ethers

Simple ethers are not found in nature. They are formed by the direct covalent bonding of one oxygen atom to two carbon atoms. The general

Fig. 17. The ethoxyethane molecule (diethyl ether).

type formula of ethers is ROR. Ethers may be simple or mixed. They are called simple ethers when the alkyl groups are the same. Mixed ethers have different alkyl groups. Ethers are *metameric* isomers of alcohols.

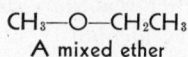

$$CH_3-O-CH_3 \qquad\qquad CH_3-O-CH_2CH_3$$
A simple ether   A mixed ether

Ethers are named in the I.U.C. system as alkyl oxides of an alkane. Thus in the two formulas given above, the first is named methoxymethane and the second is called methoxyethane. Our most important ether, the well known anesthetic, is ethoxyethane ($C_2H_5OC_2H_5$).

Although in structure ethers may be looked upon as derivatives of water with an alkyl group substituted for each hydrogen in HOH, their solubility in each other is slight. Slight solubility is attributed to a tendency for hydrogen bonding (p. 70) to the unshared electron pairs of the oxygen atom. As R— increases in size the solubility decreases. Ethers are less dense than water. Their boiling points are considerably lower than those of their isomeric alcohols with the same molecular formula. The boiling points of n-butanol and ethoxyethane (both $C_4H_{10}O$) are respectively 117.7° and 35°C.

Proof of Structure.[1] The student undoubtedly has wondered if there is a proof for the various structures assigned to organic compounds. The example just cited, namely of 1-butanol and ethoxyethane, provides an interesting proof because it illustrates the methods used by organic chemists in reaching conclusions regarding structure.

First. A quantitative analysis of both compounds yields the empirical formula $C_4H_{10}O$. (C = 64.8%; H = 13.5%, O = 21.6%) (p. 8).

Second. A determination of molecular weights yields a value of 74 for both compounds. Since 74 is the weight of the empirical formula, then the molecular formula is $C_4H_{10}O$.

Third. The following characterization reactions prove the existence of certain functional groups.

a. One mole of alcohol yields one gram-atom of hydrogen when it reacts with metallic sodium (p. 74); ether does not react with sodium. Therefore one atom of hydrogen in the alcohol must be united differently from the other nine hydrogen atoms in alcohol. This leads to the conclusion that one hydrogen atom must be joined directly to the oxygen atom in the alcohol. In ether the oxygen must be joined in some other manner. The only possibility, considering there are 10 hydrogen atoms present, is to have the oxygen joined directly to carbon atoms.

b. Alcohol reacts directly with PCl_3 (p. 59), producing C_4H_9Cl and phosphorous acid; ether does not undergo this change. This further establishes the presence of an —OH group in the alcohol and the absence of the group in ether.

c. Typical reactions of 1-butanol with sulfuric acid (p. 75) yields C_4H_8 and $C_4H_9OC_4H_9$ which are relatively large molecules; typical reactions of ether with sulfuric acid (p. 79) yield C_2H_5OH, a small molecule, showing that the ether molecule had been cleaved.

d. Further reactions show other properties, such as that the alcohol is a primary alcohol (p. 74). These chemical changes further prove the structure of the alcohol and of the ether.

Within recent years studies of many compounds with the electron microscope and with X-ray diffraction photographs (physical-chemical methods applied to the study of molecular structure) have shown the conclusions reached by early organic chemists to be correct.

---

[1] The size of this text will not permit the elaboration of the complex procedures required for the characterization of the structure of other substances. From the example given here, however, the student will gain an insight into the methods employed by chemists in proving structure and a confidence in the conclusions of organic research.

Preparation. a. Simple ethers may be prepared by the removal of one molecule of water from two molecules of alcohol using sulfuric acid as a dehydrating agent. This reaction was shown in steps on page 75

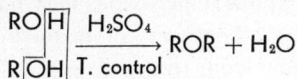

$$\begin{array}{c} RO\boxed{H} \\ R\boxed{OH} \end{array} \xrightarrow[\text{T. control}]{H_2SO_4} ROR + H_2O$$

(alcohols). Phosphoric acid may also be used as the dehydrating agent. If aluminum sulfate is present, ethers form at a lower temperature. Primary alcohols give the best results in the formation of ethers. Secondary and tertiary alcohols tend rather to the formation of alkenes. When two different alcohols are used, a mixture of several different ethers will be formed.

b. Mixed ethers (R′—O—R″) are prepared by the Williamson synthesis. Sodium alkoxide reacts with haloalkane, usually the bromide or iodide, to form the ether and sodium halide. The volatile ether is then

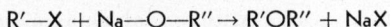

$$R'—X + Na—O—R'' \rightarrow R'OR'' + NaX$$

separated by distillation. Simple ethers may also be prepared by this method.

Reactions. a. Ethers react with sulfuric acid to form alkyl hydrogen sulfate. When the mixture is treated with water, alcohols result.

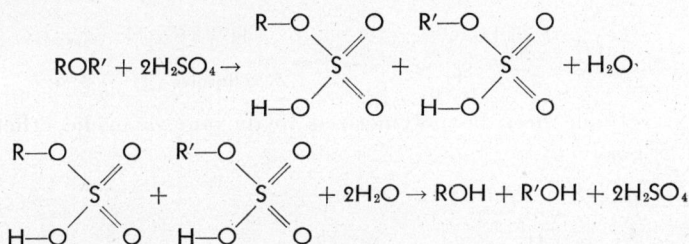

$$ROR' + 2H_2SO_4 \rightarrow \underset{H-O}{\overset{R-O}{\diagdown}}S\underset{O}{\overset{O}{\diagup}} + \underset{H-O}{\overset{R'-O}{\diagdown}}S\underset{O}{\overset{O}{\diagup}} + H_2O$$

$$\underset{H-O}{\overset{R-O}{\diagdown}}S\underset{O}{\overset{O}{\diagup}} + \underset{H-O}{\overset{R'-O}{\diagdown}}S\underset{O}{\overset{O}{\diagup}} + 2H_2O \rightarrow ROH + R'OH + 2H_2SO_4$$

The student will observe that this reaction is essentially the reverse of that by which ethers may be prepared.

b. With hydrogen iodide ethers form an alcohol and an iodoalkane. If the HI is in excess, further reaction of the ROH formed will produce more

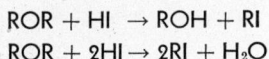

$$ROR + HI \rightarrow ROH + RI$$
$$ROR + 2HI \rightarrow 2RI + H_2O$$

alkyl iodide. Fuming hydrobromic acid will react in the same manner but higher temperatures are required.

Ethoxyethane (also called diethyl ether or simply "ether") was the first organic anesthetic used in medicine in 1842. It still retains its place as our most important general anesthetic (p. 284). In industry and in the laboratory ether finds important use as a solvent for organic substances. Its selective solvent action in relation to water leads to its use in the characterization of organic compounds. *Anhydrous ether* is the important medium for Grignard reactions. One caution should always be remembered in using ether from previously opened containers: due to air-oxidation of the ether, dangerously explosive peroxides may have formed, and the user, whether in industry, the laboratory, or in medicine, must use approved precautions against injury from this source.

Inner and Cyclic Ethers. Internal ethers are formed by the reaction of a 2-haloalkanol with a strong base. The product is called an epoxyalkane (alkene oxide).

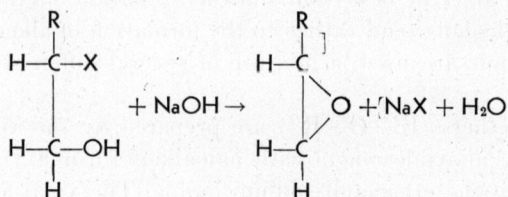

When epoxyethane reacts with an alcohol, a cellosolve is formed. These are valuable as solvents for lacquers and varnishes.

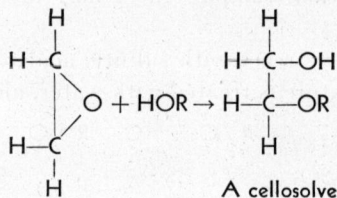

A cellosolve

Epoxyethane may be polymerized to dioxane, a cyclic ether, very useful as a solvent.

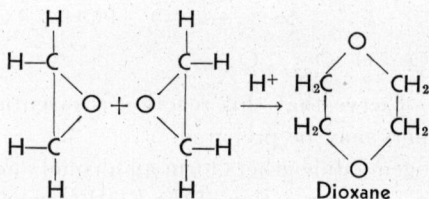

Dioxane

Reaction Summaries

Preparation ALCOHOLS **Reactions**

Preparation:

RX + HOH $\xrightarrow{\text{alkaline}}$

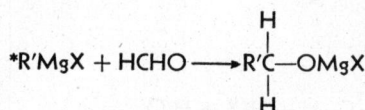

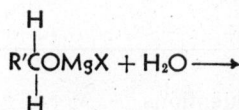

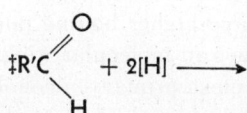

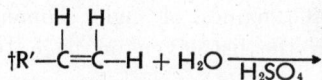

*R'MgX + HCHO ⟶ R'C̣—OMgX (with H above and below C)

R'C̣OMgX + H₂O ⟶ (with H above and below C)

‡R'C(=O)H + 2[H] ⟶

†R'—C=C—H + H₂O $\xrightarrow{H_2SO_4}$

{ ROH }

Reactions:

+ Na ⟶ R—O—Na

+ HOC̈R' (O double bond) ⟶ R—O—C̈—R' (O double bond)

+ HOSO₃H ⟶ R—OSO₃H

+ R—OSO₃H ⟶ ROR

Dehydration ⟶ R'C=C—H (with H H above)

+ [O] ⟶ R'C(=O)H §

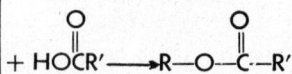

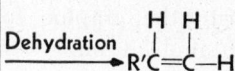

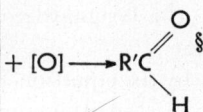

Special Preparations

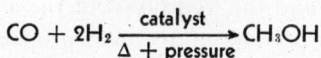

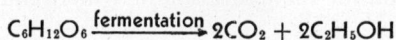

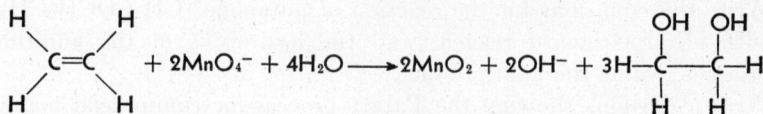

$$CO + 2H_2 \xrightarrow[\Delta + \text{pressure}]{\text{catalyst}} CH_3OH$$

$$C_6H_{12}O_6 \xrightarrow{\text{fermentation}} 2CO_2 + 2C_2H_5OH$$

C=C (ethylene) + 2MnO₄⁻ + 4H₂O ⟶ 2MnO₂ + 2OH⁻ + 3H—C̣—C̣—H (with OH OH above and H H below)

Comments

* With formaldehyde, Grignard yields primary alcohols; other aldehydes, secondary; ketones, tertiary.
† Beyond C3, alkenes tend to produce secondary and tertiary alcohols.
‡ Reduction of ketones produces secondary alcohols.
§ Primary alcohols yield aldehydes; secondary, ketones; tertiary, more complex products.

ETHERS

$$ROH + HOR \xrightarrow[150°C]{H_2SO_4}$$

$$RX + NaOR \longrightarrow$$

$$\Big\} ROR \Big\{$$

$$+ H_2SO_4 \longrightarrow ROSO_3H$$

$$+ HI \longrightarrow ROH + RI$$

A Special Ether

$$R-\underset{\underset{H}{|}}{\overset{\overset{X}{|}}{C}}-\underset{\underset{H}{|}}{\overset{\overset{OH}{|}}{C}}-H + NaOH \longrightarrow RC\overset{O}{\overbrace{}}C-H + NaX + H_2O$$

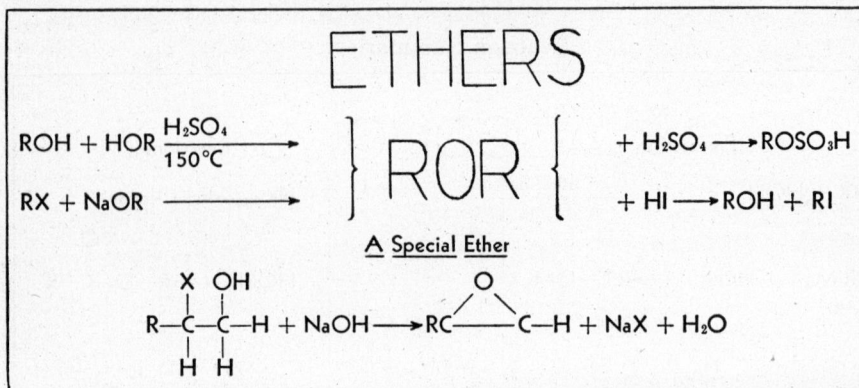

Study and Review Questions

1. Explain in some detail why alcohols have higher boiling points than other organic compounds of similar size and molecular weight.
2. Write the graphic formula of the simplest primary alcohol. Of the simplest secondary alcohol. Of the simplest tertiary alcohol.
3. Write the graphic formula and I.U.C. names of eight isomeric pentanols. Compare your results with the list given on page 169 (Chapter 12).
4. Write the Grignard reaction for preparing 1-butanol from 1-chloropropane.
5. Show by an equation the hydrolysis reaction of 1-chloropropane.
6. Write in steps the equations for the hydration of propene.
7. Write the equations for the reaction of ethyl Grignard reagent and ethanal (CH_3CHO) and the hydrolysis of the addition product.
8. Name the reduction product of acetone (propanone, CH_3COCH_3).
9. Write the equation for the hydration of 2-methylpropene. Name the product.
10. Write the equations for the reaction of butanone ($CH_3COCH_2CH_3$) with ethyl Grignard reagent and the hydrolysis of the addition product. Name the final product.
11. Write equations showing the Patart process for commercial manufacture of methanol.
12. Write in steps equations showing how a starch is fermented to ethanol. What is the alcoholic content of the fermentation product?
13. Show by equations how ethene can be hydrated to ethanol.
14. Name the products of the reaction of ethanol with sodium metal.
15. What is the order of activity with sodium of 1-butanol, 2-butanol, and 2-methyl-2-propanol?
16. Write the equation for the reaction of ethanol with propanoic acid (CH_3CH_2COOH).

17. Show in steps the dehydration reactions of ethanol with sulfuric acid at two different temperature levels. Name the products of both reactions.

18. Show in steps the dehydration reactions of 2-butanol with sulfuric acid at two different temperature levels. Name the products of both reactions.

19. If a mixture of 1 mole of 2-methyl-2-propanol, 1 mole of 2-butanol, and 1 mole of 1-butanol was treated with $\frac{1}{3}$ mole of PCl_3, name the compound most likely to be formed by the reaction.

20. Write the equation for the oxidation of 1-propanol with potassium permanganate.

21. Write the equation for the reaction of 2-propanol with potassium permanganate.

22. Show by equations two methods of preparing 1,2-ethanediol (ethylene glycol).

23. By a series of equations show how glycerol can be prepared from propene.

24. Write out a proof that methoxymethane and ethanol (molecular formula C_2H_6O for both) have different molecular structures.

25. How does the reaction of an ether with HI differ from that of an alcohol?

26. By a series of equations show how epoxyethane may be prepared from ethanol.

27. Write the equation for the preparation of methoxypropane.

Chapter 6

COMPOUNDS OF CARBON
WITH OXYGEN AND HYDROGEN
II. SECOND OXIDATION: ALDEHYDES; KETONES

CHAPTER PROLOGUE

Aldehydes are the Jekyll-Hydes of organic chemicals. They are quick-change artists, first presenting one appearance then another. Fortunately there is nothing unpleasant about aldehydes or ketones except perhaps the odor emanating from a few small members of the family. Aldehydes and ketones are loyal, active busybodies. They are like the small children of a family, always active with something during waking hours and hard to arouse when once asleep. Like children also, they are active with almost any other thing seeking action. They seldom forget the alcoholic condition of their origin. They act as individual units at times and on other occasions they unite among themselves to present a unified complex front.

Because of these fundamental characteristics aldehydes and ketones are among our most important organic chemicals. They move up and down the ladder of oxidation readily changing to alcohols or to acids. They add many kinds of strange molecules and thereby start entirely different families of compounds.

In this chapter we shall explain the nature and structure of both aldehydes and ketones, and see some of their characteristic reactions.

Names of Compounds. The second oxidation of an alkane results in the formation of the carbonyl group, $\diagdown C=O$. When the carbon of the carbonyl group is a primary carbon atom, the compound is an aldehyde. In the I.U.C. system aldehydes are called alkanals. When the carbon of the carbonyl group is a secondary carbon atom, the compound is a ketone. Ketones are called alkanones in the I.U.C. system of nomenclature. Each name is specifically derived by adding the suffix -*al* or -*one* to the root signifying the total number of carbon atoms in the compound. The location of the carbonyl group is given by a number when it is possible to have the carbonyl group located in more than one place within the molecule. Thus propanal and propanone have no number since the carbonyl group can be located in only one place within the molecule. On the other hand 2-pentanone and 3-pentanone would refer to distinct molecules and the name pentanone alone would not clearly designate either molecule accurately.

The general formula of alkanals is RCHO and for alkanones is RCOR'. In writing structural formulas the correct order of symbols is very important. The structural formula CH_3COH is *not* suitable for ethanal since the reader would interpret the OH portion as referring to an alcohol.

Structure and Bonding. Structurally the carbonyl group consists of an atom of oxygen bonded to a carbon atom by a double covalent

Fig. 18. (Left) Ethanal and (right) propanone (acetone) molecules.

bond. This makes such an oxygen atom highly negative. High negativity

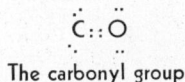

$$\ddot{C} :: \ddot{O}$$

The carbonyl group

of the oxygen atom in a molecule induces a complementary high positive polarity on the carbon atom adjacent to the carbonyl group. The effect of this high positive polarity is to lessen the bonding potential of attached hydrogen atoms, making them more active ionically than other ordinary hydrogen atoms in the alkyl group. Hence the hydrogen atoms on the 2-carbon atom (the carbon adjacent to the carbonyl group) are more labile.

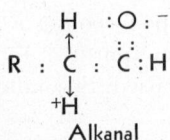

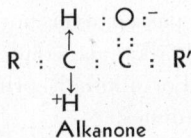

$$R : \overset{\uparrow}{\underset{^+H}{C}} : \overset{:\ddot{O}:^-}{C} : H$$

Alkanal

$$R : \overset{\uparrow}{\underset{^+H}{C}} : \overset{:\ddot{O}:^-}{C} : R'$$

Alkanone

The effect of the oxygen's negativity and the weakened bonding of hydrogens on an adjacent carbon atom provides the opportunity for the tautomeric existence of the carbonyl group. Tautomerism is an ability of a compound to shift its structure to an isomeric form, hence under certain conditions it may exist in two different forms. In the case of the aldehydes, the shift in structure is greatly aided by the presence of any base. At the beginning of the tautomeric cycle the highly negative oxygen

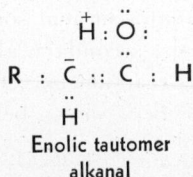

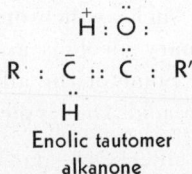

$$R : \overset{}{\underset{H}{\bar{C}}} :: C : H \qquad \overset{^+\dot{H}:\ddot{O}:}{}$$

Enolic tautomer
alkanal

$$R : \overset{}{\underset{H}{\bar{C}}} :: C : R' \qquad \overset{^+\dot{H}:\ddot{O}:}{}$$

Enolic tautomer
alkanone

attracts one hydrogen nucleus to it, forming an hydroxyl group and a double bond between the two carbon atoms. In this condition the compound becomes an *alkenol*. It is called the *enol* tautomer of a carbonyl group and is often shown as

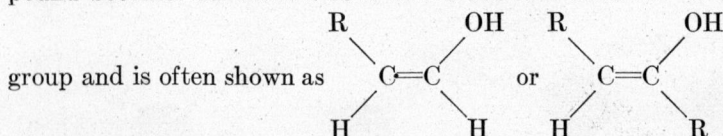

In the enol form, polarity is reversed. Under appropriate conditions the keto structure is reformed. Hence, normally there exists an equilibrium mixture containing both the keto and the enol forms:

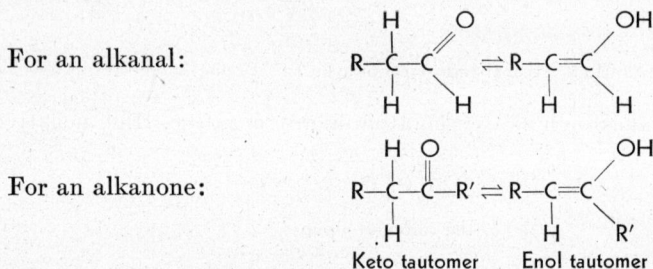

For an alkanal:

For an alkanone:

Keto tautomer   Enol tautomer

From the bond energies in Table 2, p. 37, it will be observed that alkanones require a slightly higher energy (152 to 149) to break the bond than do the alkanals. Hence the alkanals are slightly more reactive.

Preparation.

a. The commercial methods of preparing alkanals is by the catalytic dehydrogenation of suitable alcohols in the vapor phase. Hot vapors of alcohols are passed over hot copper gauze. Upon cooling a mixture of alcohols and alkanals condenses which is then purified by fractional distillation. For alkanals primary alcohols are used. Secondary alcohols produce alkanones.

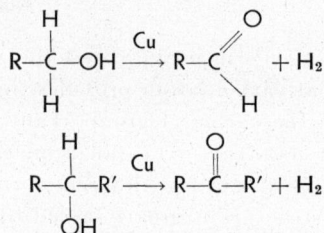

b. In laboratory preparation of alkanals and alkanones strong oxidizing agents such as dichromates and permanganates in acid solution are used. Primary alcohols are oxidized to alkanals, secondary alcohols to alkanones. Some of the alkanal formed by this reaction may be further oxidized to acid. One typical reaction of each type is shown below.

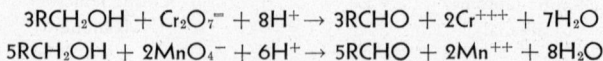

$$3RCH_2OH + Cr_2O_7^{--} + 8H^+ \rightarrow 3RCHO + 2Cr^{+++} + 7H_2O$$
$$5RCH_2OH + 2MnO_4^- + 6H^+ \rightarrow 5RCHO + 2Mn^{++} + 8H_2O$$

Another laboratory method consists of the action of alkali on 1,1-dihaloalkane for alkanals and secondary dihaloalkanes for ketones.

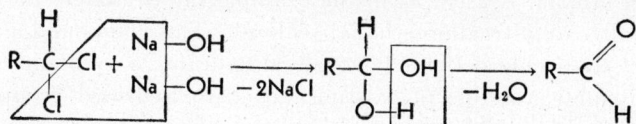

A common laboratory method of preparing certain specific alkanones is by the dry distillation of suitable calcium, barium, or thorium salts of organic acids.

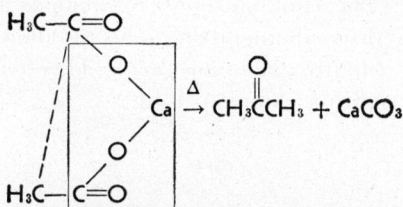

General Reactions of Alkanals and Alkanones: A. WITH GRIGNARD REAGENT. Grignard reagents add to methanal, alkanals, and alkanones and then hydrolize to form primary, secondary, and tertiary alkanols respectively (p. 72).

B. WITH HALOGENS. Halogens add to the enol tautomer of alkanals or alkanones forming haloalkanals or haloalkanones with hydrogen halide. In an excess of halogen, enolization will continue with addition of the

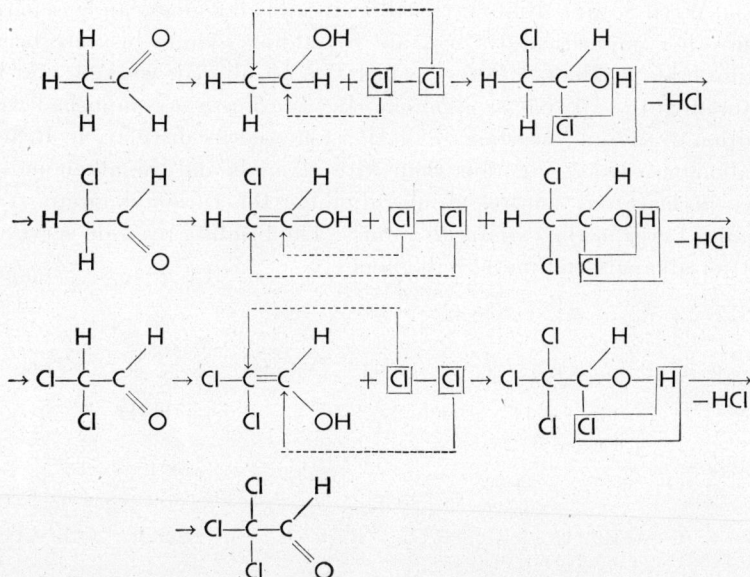

halogen until the carbon nearest the carbonyl group is saturated with halogen. The reaction probably occurs in steps.

When ethanal is used, as in the example shown above, the resulting compound is 2,2,2-trichloroethanal (chloral). This compound adds water to form 2,2,2-trichloroethanediol (chloral hydrate), a solid and one of the few compounds capable of existing with two hydroxyl groups on the same carbon atom. When methyl ketones, ethanol, or ethanal are halogenated using sodium hypoiodite (NaOI) as the halogenating agent, triiodomethane is produced. This reaction, known as the haloform reaction (p. 63) constitutes an analytical test for the compounds indicated.

c. WITH HCN. The enol tautomer of alkanals and alkanones adds HCN to form 2-hydroxyalkanenitrile. This product may then be hydrolyzed to form 2-hydroxyalkanoic acid. This reaction is useful in adding a carbon atom to a chain (p. 148).

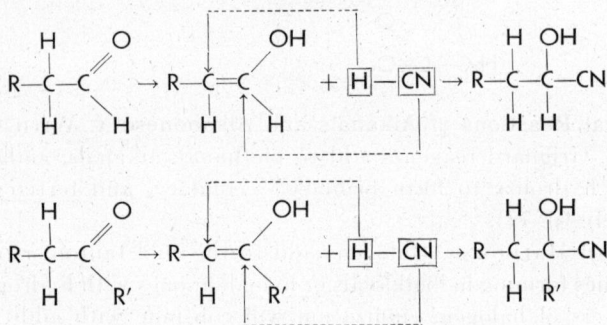

d. WITH SODIUM BISULFITE. Alkanals and alkanones can be separated from other impurities in a mixture by adding sodium bisulfite because an insoluble product is formed which can be filtered, washed, and later restored to the alkanal or alkanone. Restoration is accomplished with a solution of sodium bicarbonate or HCl followed by distillation. Reaction of alkanones is less vigorous than with alkanals and the alkanone must have at least one methyl group adjoining the carbonyl group. Yields are also lower in the case of alkanones. The bisulfite reaction serves as a test for alkanals and methyl alkanones.

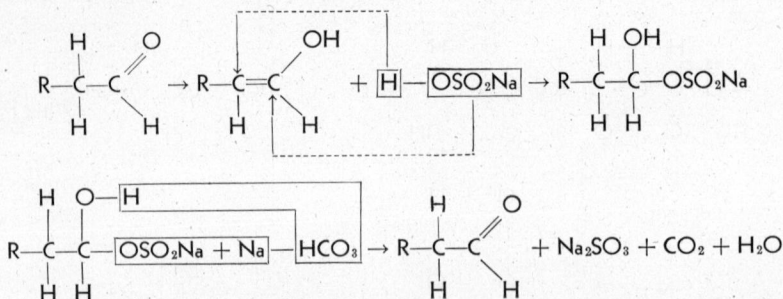

E. Aldol Condensation. In basic solution alkanals add other molecules of alkanals forming hydroxyalkanals. In the simplest addition of one alkanal molecule, as one molecule of ethanal by another molecule of ethanal, the product is called an aldol and the type of reaction is called an aldol condensation. When an aldol dimer is heated it will lose water, forming an alkenal. If several molecules are united by successive aldol

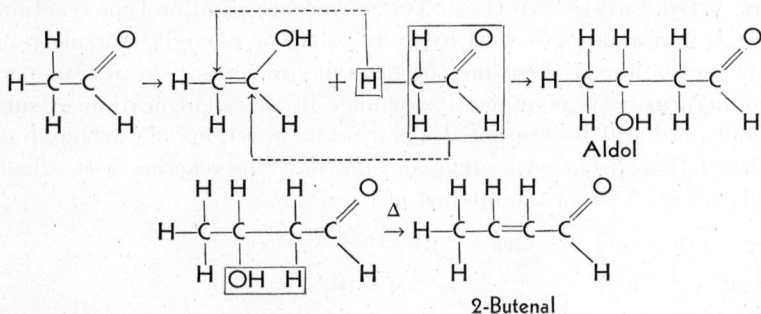

Aldol

2-Butenal

condensations, the product may be dehydrated, resulting in the formation

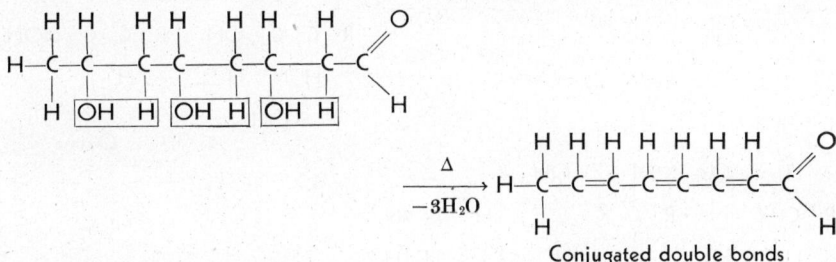

Conjugated double bonds

of a system of conjugate double bonds (p. 46). The name "aldol condensation" refers to the two different functional groups that form in the molecule, "al-" or "ald-" referring to alkanal or aldehyde and "-ol" referring to alkanol. Alkanones also undergo condensation reactions in much the same manner as alkanals, but less readily.

F. Hemiacetal and Acetal Formation. An alkanal can add ethanol in the presence of dry hydrogen chloride to form hemiacetal and acetal.

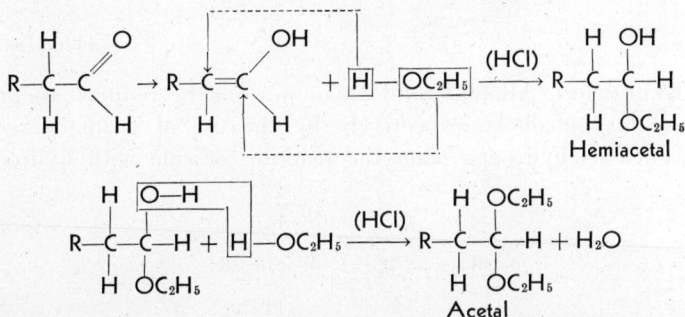

Hemiacetal

Acetal

The acetal is more stable than the hemiacetal. The reaction is easily reversed in mineral acid solution. Alkaline solution promotes condensation of alkanals but acetal is fairly stable to alkaline reaction. To preserve an alkanal function, alkanals are sometimes converted to the acetal for alkaline reaction of some other functional group, and the acetal is later reconverted to the alkanal.

G. WITH AMINES AND HYDRAZINES. Another addition type reaction of alkanals and alkanones with hydroxylamine or phenylhydrazine results in the formation of aldoximes or phenylhydrazones. The reaction with phenylhydrazine is of major importance in the identification of sugars (p. 159) and will be presented again under reactions of carbohydrates. Both reactions occur with alkanones but only the reaction with alkanals will be shown here in the interest of brevity.

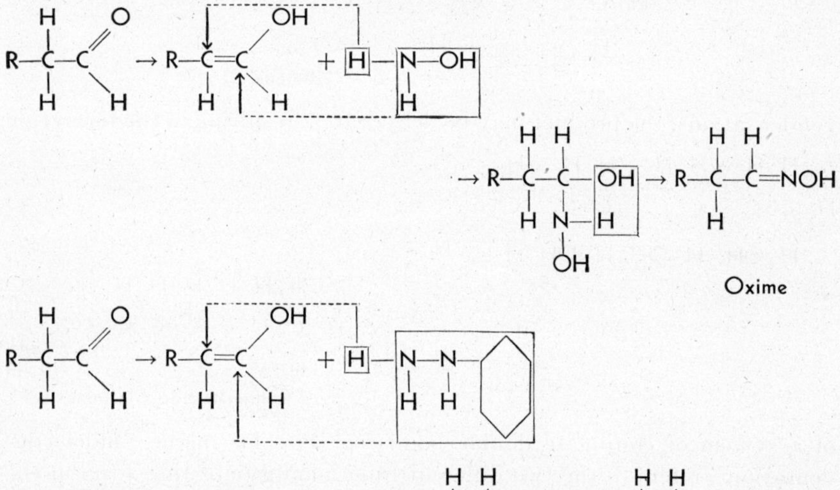

Oxime

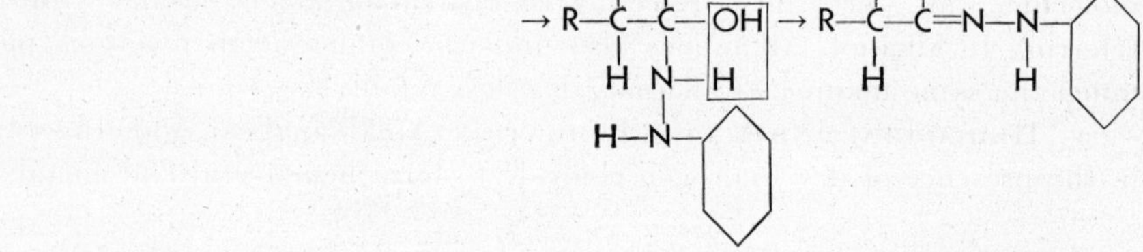

Hydrazone

H. REDUCTION. Alkanals and alkanones can be reduced to primary and secondary alcohols respectively by the use of ordinary reducing agents, such as hydrogen from the reaction of zinc with hydrochloric acid.

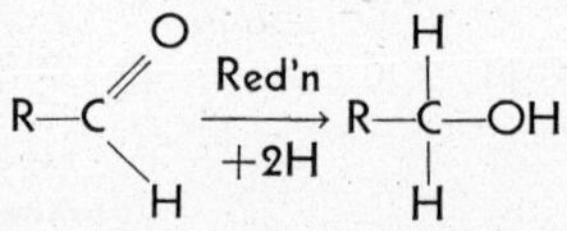

Meerwein, Ponndorf, and Verley discovered that aluminum 2-propoxide is a specific reducing agent for alkanals and alkanones. Reduction beyond the alcohol does not occur. The oxidation-reduction reaction is really one of aluminum exchange. The aluminum alkoxide is converted to the alcohol and aluminum chloride by the addition of dilute hydrochloric acid.

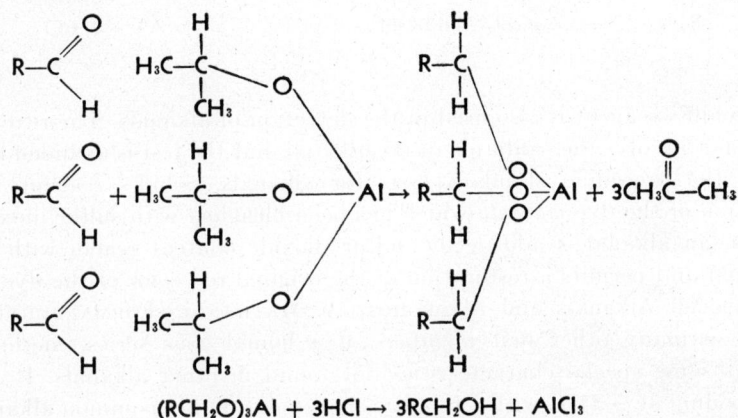

$$(RCH_2O)_3Al + 3HCl \rightarrow 3RCH_2OH + AlCl_3$$

1. OXIDATION. Alkanoic acids are formed by the oxidation of alkanals with vigorous oxidizing agents such as dichromate ion in acid solution. Permanganates in acid solution may also be used. Alkanone molecules are split by strong oxidizing solutions and may oxidize to form acids from both portions of the molecule.

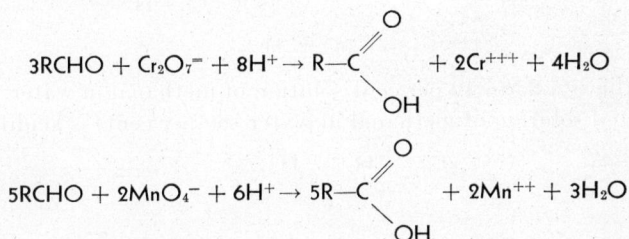

$$3RCHO + Cr_2O_7^= + 8H^+ \rightarrow R-C\underset{OH}{\overset{O}{\lessgtr}} + 2Cr^{+++} + 4H_2O$$

$$5RCHO + 2MnO_4^- + 6H^+ \rightarrow 5R-C\underset{OH}{\overset{O}{\lessgtr}} + 2Mn^{++} + 3H_2O$$

Fehling's solution or Benedict's solution are oxidizing agents used in testing for alkanals. They contain cupric-complex ions in basic solution and they oxidize the alkanal to an alkanoic acid. The cupric-complex ion (blue) is reduced to cuprous ion which precipitates in reddish brown to yellow cuprous oxide when the mixture is boiled.

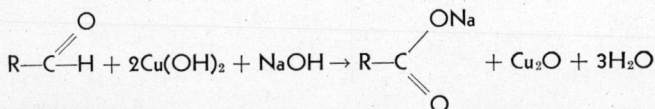

$$R-\overset{O}{\overset{\|}{C}}-H + 2Cu(OH)_2 + NaOH \rightarrow R-C\underset{O}{\overset{ONa}{\lessgtr}} + Cu_2O + 3H_2O$$

Tollen's solution also oxidizes alkanals to alkanoic acids. Tollen's solution contains silver ammonia ion in alkaline solution. The silver complex is reduced to free silver, leaving a mirror deposit on clean glassware. Tollen's solution will also oxidize alkanones and thus, with Fehling's, serves as a means of distinguishing between alkanals and alkanones.

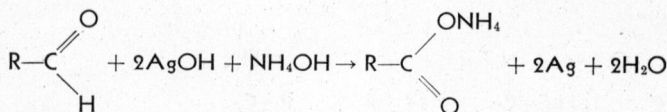

$$R-C\underset{H}{\overset{O}{\diagup}} + 2AgOH + NH_4OH \rightarrow R-C\underset{O}{\overset{ONH_4}{\diagup}} + 2Ag + 2H_2O$$

Schiff's reagent is also used in the detection of alkanals. The reaction does not involve the oxidation of the alkanal, but the test is included here with the preceding standard tests for alkanals. Schiff's reagent is a solution of the dye *magenta* which has been bleached with sulfur dioxide. When an alkanal is added, the sulfur dioxide content reacts with the alkanal and permits a restoration of the original red color of the dye.

Special Alkanals and Alkanones: A. METHANAL (FORMALDEHYDE). Like so many other first members of a homologous series, methanal shows some special characteristics not found in other alkanals. It is a gas boiling at −21°C with a familiar sharp odor. Other common alkanals are liquids.

1. PREPARATION. Methanal is prepared commercially by passing hot methanol vapors over heated copper gauze. The combustion of the hydrogen with added oxygen provides the heat energy required.

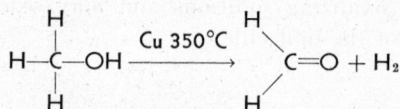

$$H-\underset{H}{\overset{H}{\underset{|}{C}}}-OH \xrightarrow{Cu\ 350°C} \overset{H}{\underset{H}{\diagdown}}C=O + H_2$$

Formalin is a 35 to 40 per cent solution of methanal in water. When a concentrated solution of methanal in water (60 per cent) is acidified with

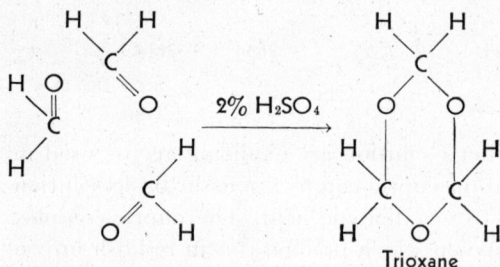

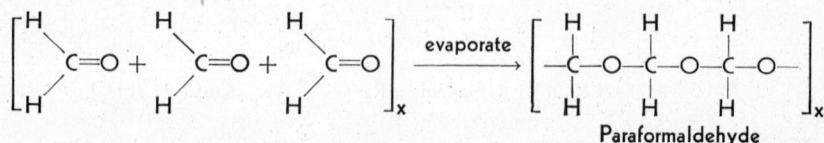

Trioxane

Paraformaldehyde

2 per cent sulfuric acid and evaporated, a solid cyclic trimer of methanal called trioxane is produced. Methanal is regenerated from trioxane by heating. Paraformaldehyde, a solid linear polymer of indefinite length is made by slowly evaporating a solution of methanal. Paraformaldehyde candles are used in fumigating. Methanal is liberated from paraformaldehyde by heat.

2. REACTIONS OF METHANAL: *a. Auto-oxidation-reduction.* Methanal undergoes auto-oxidation and reduction in the presence of sodium hydroxide. One molecule of methanal is oxidized to methanoic acid at the expense of another molecule reduced to methanol. It is called the Cannizzaro reaction. This reaction is of greater importance in aromatic aldehydes (p. 222) and will be introduced in more detail later.

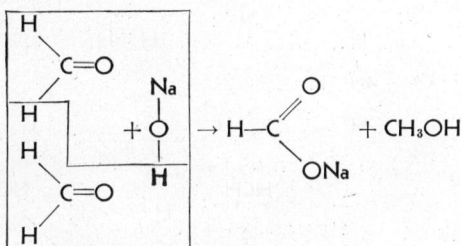

b. Condensation with Other Alkanals. Methanal cannot condense by aldol condensation because it has only one carbon atom. It utilizes its additive power to unite with the active hydrogen of other alkanals forming polyhydric alcohols. With ethanal, by successive reaction with the

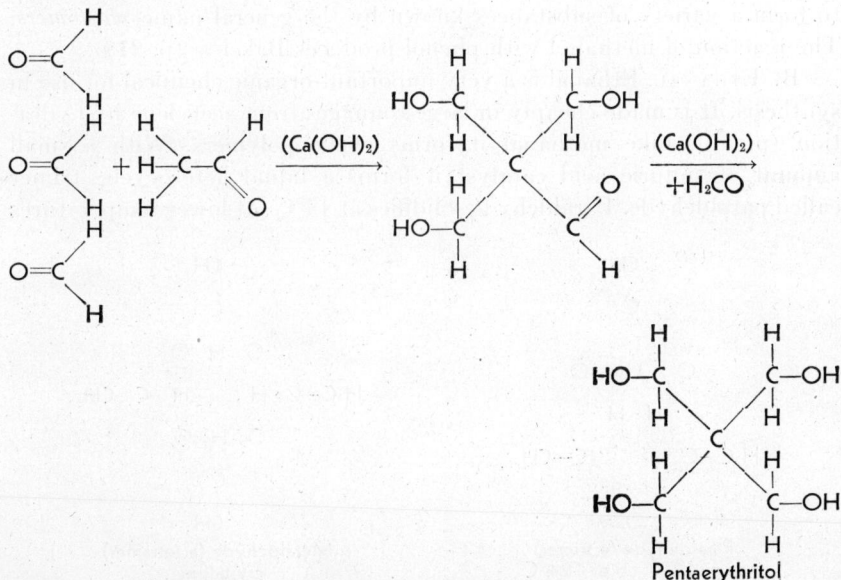

Pentaerythritol

three activated hydrogen atoms, methanal forms pentaerythritol. Calcium hydroxide serves as a catalyst. The high explosive, pentaerythritoltetranitrate PETN, is the nitric acid ester, of this compound.

c. With Ammonia. An aqueous solution of methanal and ammonia, when evaporated forms a crystalline solid, urotropin, also called hexamethylenetetramine. The high explosive RDX cyclonite is made from urotropin.

$$6 \quad \overset{H}{\underset{H}{>}}C{=}O + 4N{-}H \rightarrow$$

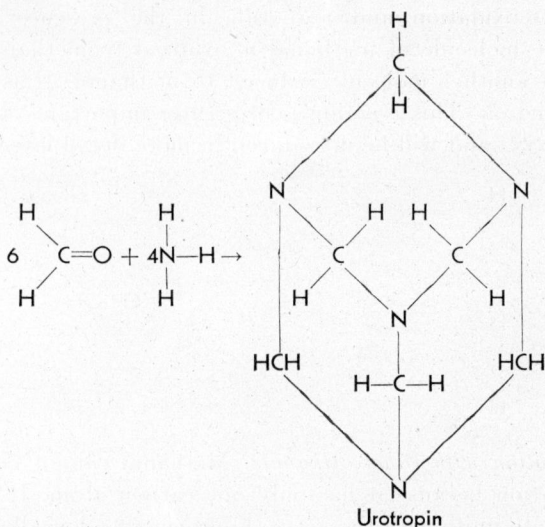

Urotropin

d. Methanal copolymerizes with a number of other organic compounds to form a variety of substances known by the general name *plastomers.* The reaction of methanol with phenol produces Bakelite (p. 218).

B. ETHANAL. Ethanal is a very important organic chemical for use in synthesis. It is made cheaply in large tonnages from acetylene by hydration (p. 50). Like methanal it forms cyclic polymers. With a small amount of sulfuric acid catalyst it forms a liquid heterocyclic trimer called paraldehyde. Paraldehyde solidifies at 12°C. At lower temperatures

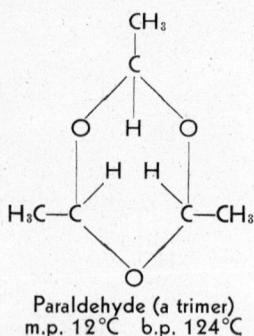

Paraldehyde (a trimer)
m.p. 12°C   b.p. 124°C

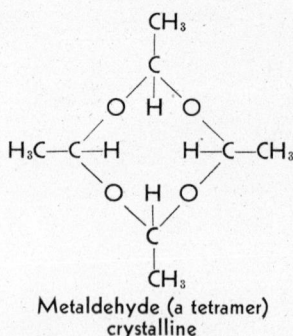

Metaldehyde (a tetramer)
crystalline

an eight-membered ring similar to paraldehyde is formed. This crystalline compound is called metaldehyde. Ethanal may be regenerated from either paraldehyde or metaldehyde by distillation after adding a small amount of dilute sulfuric acid.

C. 2-BUTENAL (CROTONALDEHYDE). 2-Butenal is formed by heating the product of an aldol condensation of ethanal. It is an unsaturated compound.

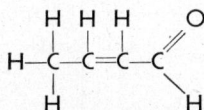

$$H-\overset{\overset{\displaystyle H}{|}}{\underset{\underset{\displaystyle H}{|}}{C}}-\overset{\overset{\displaystyle H}{|}}{C}=\overset{\overset{\displaystyle H}{|}}{C}-C\overset{\displaystyle O}{\underset{\displaystyle H}{}}$$

D. PROPANONE (ACETONE). Propanone is the most used member of the alkanone family. It was formerly made from pyroligneous acid (p. 73). It is now made synthetically from acetylene and also by the fermentation of certain starches with *Clostridium acetobutylicum* (p. 74).

ALDEHYDES †

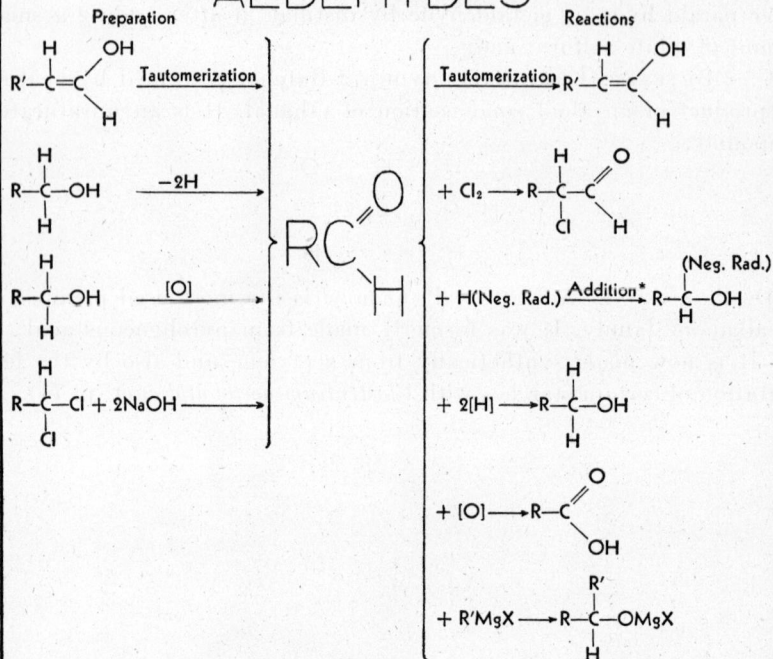

Preparation

Reactions

*Summary of Addition Reactions with Ethanal

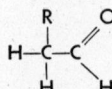

H—Neg. Radical

H—CN

H—OSO₂Na

H—OC₂H₅

H—NH₂

H—NHOH

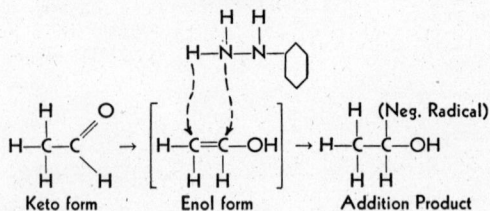

Keto form   Enol form   Addition Product

† Replacing —H of carbonyl group by R' provides most ketone reactions.

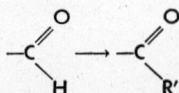

Study and Review Questions

1. Show how a carbonyl group is a dehydrated -diol.
2. Explain keto-enol tautomerism of aldehydes and ketones.
3. Name the compounds whose structural formulas are the following: (a) $CH_3CH_2CH_2CH_2CHO$, (b) $(CH_3)_2CH_2CHO$, (c) $CH_3CH(CH_3)$-CHO, (d) $(CH_3)_3CCHO$, (e) $CH_3COCH_2CH_3$, (f) $(CH_3)_2CHCOCH_3$, (g) $CH_3CH_2COCH_2CH_3$, (h) $(CH_3)_3CCOCH_3$, and (i) CH_3CH_2-$COCH_3$.
4. Write the equations for a commercial method of preparing butanal and butanone.
5. Starting with chloroethane, show by a series of equations how to prepare ethanal.
6. Starting with 2-chloropropane, show by a series of equations how to prepare (a) propanone and (b) propanal.
7. Show in steps the reaction of ethanal with chlorine.
8. By a series of equations show in steps what happens when propanal reacts with hydrogen cyanide.
9. Write a series of reactions showing how propanone (acetone) may be separated from impurities in a mixture.
10. In the aldol condensation reaction of ethanal why is 2-butenal formed on dehydration rather than 3-butenal?
11. Aldols are formed in basic solution. What is produced by the reaction of ethanal with ethanol in the presence of hydrogen chloride? Write equations for the reaction.
12. Propanal and propanone both react with either phenylhydrazine or hydroxylamine. Write the equations for all four reactions.
13. What is the chief difficulty encountered in trying to reduce propanal or propanone to an alcohol using the zinc and hydrochloric acid method?
14. Write an equation for reducing propanal and propanone to their proper alcohols that will avoid the difficulty you have noted in Question 13.
15. Prepare a small chart showing the reagents used and the color changes involved in the use of Fehling's, Tollen's, and Schiff's reagents with an aldehyde.
16. Contrast the methods of preparing trioxane and paraformaldehyde. How is methanal liberated from each polymer?
17. Outline a method of preparing methanoic (formic) acid from methanol, using only reactions shown in this chapter.
18. Write the equation for the preparation of ethanal from acetylene.
19. Contrast paraldehyde and paraformaldehyde from the point of view of preparation, structure of compounds, phase condition at 20°C (room temperature), and method of regenerating the proper aldehyde.

Chapter 7

COMPOUNDS OF CARBON
WITH OXYGEN AND HYDROGEN
III. THIRD OXIDATION: ORGANIC ACIDS

CHAPTER PROLOGUE

Acids are the "goofiest" things in organic chemistry. They are relatively simple compounds but they react in a complex way. As acids they yield hydrogen ion, but they are very stingy in their yield. With an hydroxyl group attached to a carbon atom, we might expect them to react like alcohols, but they don't. They have an oxygen atom double-bonded to a carbon atom, one of the requirements of a carbonyl group, yet they do not react like aldehydes or ketones. Other than the very simple acids, they form their alkyl chain in nature by avoiding odd numbers of carbon atoms in the molecule. The most common of all long-chain acid molecules is stearic acid with eighteen saturated carbon atoms. It is a solid at room temperature. When two little atoms of hydrogen are removed, one each from the ninth and tenth carbon atoms, the resulting acid, oleic, is a liquid. Remove a few more hydrogen atoms from oleic acid, and the resulting acid is again a solid. In water most of these acids are insoluble especially those with a long alkyl group, yet if sodium or potassium replaces the hydrogen of the carboxyl group, even the long alkyl group dissolves in water. The sodium salt dissolves not only in water, it also dissolves in alcohol, ether, and even kerosene (hydrocarbon). Many of the organic acids are odorless and some of them have a perfectly disgusting smell. Those with the disgusting smell can be altered by a simple reaction to the inviting odor of pineapples, or peaches, or apricots.

Here then are the acids, the "goofiest" compounds in organic chemistry. This chapter will tell you about their structure and behavior.

The distinguishing structural characteristic of the organic acids is the carboxyl group, —COOH. Since all but one of the bonds of the group are required to hold the oxygen and hydroxyl groups, the carbon atom is always a primary carbon atom. In this chapter we shall consider the characteristics of this carboxyl group in detail, particular attention being given to those acids having but one such group. Certain alkanedioic acids (those with two carboxyl groups) will be briefly discussed in the latter part of the chapter.

Of great importance are the compounds derived from acids in which part of the carboxyl group is replaced by some other atom or group, and also those compounds in which other functional groups have been substituted for hydrogen atoms on the aliphatic portion of the molecule. Very significant influence is caused by these derivatives and substituted products. They will be studied in detail in Chapters 8 (Acid Derivatives) and 9 (Substituted Acids) respectively.

Naming of Acids. Carboxylic acids are named in the I.U.C. system by substituting the suffix -oic for the e of the particular alkane having the same number of carbon atoms. The separate word "acid" is always added. Thus the carboxylic acid whose formula is CH_3CH_2COOH is

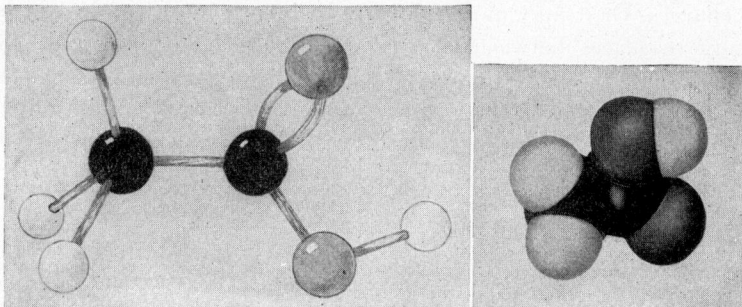

Fig. 19. The acetic (ethanoic) acid molecule.

called propanoic acid. Acids having only one carboxyl group are known by the general name, alkanoic acids. The general type formula for alkanoic acids is RCOOH or RCO_2H.

Structure of Carboxyl Group. The exact structure of the carboxyl group is still undetermined and the formula may be correctly written as

$$\overset{O}{\underset{}{\overset{\|}{-}C}-OH}$$

—C—OH or as —CO_2H. There is evidence to support both formulas and the structures they imply. In its effect upon the hydrogen atoms attached to the adjoining carbon atom the carboxyl group is similar to the carbonyl group. These hydrogens are activated by the induced polarity of the highly negative oxygens of the carboxyl group. Also the —OH group may be replaced with such agents as PCl_5. This evidence

favors the formula —C—OH.

Evidence for the acceptance of the second formula is based upon the fact that the attributed carbonyl structure is not truly a carbonyl structure. True carbonyl groups always react with hydroxylamine to form oximes (p. 90). Carboxyl groups do not react with hydroxylamine. This failure to form oximes is explained by the negative character of oxygen which attracts the hydrogen in the neighboring hydroxyl of the carboxyl group. Thus we can say that the carbonyl function of the acid group is significantly modified by the presence of the second oxygen atom. The oxygen temporarily holding the hydrogen has been rendered less intensely negative by having the hydrogen attached to it. Under these circumstances the hydrogen follows the stronger attraction and tends to migrate from the hydroxyl group to the lone oxygen, transforming it into a hydroxyl group, thereby reversing the previous electro-

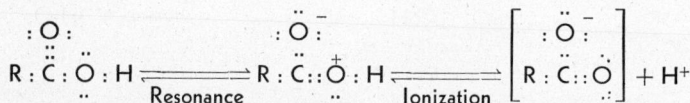

$$R : \overset{..}{\underset{..}{C}} : \overset{..}{\underset{..}{O}} : H \underset{\text{Resonance}}{\rightleftharpoons} R : \overset{..}{C} :: \overset{..}{O} : H \underset{\text{Ionization}}{\rightleftharpoons} \left[R : \overset{..}{C} :: \overset{..}{O} \right]^- + H^+$$

static charges. Or it may more correctly be interpreted that the electron structure resonates between the carbon and its bonded oxygens with lightning rapidity. The ultimate state of the hydrogen then becomes that of a proton held between two oxygen atoms which are alternately highly negative and slightly less negative. The ionized state of the carboxyl

group may be expressed as $\left[\begin{matrix} :\ddot{O}: \\ R:\ddot{C}:\ddot{O}: \end{matrix} \right]^- + H^+.$

Physical Properties. Alkanoic acids from C 1 through C 9 are liquids at room temperatures. The melting points of methanoic acid (8°) and ethanoic acid (16.7°) are disproportionately high. Lower homologues (C 1–C 4) are miscible with water in all proportions and have sharp odors. Acids in the range C 5–C 10 are less soluble. Those from C 5 to C 10 also have disgusting animal odors. Higher homologues are practically odorless. The predominance of the alkyl part of the molecule gives these higher homologues a waxlike appearance and makes them feel greasy. Alkanoic acids are all weakly but uniformally ionized. Formic acid is the strongest acid $K = 2.1 \times 10^{-4}$; the rest have ionization constants near 1.6×10^{-5}.

The alkanoic acids with an even number of carbon atoms in the chain are commonly found in fats and oils as esters of glycerol. Most abundant of all are those with 18 carbon atoms. The homologous series may contain unsaturated structure in the aliphatic portion of the molecule. The specific acids of this type and their relation to the properties and uses of the resulting fats and oils will be discussed under esters in the chapter on Acid Derivatives (Chapter 8).

Preparation. The commercial method of preparing acids is usually specific for each acid considered. The preparation of methanoic (formic) acid and ethanoic (acetic) acid will be discussed specifically later in this chapter.

General methods of preparing acids are as follows:

1. OXIDATION OF AN ALDEHYDE. Equivalent weights of strong oxidizing agents such as potassium dichromate or potassium permanganate are used with dilute sulfuric acid. Concentrated sulfuric acid tends to produce esters.

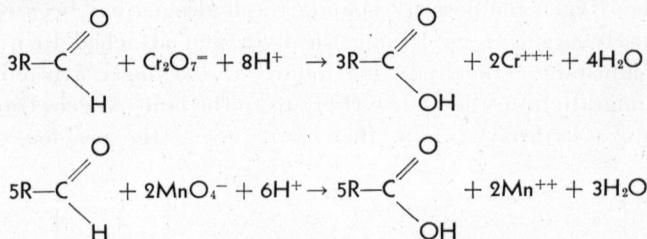

$$3R-\overset{\displaystyle O}{\underset{\displaystyle H}{C}} + Cr_2O_7^- + 8H^+ \rightarrow 3R-\overset{\displaystyle O}{\underset{\displaystyle OH}{C}} + 2Cr^{+++} + 4H_2O$$

$$5R-\overset{\displaystyle O}{\underset{\displaystyle H}{C}} + 2MnO_4^- + 6H^+ \rightarrow 5R-\overset{\displaystyle O}{\underset{\displaystyle OH}{C}} + 2Mn^{++} + 3H_2O$$

2. Hydrolysis of a Nitrile. Nitriles may be hydrolyzed in either acidic or basic solution. Thus by converting a compound to the nitrile

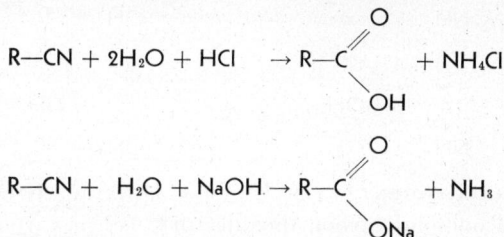

$$R\text{—}CN + 2H_2O + HCl \rightarrow R\text{—}C\overset{\displaystyle O}{\underset{\displaystyle OH}{\big\backslash}} + NH_4Cl$$

$$R\text{—}CN + H_2O + NaOH \rightarrow R\text{—}C\overset{\displaystyle O}{\underset{\displaystyle ONa}{\big\backslash}} + NH_3$$

and oxidizing it we have a convenient method of increasing the carbon atoms in a chain.

3. By Action of a Strong Acid on a Salt. Mineral acids convert salts of organic acids to the acid state.

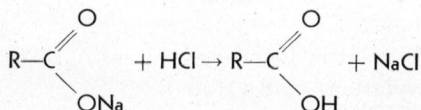

$$R\text{—}C\overset{\displaystyle O}{\underset{\displaystyle ONa}{\big\backslash}} + HCl \rightarrow R\text{—}C\overset{\displaystyle O}{\underset{\displaystyle OH}{\big\backslash}} + NaCl$$

4. By Hydrolysis of Acid Derivatives:[1] **A. hydrolysis of esters.** Sodium hydroxide with esters liberates an alcohol and produces the salt of an organic acid. The organic acid is then formed by adding mineral acid.

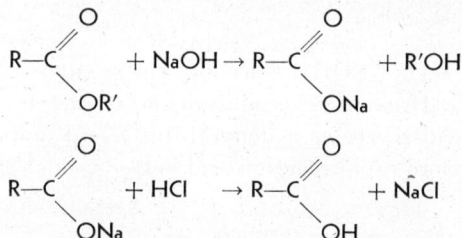

$$R\text{—}C\overset{\displaystyle O}{\underset{\displaystyle OR'}{\big\backslash}} + NaOH \rightarrow R\text{—}C\overset{\displaystyle O}{\underset{\displaystyle ONa}{\big\backslash}} + R'OH$$

$$R\text{—}C\overset{\displaystyle O}{\underset{\displaystyle ONa}{\big\backslash}} + HCl \rightarrow R\text{—}C\overset{\displaystyle O}{\underset{\displaystyle OH}{\big\backslash}} + NaCl$$

B. hydrolysis of an acid anhydride. The addition of water produces acids.

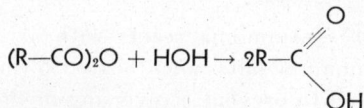

$$(R\text{—}CO)_2O + HOH \rightarrow 2R\text{—}C\overset{\displaystyle O}{\underset{\displaystyle OH}{\big\backslash}}$$

C. hydrolysis of amides. Amides boiled with water containing a mineral acid or sodium hydroxide will be converted to an organic acid or the salt of such an acid.

---

[1] More complete information about these derivatives will be presented in Chapter 8.

$$R-\underset{NH_2}{\overset{O}{\underset{|}{\overset{\|}{C}}}} + H_2O + H_2SO_4 \rightarrow R-\underset{OH}{\overset{O}{\underset{|}{\overset{\|}{C}}}} + NH_4HSO_4$$

$$R-\underset{NH_2}{\overset{O}{\underset{|}{\overset{\|}{C}}}} + NaOH \rightarrow R-\underset{ONa}{\overset{O}{\underset{|}{\overset{\|}{C}}}} + NH_3$$

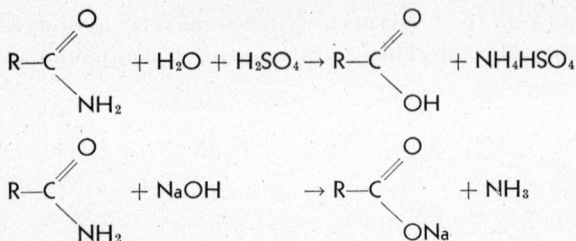

5. GRIGNARD REAGENT WITH CARBON DIOXIDE. Grignard reagent may be poured on solid carbon dioxide (dry ice) or carbon dioxide gas may be bubbled through Grignard reagent to produce an acid upon acid hydrolysis of the addition product.

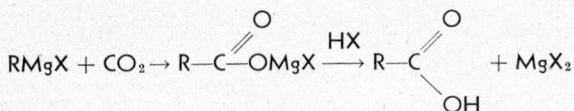

$$RMgX + CO_2 \rightarrow R-\overset{O}{\overset{\|}{C}}-OMgX \overset{HX}{\longrightarrow} R-\underset{OH}{\overset{O}{\underset{|}{\overset{\|}{C}}}} + MgX_2$$

Reactions. Acids react in four ways: (1) as —H, (2) as —OH, (3) as —COOH (decarboxylation), and (4) as R—.

1. REACTIONS OF THE —H. Being a true acid the —H portion of the carboxyl radical can be neutralized by a base to form a salt and water.

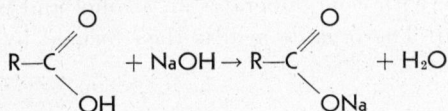

$$R-\underset{OH}{\overset{O}{\underset{|}{\overset{\|}{C}}}} + NaOH \rightarrow R-\underset{ONa}{\overset{O}{\underset{|}{\overset{\|}{C}}}} + H_2O$$

2. THE HYDROXYL (—OH) PART OF THE CARBOXYL GROUP reacts (a) *by Alcoholysis.* Organic acids and alcohols react to produce esters. A little sulfuric acid serves as a dehydrating agent and to some extent as a catalyst for more rapid reaction and better yields.

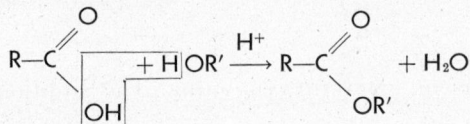

$$R-\underset{OH}{\overset{O}{\underset{|}{\overset{\|}{C}}}} + H \big| OR' \overset{H^+}{\longrightarrow} R-\underset{OR'}{\overset{O}{\underset{|}{\overset{\|}{C}}}} + H_2O$$

(b) *By Ammonolysis.* Ammonia reacts with an acid to produce an ammonium salt. Ammonium salts when heated form amides and water. Water must be removed to prevent a reversing of the reaction.

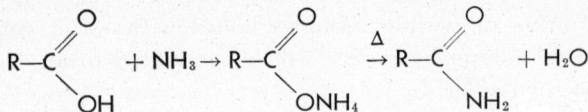

$$R-\underset{OH}{\overset{O}{\underset{|}{\overset{\|}{C}}}} + NH_3 \rightarrow R-\underset{ONH_4}{\overset{O}{\underset{|}{\overset{\|}{C}}}} \overset{\Delta}{\rightarrow} R-\underset{NH_2}{\overset{O}{\underset{|}{\overset{\|}{C}}}} + H_2O$$

(c) *By Replacement with a Halogen.* Phosphorus trihalide reacts with acids to form acyl halide and phosphorous acid.

$$3R-\overset{O}{\underset{OH}{C}} + PX_3 \rightarrow 3R-\overset{O}{\underset{X}{C}} + H_3PO_3$$

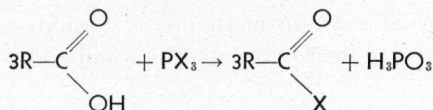

Since it is somewhat more difficult to replace the —OH group of the acid with halogen than it is to replace the —OH group of an alcohol, phosphorus pentahalide, as PCl_5, gives much better yields than the trihalide. The use of thionyl chloride $SOCl_2$ is also important since it gives an excellent yield and purification of the acyl chloride is easier. It will be noted that the other products of the reaction in the following example are gases.

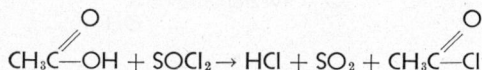

$$CH_3\overset{O}{C}-OH + SOCl_2 \rightarrow HCl + SO_2 + CH_3\overset{O}{C}-Cl$$

3. REACTION OF THE —COOH. The removal of the carboxyl group from an acid is known as decarboxylation. Although not of particular importance in the behavior of simple aliphatic acids, the removal of a carboxyl group from complex acids has many significant applications in physiology and industry. There are several examples cited in this text: (a) the removal of carbon dioxide from a salt in the formation of a hydrocarbon with one less carbon atom (p. 22); (b) the synthesis of a vital amino acid through a decarboxylation reaction is shown on p. 107; (c) the decarboxylation of amino acids within the body is described on p. 296; and (d) preparation of methanoic acid by decarboxylation of oxalic acid is shown on p. 104.

4. THE ALKYL PART OF AN ACID (R—) may be modified by certain replacements for hydrogen atoms thereby producing substituted acids. For example, a halogen may react with the alkyl part of an acid by virtue of the special activity of those hydrogen atoms attached to the carbon atom nearest the carboxyl group. These are discussed in greater detail in Chapter 9.

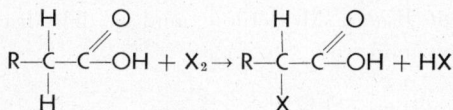

$$R-\overset{H}{\underset{H}{C}}-\overset{O}{C}-OH + X_2 \rightarrow R-\overset{H}{\underset{X}{C}}-\overset{O}{C}-OH + HX$$

Specific Acids: A. METHANOIC (FORMIC) ACID, like the first member of most homologous series, is unique among the acids. By its structure it is both an aldehyde and an acid, $HO-\overset{O}{C}-H$ and $H-\overset{O}{C}-OH$. Its common name, formic acid, is derived from the Latin *formica*, ant, because it was formerly prepared by the distillation of red ants.

PREPARATION. 1. Methanoic acid is produced commercially in very large quantities by the reaction of carbon monoxide under pressure on hot sodium hydroxide. Sodium methanoate (formate) produced by this reaction is converted to methanoic acid by a mineral acid.

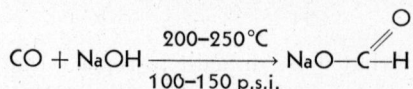

$$CO + NaOH \xrightarrow[\text{100–150 p.s.i.}]{\text{200–250 °C}} NaO-C\!\!\begin{smallmatrix}O\\||\\\ \end{smallmatrix}\!\!H$$

2. In the laboratory methanoic acid is prepared by the decarboxylation of oxalic acid in glycerol.

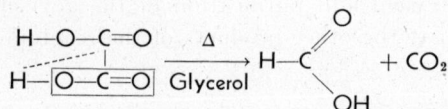

$$\begin{array}{c} H-O-C=O \\ H-\boxed{O-C=O} \end{array} \xrightarrow{\Delta} \text{Glycerol} \quad H-C\!\!\begin{smallmatrix}O\\\\OH\end{smallmatrix} + CO_2$$

REACTIONS: 1. *As a Reducing Agent.* Methanoic acid, because it is also an aldehyde, is one of the few organic acids capable of acting as a reducing agent. Water and carbon dioxide form when formic acid is used as a reducing agent.

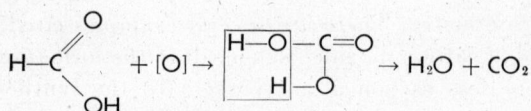

$$H-C\!\!\begin{smallmatrix}O\\\\OH\end{smallmatrix} + [O] \rightarrow \boxed{\begin{array}{c} H-O-C=O \\ H-O \end{array}} \rightarrow H_2O + CO_2$$

Certain metal salts are reduced to lower oxidation states with methanoic acid. The reduction of soluble mercuric chloride to insoluble mercurous chloride is a test for methanoic acid.

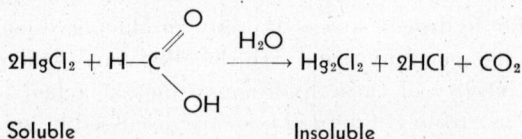

$$2HgCl_2 + H-C\!\!\begin{smallmatrix}O\\\\OH\end{smallmatrix} \xrightarrow{H_2O} Hg_2Cl_2 + 2HCl + CO_2$$

Soluble   Insoluble

2. *With Loss of Water.* Methanoic acid is dehydrated by sulfuric acid to form carbon monoxide.

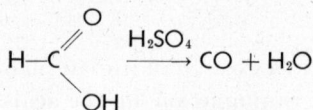

$$H-C\!\!\begin{smallmatrix}O\\\\OH\end{smallmatrix} \xrightarrow{H_2SO_4} CO + H_2O$$

3. *By Heating.* When heated in an autoclave methanoic acid decomposes to form hydrogen and carbon dioxide.

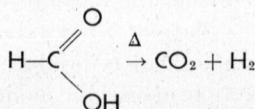

$$H-C\!\!\begin{smallmatrix}O\\\\OH\end{smallmatrix} \xrightarrow{\Delta} CO_2 + H_2$$

4. *By Heating a Salt.* Sodium methanoate when heated strongly forms hydrogen gas and sodium oxalate. This method is used in the manufacture of oxalic acid.

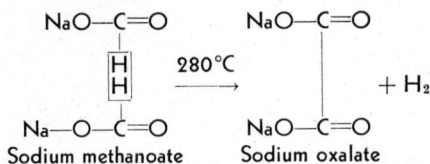

$$
\underset{\text{Sodium methanoate}}{
\begin{array}{c}
\text{NaO—C}{=}\text{O} \\
\boxed{\begin{array}{c}\text{H}\\ \text{H}\end{array}} \\
\text{Na—O—C}{=}\text{O}
\end{array}}
\quad \xrightarrow{280\,°\text{C}} \quad
\underset{\text{Sodium oxalate}}{
\begin{array}{c}
\text{NaO—C}{=}\text{O} \\
| \\
\text{NaO—C}{=}\text{O}
\end{array}}
\; + \text{H}_2
$$

B. Ethanoic (Acetic) Acid. Acetic acid can be prepared in several ways. Until the development of a synthetic process in Canada during World War I, most concentrated acetic acid was prepared from a wood tar known as pyroligneous acid obtained from the destructive distillation of wood (p. 73). Pure acetic acid is called "Glacial Acetic Acid" because it so readily solidifies (without expanding). Its melting point is 16.6°C.

Vinegar contains from 3 to 6 per cent acetic acid. Dilute solutions of ethanol are catalyzed by the air-borne bacteria *Acetobacter*. Several species of this bacteria (called "Mother of Vinegar") are capable of causing the oxidation of ethanol with the aid of oxygen from air to acetic acid. Acetaldehyde is an intermediate product.

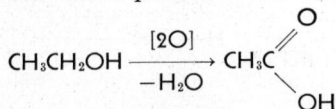

$$
\text{CH}_3\text{CH}_2\text{OH} \xrightarrow[-\text{H}_2\text{O}]{[2\text{O}]} \text{CH}_3\text{C}{\overset{\displaystyle\text{O}}{\underset{\displaystyle\text{OH}}{\Big\langle}}}
$$

Acetic acid is prepared commercially by the catalytic oxidation of acetaldehyde produced from acetylene (p. 50). Manganese acetate is used as the catalyst and oxygen is taken from the air.

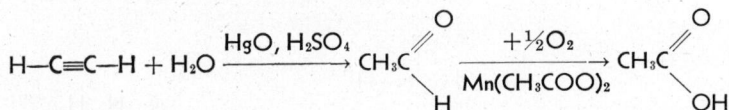

$$
\text{H—C}{\equiv}\text{C—H} + \text{H}_2\text{O} \xrightarrow{\text{HgO, H}_2\text{SO}_4} \text{CH}_3\text{C}{\overset{\displaystyle\text{O}}{\underset{\displaystyle\text{H}}{\Big\langle}}} \xrightarrow[\text{Mn(CH}_3\text{COO)}_2]{+\tfrac{1}{2}\text{O}_2} \text{CH}_3\text{C}{\overset{\displaystyle\text{O}}{\underset{\displaystyle\text{OH}}{\Big\langle}}}
$$

Alkanedioic Acids. Type formula $C_nH_{2n}(CO_2H)_2$. Alkanedioic acids are solid crystalline substances. Each carboxyl radical acts as a unit in undergoing the typical reactions described under acids. Two additional outstanding reactions of alkanedioic acids are (a) their ability, with a few exceptions, to form internal anhydrides and (b) the decarboxylation of the acid in the case of lower members of the series to form alkanoic acids. Neither oxalic nor malonic acids form anhydrides. These reactions will be illustrated during the discussion of specific acids which follows.

Oxalic Acid. Ethanedioic Acid is also the I.U.C. name for oxalic acid. It is produced commercially as previously shown in this chapter (p. 105). It is produced as a hydrated acid, $(CO_2H)_2 \cdot 2H_2O$. The water of hydration is easily removed by heating for a time at temperatures slightly above **100°C.** **The** hydrated salt suggests the interesting possibility of its existence as a hexahydroxyethane, one of the few cases in which more than one —OH group is joined to a carbon atom.

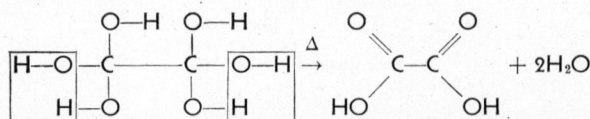

$$H-O-\overset{\overset{\displaystyle O-H}{|}}{\underset{\underset{\displaystyle H-O}{|}}{C}}-\overset{\overset{\displaystyle O-H}{|}}{\underset{\underset{\displaystyle O-H}{|}}{C}}-O-H \xrightarrow{\Delta} \underset{HO}{\overset{O}{\diagup}}C-C\underset{OH}{\overset{O}{\diagdown}} + 2H_2O$$

Like methanoic acid, the lowest homologue in its series, this lowest homologue of the alkanedioic acid series is also a reducing agent. It is oxidized by powerful oxidizing agents to CO_2 and H_2O. Sodium oxalate in acid solution $(H_2C_2O_4)$ is used in the standardizing of permanganate solutions. Oxalic acid in its first stage of ionization is a strong organic

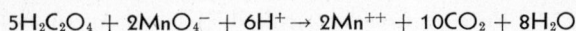

$$5H_2C_2O_4 + 2MnO_4^- + 6H^+ \rightarrow 2Mn^{++} + 10CO_2 + 8H_2O$$

acid, ranking next among the acids in acidity to trichloroacetic acid.

Malonic Acid. Propanedioic acid is also the I.U.C. name for malonic acid. It is usually prepared by converting monochloroacetic acid to the nitrile with hydrolysis of the product to form the malonic acid.

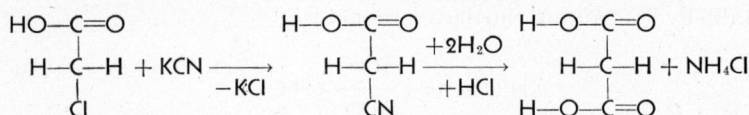

$$\underset{Cl}{\overset{HO-C=O}{\underset{|}{H-C-H}}} + KCN \xrightarrow[-KCl]{} \underset{CN}{\overset{H-O-C=O}{\underset{|}{H-C-H}}} \xrightarrow[+HCl]{+2H_2O} \underset{H-O-C=O}{\overset{H-O-C=O}{\underset{|}{H-C-H}}} + NH_4Cl$$

The ethyl ester of malonic acid is one of our most important agents of synthesis. As an example the reaction with urea in the presence of sodium ethoxide to form barbituric acid is cited. Modifications of barbituric acid are used as "sleeping pills."

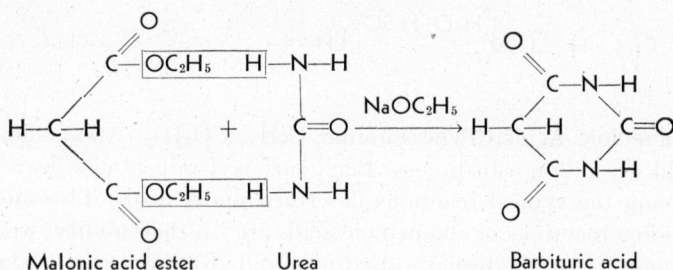

$$\underset{\text{Malonic acid ester}}{\overset{C-\boxed{OC_2H_5}}{\underset{C-\boxed{OC_2H_5}}{H-C-H}}} \quad + \quad \underset{\text{Urea}}{\overset{H-N-H}{\underset{H-N-H}{C=O}}} \xrightarrow{NaOC_2H_5} \underset{\text{Barbituric acid}}{\overset{C-N-H}{\underset{C-N-H}{H-C-H}}C=O}$$

Another example of the use of malonic ester is in the synthesis of the amino acid, alinine. In this synthesis, decarboxylation of an acid is also

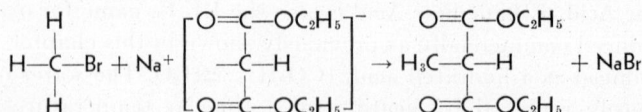

$$\underset{H}{\overset{H}{\underset{|}{H-C-Br}}} + Na^+ \left[\underset{O=C-OC_2H_5}{\overset{O=C-OC_2H_5}{-C-H}}\right]^- \rightarrow \underset{O=C-OC_2H_5}{\overset{O=C-OC_2H_5}{H_3C-C-H}} + NaBr$$

illustrated. The 2-hydrogen of the diethyl ester of malonic acid is sufficiently activated by two adjacent carbonyl groups to be replaced by
sodium, forming sodiomalonic ester. This reacts with bromomethane
to form the diethyl ester of 2-methylpropanedioic acid. After hydrolyzing
this ester, bromine replaces the remaining activated hydrogen atom.

$$
\begin{array}{ccc}
\text{O=C—OC}_2\text{H}_5 & \text{O=C—OH} & \text{O=C—OH} \\
| & | & | \\
\text{H}_3\text{C—C—H} \xrightarrow{+2\text{H}_2\text{O}} & \text{H}_3\text{C—C—H} \xrightarrow{+\text{Br}_2} & \text{H}_3\text{C—C—Br} + \text{HBr} \\
| & | & | \\
\text{O=C—OC}_2\text{H}_5 & \text{O=C—OH} & \text{O=C—OH}
\end{array}
$$

Decarboxylation occurs when this 2-methyl-2-bromopropanedioic acid
is heated, and the product with ammonia will form alinine.

$$
\begin{array}{ccc}
\text{O=C—OH} & \text{O=C—OH} & \text{H} \quad \text{O} \\
| & | & | \quad \diagup\diagup \\
\text{H}_3\text{C—C—Br} \xrightarrow[-\text{CO}_2]{\Delta} & \text{H}_3\text{C—C—Br} \xrightarrow{+2\text{NH}_3} & \text{H}_3\text{C—C—C} \quad + \text{NH}_4\text{Br} \\
| & | & | \quad \diagdown \\
\boxed{\text{O=C—O}}\text{H} & \text{H} & \text{NH}_2 \quad \text{OH}
\end{array}
$$

Alanine

It should not be necessary to learn the foregoing synthesis. It is
included primarily to illustrate the use and reactions of these acids.

Succinic Acid. Succinic acid in the I.U.C. system is butanedioic acid.
It can be prepared from ethene by chlorination, treating the product
with KCN and hydrolyzing the resulting nitrile.

$$
\begin{array}{cccc}
& \text{H} & \text{H} & \text{H} \quad \text{O} \\
& | & | & | \quad \diagup\diagup \\
\text{H—C—H} & \text{H—C—Cl} & \text{H—C—CN} & \text{H—C—C} \\
\| \quad + \text{Cl}_2 & | \xrightarrow[-2\text{KCl}]{+2\text{KCN}} & | \xrightarrow[+2\text{HCl}]{+4\text{H}_2\text{O}} & \quad\quad \diagdown\text{OH} \quad + 2\text{NH}_4\text{Cl} \\
\text{H—C—H} & \text{H—C—Cl} & \text{H—C—CN} & \text{H—C—C} \diagup\text{OH} \\
& | & | & | \quad \diagdown \\
& \text{H} & \text{H} & \text{H} \quad \text{O}
\end{array}
$$

Succinic acid and its next higher homologue, glutaric acid (pentane-
dioic acid) form cyclic anhydrides when heated alone or with dehydrating
agents. Five- and six-membered ring structures, such as these two
anhydrides possess, are the most stable ring compounds.

$$
\begin{array}{cc}
\text{H} & \text{H} \\
| & | \\
\text{H—C—C=O} & \text{H—C—C=O} \\
\quad \diagdown\text{O—H} \xrightarrow{\Delta} & \quad\quad \diagdown\text{O} \\
\quad \diagup\text{O—H} \quad {-\text{H}_2\text{O}} & \text{H—C—C=O} \\
\text{H—C—C=O} & | \\
| & \text{H} \\
\text{H} &
\end{array}
$$

Adipic Acid. Adipic acid (hexanedioic acid) is prepared by hydrogenating phenol and oxidizing the resulting cyclohexanol. It is used in the manufacture of Nylon (p. 170).

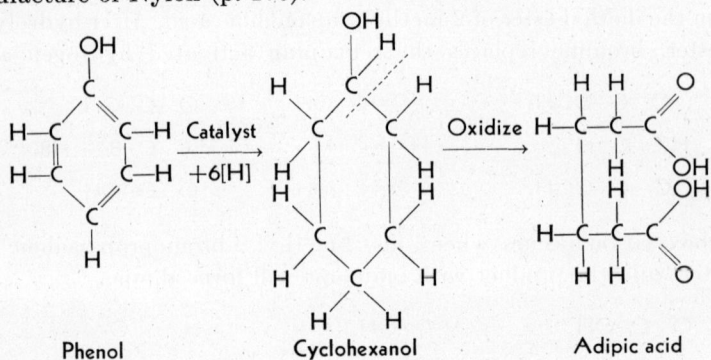

Phenol   Cyclohexanol   Adipic acid

ACIDS

Preparation

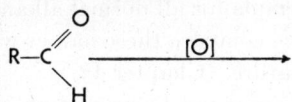

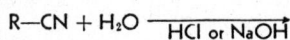

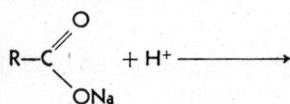

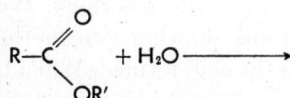

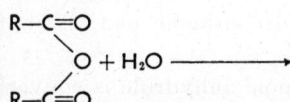

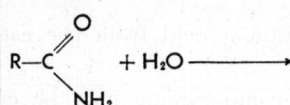

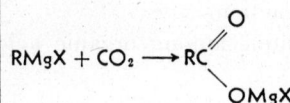

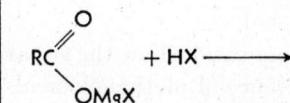

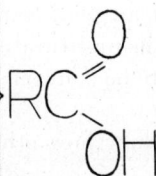

Reactions

Reactions of H+

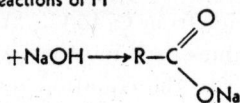

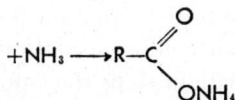

Reaction of —OH

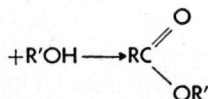

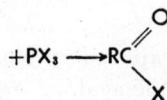

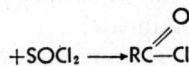

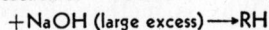

Reaction of —COOH
+NaOH (large excess) ⟶ RH

Reaction of R—

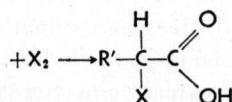

SPECIFIC ACIDS

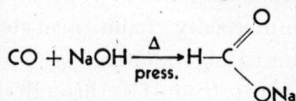

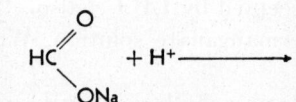

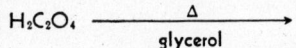

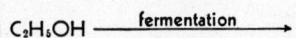

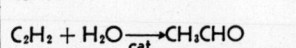

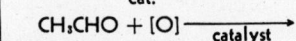

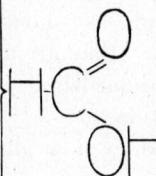

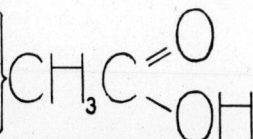

Study and Review Questions

1. Write the I.U.C. name and structural formula for all normal alkanoic acids from C_1 to C_{10}. As a useful reference, compare these names with names used in other systems of nomenclature. (Chapter 12.)

2. Write the equation for the oxidation of propanal to propanoic acid with permanganate solution.

3. Write the equation for the oxidation of propanal to propanoic acid with Fehling's solution.

4. Write in steps the equations for the oxidation of 1-propanol to propanoic acid using potassium dichromate as an oxidant.

5. When propanal reacts with HCN, a type of nitrile is formed. Write the equation for the reaction and equations showing two methods of converting this nitrile to an acid. Name the acid formed. When this acid is dehydrated, name the unsaturated acid formed by the reaction.

6. How can sodium butanoate be converted to butanoic acid? Write the equation.

7. Using graphic formulas, show how ethanoic anhydride is converted to ethanoic acid.

8. Write the equations for preparing butanoic acid from the ester, ethylbutanoate.

9. By a series of equations show how 1-bromopropane may be converted to butanoic acid by a Grignard reaction.

10. Why is methanoic acid structurally unique among organic acids? Give contrasting examples.

11. Write the equations for the commercial manufacture of methanoic acid, showing its final conversion to the acid.

12. Based upon the reducing action of methanoic acid, show the equation and explain how you can test for the presence of methanoic acid.

13. Write a series of equations showing how oxalic acid can be prepared by starting with carbon monoxide and other reagents of your choice.

14. Ethanoic acid is being prepared commercially from acetylene. Write equations for the reactions showing catalysts used.

15. Name all alkanedioic acids as -dioic acids from C_2 through C_6. Write their common names (they are accepted by I.U.C.) also.

16. Oxalic acid is used in standardizing permanganate solution. Write the equation for this reaction.

17. By a series of equations show how glutaric anhydride may be prepared from 3-chloro-1-propene.

Chapter 8

ACID DERIVATIVES
SALTS, ESTERS, ACID ANHYDRIDES, ACYL HALIDES, AND AMIDES

CHAPTER PROLOGUE

Modified functional units of the carboxyl radical of organic acids are versatile combinations of latent power. The acid itself is persistently powerful yet mild in comparison with some inorganic acids. It is like an engine with a speed governor traveling at neither too rapid nor too slow a rate. Man in his impatience to annihilate time has modified the functional group of acids to give him two rapid reagents of synthesis, the acid anhydrides and the acyl halides, neither of which occurs free in nature.

Contrariwise, man has used another modification of acids, their metallic salts, to insure by their hydrolysis, a persistent, uniform basic concentration. Nature, herself uses this hydrolysis of metallic salts of organic acids in the bodies of men and animals for controlling the concentration of vital fluid such as blood. On the other hand nature also uses an acid derivative that is insoluble in water as a reserve supply of energy material and as a protective cushion for vital organs; it is the glyceryl ester of the acids known as fats that serve these functions. Other esters, those of the acids with simple alcohols, are used by nature to give flavor to fruit and odor to flowers.

Still another acid derivative plays an important part in physiological processes, the diamide of carbonic acid, urea. This acid derivative is the chief form in which products of protein metabolism within the body are excreted. And uniquely enough it was this simple compound that Woehler synthesized from inorganic material, giving the breath of life to modern organic chemistry.

In this chapter we shall investigate these important acid derivatives to learn how they are made, how they react, and why they are so vital to man.

Acid derivatives are compounds that are formed by substitution in the carboxyl radical. It follows that, whether the substitution be for the hydrogen ion or the hydroxyl group these acid derivatives have one thing in common, the acyl group, $R-C\overset{\displaystyle O}{\overset{\|}{}}-$. The acyl group differs from the carbonyl group in that one of its bonds is never attached directly to either a carbon or a hydrogen atom. The similarity of the acid derivatives is illustrated in Fig. 20.

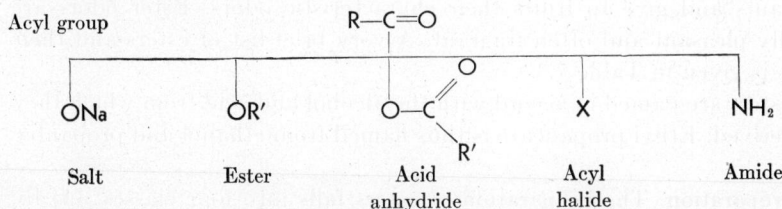

Fig. 20. Interrelationship of acid derivatives.

111

Salts

As in inorganic chemistry salts are formed by the neutralization reaction of organic acids with bases or by the direct action of the acid upon a metal. Only hydrogen in the carboxyl group reacts. Neutralization of the acid results in the formation of a salt and water.

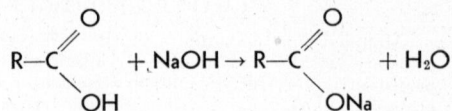

$$R-C\!\!\underset{OH}{\overset{O}{\diagup}} + NaOH \rightarrow R-C\!\!\underset{ONa}{\overset{O}{\diagup}} + H_2O$$

Salts of organic acids are named in the same manner as salts of inorganic acids, by dropping the suffix *-ic* from the name of the acid and adding in its place the suffix *-ate*. The general name of the salt shown above is sodium alkanoate. With strong mineral acids, salts react to form the organic acid and the salt of the mineral acid. The relative ionization

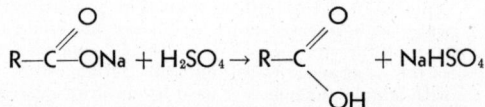

$$R-C\!\!\overset{O}{\diagup}-ONa + H_2SO_4 \rightarrow R-C\!\!\underset{OH}{\overset{O}{\diagup}} + NaHSO_4$$

of acids, bases, and salts will be recalled from studies in general chemistry. Organic acids are practically all weak electrolytes but their soluble salts are strong.

Again as was true in inorganic chemistry, the salts of the alkali metals are soluble and are highly ionized forming strong electrolytes. This holds true even for the salts of very long acids such as octadecanoic acid (stearic acid). The sodium salt of this acid ($C_{17}H_{35}COONa$) is the principal constituent of soap.

Esters

Two kinds of esters exist, those formed by the reaction of an alkanoic acid with an alcohol and those formed by the reaction of a mineral acid with an alcohol. Only the former esters of alkanoic acids will be discussed in this chapter. The latter, known as inorganic esters, have been introduced in previous chapters, pp. 42 and 75.

Esters are very common in nature. They are found in essential oils of plants and give to fruits their characteristic odors. Ester odors are usually pleasant and often fragrant. A very brief list of esters and their odors is given in Table 7.

Esters are named in accord with the alcohol and acid from which they are derived. Ethyl propanoate is thus named from ethanol and propanoic acid.

Preparation. The preparation of esters falls into four classes: (1) by replacement of the hydroxyl part of the carboxyl group, (2) by reaction

Table 7

ESTERS AND THEIR ODORS OR FLAVORS

Ester	*Found in:*
Ethyl methanoate	Artificial rum flavors
Ethyl ethanoate	Apple blossoms
3-Methylbutyl ethanoate	Banana oil
Pentyl ethanoate	Pears
2-Octyl ethanoate	Oranges
Methyl butanoate	Pineapples
Ethyl butanoate	Peaches
Pentyl butanoate	Apricots
3-Methylbutyl 3-methylbutanoate	Apples
Ethyl heptanoate	Synthetic cognac
Ethyl hendecanoate	Ylang odor
Ethyl dodecanoate	Begonias
Methyl tetradecanoate	Iris
Benzyl ethanoate	Jasmine
Methyl salicylate	Wintergreen

with another acid derivative, (3) by the direct addition of an unsaturated hydrocarbon to an acid, and (4) by double decomposition reaction.

1. BY REPLACEMENT OF —OH. Organic acids react with alcohols to form esters and water. If a small amount of gaseous HCl or concentrated

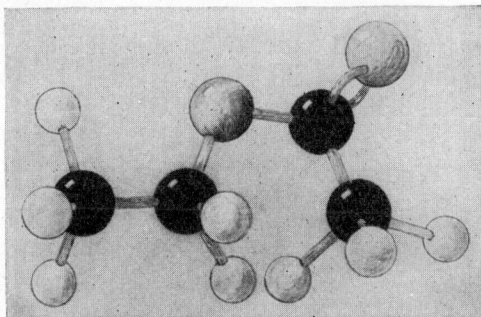

Fig. 21. The ethyl acetate (ethanoate) molecule.

H_2SO_4 is present in the solution, it serves as a catalyst and dehydrating agent, shortening the time of reaction and giving better yields. The reaction is readily reversible particularly in basic solution.

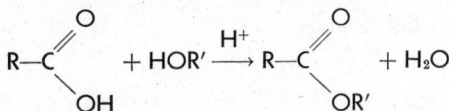

$$R-C \overset{O}{\underset{OH}{\big\langle}} + HOR' \xrightarrow{H^+} R-C \overset{O}{\underset{OR'}{\big\langle}} + H_2O$$

2. BY ACTION OF ACID DERIVATIVE. a. Acid anhydrides react with alcohols to form an ester and an acid.

$$R-C=O$$
$$|$$
$$O \qquad + HOR' \rightarrow R-C \overset{O}{\underset{OR'}{\big\langle}} \quad + R-C \overset{O}{\underset{OH}{\big\langle}}$$
$$R-C=O$$

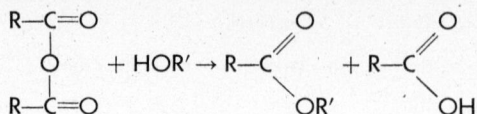

b. Acyl halides and alcohol form an ester and hydrogen halide. A

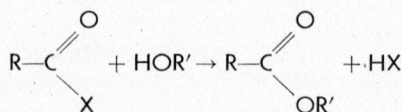

$$R-C \overset{O}{\underset{X}{\big\langle}} + HOR' \rightarrow R-C \overset{O}{\underset{OR'}{\big\langle}} + HX$$

better yield is obtained by using the sodium alcoholate, RONa.

3. By Direct Addition. It is interesting to discover that unsaturated hydrocarbons will add directly to the acid producing an ester. An acid medium and the presence of a special catalyst favor these additions. One such reaction has assumed great importance in the plastics industry, for it is by the addition of acetylene to acetic acid that vinyl acetate is produced.

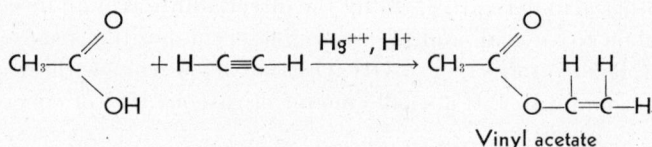

$$CH_3-C \overset{O}{\underset{OH}{\big\langle}} + H-C\equiv C-H \xrightarrow{Hg^{++}, H^+} CH_3-C \overset{O}{\underset{O-C=C-H}{\big\langle}} \begin{array}{c} H \ H \\ | \ | \end{array}$$

Vinyl acetate

4. Salt with Alkyl Halide. With haloalkanes organic salts of silver form insoluble silver halide and an ester. The best results are obtained with the alkyl iodide.

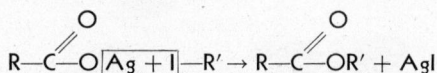

$$R-C \overset{O}{\underset{}{\big\langle}} O\boxed{Ag+I}-R' \rightarrow R-C \overset{O}{\underset{}{\big\langle}} OR' + AgI$$

Reactions. Esters react in many ways. The five reactions given below are characteristic of most esters. It will be noted that the changes are similar to the double decomposition reactions of general chemistry.

1. Hydrolysis. Esters hydrolyze slowly forming an acid and alcohol. With strong mineral acid an organic acid and the ester of a mineral acid are formed. With strong base, esters produce a salt of an organic acid and an alcohol. This latter reaction is called *saponification*.

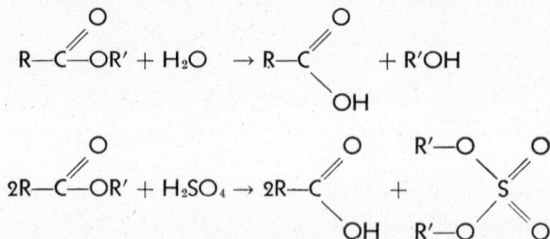

$$R-C \overset{O}{\underset{}{\big\langle}} OR' + H_2O \rightarrow R-C \overset{O}{\underset{OH}{\big\langle}} + R'OH$$

$$2R-C \overset{O}{\underset{}{\big\langle}} OR' + H_2SO_4 \rightarrow 2R-C \overset{O}{\underset{OH}{\big\langle}} + \begin{array}{c} R'-O \\ \diagdown \\ R'-O \end{array} S \overset{O}{\underset{O}{\big\langle}}$$

$$R-\overset{\displaystyle O}{\underset{\displaystyle OR'}{C}} + NaOH \rightarrow R-\overset{\displaystyle O}{\underset{\displaystyle ONa}{C}} + R'OH$$

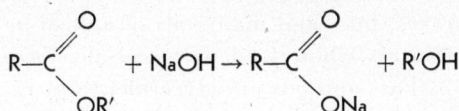

2. AMMONOLYSIS. An amide and an alcohol are formed by the reaction of an ester with ammonia gas.

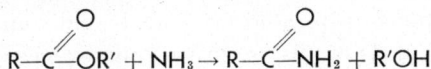

$$R-\overset{\displaystyle O}{C}-OR' + NH_3 \rightarrow R-\overset{\displaystyle O}{C}-NH_2 + R'OH$$

3. ALCOHOLYSIS. The alkyl part of an ester may be exchanged for a different alkyl group by mixing a desired alcohol in excess with an ester. Without a catalyst the reaction is very slow. A small amount of sodium alkoxide catalyzes the reaction. For best results one of the end products should be removed during the course of the reaction.

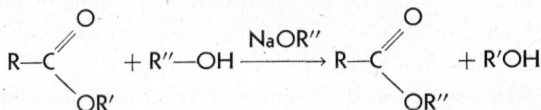

$$R-\overset{\displaystyle O}{\underset{\displaystyle OR'}{C}} + R''-OH \xrightarrow{\quad NaOR'' \quad} R-\overset{\displaystyle O}{\underset{\displaystyle OR''}{C}} + R'OH$$

4. REDUCTION OF AN ESTER. Esters may be reduced to two primary alcohols by direct hydrogenation at temperatures of 250–300°C in the presence of copper chromite catalyst. This method is used for the commercial production of some alcohols. A pressure of 15 to 20 atmospheres is required for this reaction.

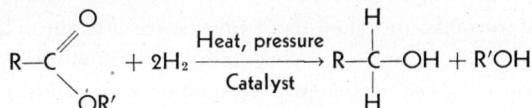

$$R-\overset{\displaystyle O}{\underset{\displaystyle OR'}{C}} + 2H_2 \xrightarrow[\text{Catalyst}]{\text{Heat, pressure}} R-\overset{\displaystyle H}{\underset{\displaystyle H}{C}}-OH + R'OH$$

In the laboratory the hydrogen produced by the action of sodium on ethanol will reduce an ester to two alcohols.

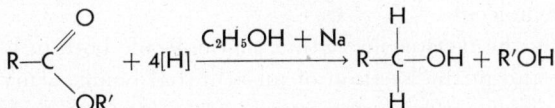

$$R-\overset{\displaystyle O}{\underset{\displaystyle OR'}{C}} + 4[H] \xrightarrow{\quad C_2H_5OH + Na \quad} R-\overset{\displaystyle H}{\underset{\displaystyle H}{C}}-OH + R'OH$$

5. GRIGNARD REACTION. Grignard reagent with an ester will form a ketone. If the Grignard reagent is in excess, the excess will react with the ketone formed and produce a tertiary alcohol.

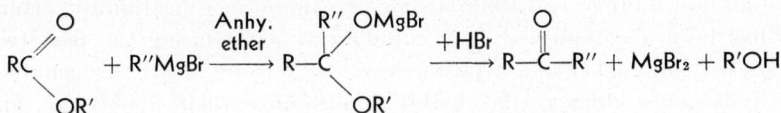

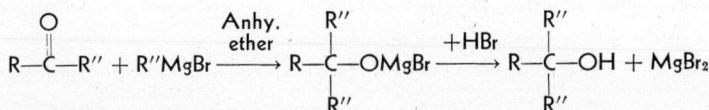

$$RC\overset{\displaystyle O}{\underset{\displaystyle OR'}{}} + R''MgBr \xrightarrow[\text{ether}]{\text{Anhy.}} R-\overset{\displaystyle R''}{\underset{\displaystyle OR'}{C}}\overset{\displaystyle OMgBr}{} \xrightarrow{+HBr} R-\overset{\displaystyle O}{C}-R'' + MgBr_2 + R'OH$$

$$R-\overset{\displaystyle O}{C}-R'' + R''MgBr \xrightarrow[\text{ether}]{\text{Anhy.}} R-\overset{\displaystyle R''}{\underset{\displaystyle R''}{C}}-OMgBr \xrightarrow{+HBr} R-\overset{\displaystyle R''}{\underset{\displaystyle R''}{C}}-OH + MgBr_2$$

Special Esters. Waxes, fats, and many oils of animal or plant origin are esters. Waxes are esters whose acid part and alcohol part are both long chain compounds. Fats and oils are glycerides, that is, esters of the alkanoic acids with the trihydric alcohol, glycerol. Because of their relationship to fats and oils and from the fact that the longer chain acids have a greasy, fatlike feel, the group is frequently spoken of as *fatty acids*. In most instances saturated fatty acids make up the acid part of solid fats; the acid parts of oils are in varying degrees from unsaturated fatty acids. A list of common names of some fatty acids is given on p. 170 (Chapter 12). Each molecule of glycerol requires three molecules of fatty acids to make one molecule of a fat or oil. A simple glyceride has three molecules of the same fatty acid united with the glycerol. A mixed glyceride may have as many as three different kinds of acids in the molecule of fat or oil. The predominant fatty acid in most fats and oils contains 18 carbon atoms; those with one point of unsaturation (one double bond like oleic acid) are the most abundant.

Fats and oils are often distinguished by their physical state, the solids being called fats and the liquids oils. Another distinction is based upon the degree of unsaturation of the fatty acids making up the glyceride. When the fatty acids are more completely saturated the glyceride is a fat. Oils have varying degrees of unsaturation. Both methods of distinguishing fats from oils are unsatisfactory as technical definitions. The terms were in common use long before technical differences were understood. Moreover the fats of animals and the oils of plants are not pure chemical compounds but rather a mixture of several similar compounds.

A typical pure fat is tristearin, the tristearic acid ester of glycerol. Its formula is $(C_{17}H_{35}COO)_3C_3H_5$. A typical pure oil is the simple glyceride, triolein, glyceryl trioleate. Its formula is $(C_{17}H_{33}COO)_3C_3H_5$. The only difference between this typical fat and this typical oil is six atoms of hydrogen per molecule.

By varying the fatty acids within the molecule both in length of the carbon chain and in the location of unsaturated bonds, a myriad of fats and oils becomes possible.

Nature has been bounteous in her supply of fats and oils. Synthetic processes of producing fats and oils from simple compounds have been unnecessary except in the production of those compounds whose acids have an odd number of carbon atoms. Synthetic modification of certain oils has been accomplished on a commercial scale during the past two decades. Using nickel as a catalyst vegetable oils are now hydrogenated, the hydrogen adding at the double bonds, until their consistency approximates that of lard. Products of this type are sold as cooking compounds under such trade names as "Crisco."

From the technical viewpoint fats and oils are classified in various

Special Esters. Waxes, fats, and many oils of animal or plant origin are esters. Waxes are esters whose acid part and alcohol part are both long chain compounds. Fats and oils are glycerides, that is, esters of the alkanoic acids with the trihydric alcohol, glycerol. Because of their relationship to fats and oils and from the fact that the longer chain acids have a greasy, fatlike feel, the group is frequently spoken of as *fatty acids*. In most instances saturated fatty acids make up the acid part of solid fats; the acid parts of oils are in varying degrees from unsaturated fatty acids. A list of common names of some fatty acids is given on p. 170 (Chapter 12). Each molecule of glycerol requires three molecules of fatty acids to make one molecule of a fat or oil. A simple glyceride has three molecules of the same fatty acid united with the glycerol. A mixed glyceride may have as many as three different kinds of acids in the molecule of fat or oil. The predominant fatty acid in most fats and oils contains 18 carbon atoms; those with one point of unsaturation (one double bond like oleic acid) are the most abundant.

Fats and oils are often distinguished by their physical state, the solids being called fats and the liquids oils. Another distinction is based upon the degree of unsaturation of the fatty acids making up the glyceride. When the fatty acids are more completely saturated the glyceride is a fat. Oils have varying degrees of unsaturation. Both methods of distinguishing fats from oils are unsatisfactory as technical definitions. The terms were in common use long before technical differences were understood. Moreover the fats of animals and the oils of plants are not pure chemical compounds but rather a mixture of several similar compounds.

A typical pure fat is tristearin, the tristearic acid ester of glycerol. Its formula is $(C_{17}H_{35}COO)_3C_3H_5$. A typical pure oil is the simple glyceride, triolein, glyceryl trioleate. Its formula is $(C_{17}H_{33}COO)_3C_3H_5$. The only difference between this typical fat and this typical oil is six atoms of hydrogen per molecule.

By varying the fatty acids within the molecule both in length of the carbon chain and in the location of unsaturated bonds, a myriad of fats and oils becomes possible.

Nature has been bounteous in her supply of fats and oils. Synthetic processes of producing fats and oils from simple compounds have been unnecessary except in the production of those compounds whose acids have an odd number of carbon atoms. Synthetic modification of certain oils has been accomplished on a commercial scale during the past two decades. Using nickel as a catalyst vegetable oils are now hydrogenated, the hydrogen adding at the double bonds, until their consistency approximates that of lard. Products of this type are sold as cooking compounds under such trade names as "Crisco."

From the technical viewpoint fats and oils are classified in various

$$R—C\overset{O}{\underset{OR'}{\big\langle}} + NaOH \rightarrow R—C\overset{O}{\underset{ONa}{\big\langle}} + R'OH$$

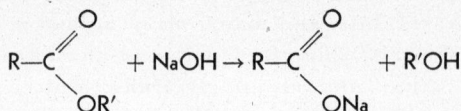

2. AMMONOLYSIS. An amide and an alcohol are formed by the reaction of an ester with ammonia gas.

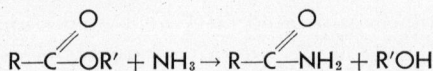

$$R—\overset{O}{\overset{\|}{C}}—OR' + NH_3 \rightarrow R—\overset{O}{\overset{\|}{C}}—NH_2 + R'OH$$

3. ALCOHOLYSIS. The alkyl part of an ester may be exchanged for a different alkyl group by mixing a desired alcohol in excess with an ester. Without a catalyst the reaction is very slow. A small amount of sodium alkoxide catalyzes the reaction. For best results one of the end products should be removed during the course of the reaction.

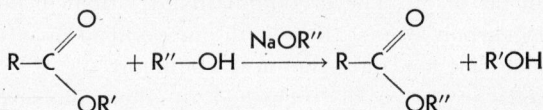

$$R—C\overset{O}{\underset{OR'}{\big\langle}} + R''—OH \xrightarrow{\text{NaOR}''} R—C\overset{O}{\underset{OR''}{\big\langle}} + R'OH$$

4. REDUCTION OF AN ESTER. Esters may be reduced to two primary alcohols by direct hydrogenation at temperatures of 250–300°C in the presence of copper chromite catalyst. This method is used for the commercial production of some alcohols. A pressure of 15 to 20 atmospheres is required for this reaction.

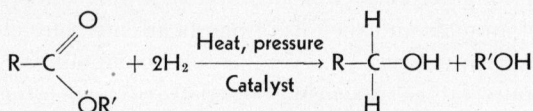

$$R—C\overset{O}{\underset{OR'}{\big\langle}} + 2H_2 \xrightarrow[\text{Catalyst}]{\text{Heat, pressure}} R—\overset{H}{\underset{H}{\overset{|}{C}}}—OH + R'OH$$

In the laboratory the hydrogen produced by the action of sodium on ethanol will reduce an ester to two alcohols.

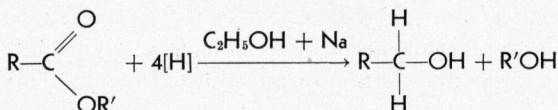

$$R—C\overset{O}{\underset{OR'}{\big\langle}} + 4[H] \xrightarrow{C_2H_5OH + Na} R—\overset{H}{\underset{H}{\overset{|}{C}}}—OH + R'OH$$

5. GRIGNARD REACTION. Grignard reagent with an ester will form a ketone. If the Grignard reagent is in excess, the excess will react with the ketone formed and produce a tertiary alcohol.

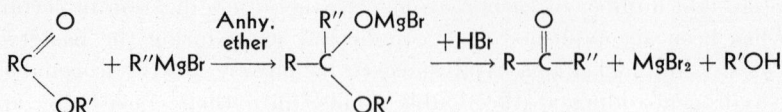

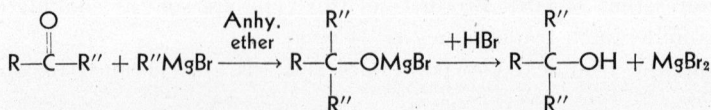

$$RC\overset{O}{\underset{OR'}{\big\langle}} + R''MgBr \xrightarrow[\text{ether}]{\text{Anhy.}} R—\overset{R''}{\underset{OR'}{\overset{|}{C}}}\overset{OMgBr}{} \xrightarrow{+HBr} R—\overset{O}{\overset{\|}{C}}—R'' + MgBr_2 + R'OH$$

$$R—\overset{O}{\overset{\|}{C}}—R'' + R''MgBr \xrightarrow[\text{ether}]{\text{Anhy.}} R—\overset{R''}{\underset{R''}{\overset{|}{C}}}—OMgBr \xrightarrow{+HBr} R—\overset{R''}{\underset{R''}{\overset{|}{C}}}—OH + MgBr_2$$

ways, two of the most common being by their saponification number and their iodine number. The reaction of saponification is shown on p. 114. The saponification number is the number of milligrams of potassium hydroxide required to saponify one gram of fat or oil. The iodine number of a fat or oil measures the degree of unsaturation existing in the fatty acid parts of the molecule. The iodine number is defined as the number of grams of iodine that will add to 100 grams of a fat or oil. Oils having an iodine number lower than 100 are called nondrying oils. When the iodine number lies between 100 and 130, such oils are called semidrying. Drying oils have an iodine number higher than 130. Many drying oils are used as the vehicle for organic and inorganic pigments in the form of paints. When a surface is painted, oxygen from the air reacts with the unsaturated structure of the drying oil. This oxidation causes a cross-linking between molecules and results in the formation of a tough wear-resisting film that clings to the painted surface.

Like all other esters, fats and oils react with inorganic bases to form glycerol and a mineral salt of fatty acids. The mineral salt of a fatty acid is a soap. In the manufacture of soap, animal fats and vegetable oils are heated usually with strong lye (sodium hydroxide) solution. Ordinary salt, sodium chloride, is added when hydrolysis is completed because an excess of sodium ion due to common ion effect "salts out" the organic salt, which then floats to the surface of the reacting mixture as soap. Glycerol, excess sodium chloride, and excess sodium hydroxide are dissolved in the lower water layer. The upper layer of soap is drained off for mixing with other ingredients to give body, color, odor, and other desirable qualities to the finished product. When potassium hydroxide is used, soft soap is produced. Hard soaps are made from sodium hydroxide. Soaps are sold in many forms such as chips, flakes, granules, powders, bars, and as liquids, i.e. solutions or suspensions of soap in water. Soaps of alkaline earth metals and most of the heavy metals are practically insoluble in water. The scumlike film that forms when ordinary soap is dissolved in "hard water," which contains calcium or magnesium ions, consists of the precipitated salts of these metals. Many soaps find interesting and unexpected uses: aluminum soaps serve as a base for automobile grease; zinc soaps are used to treat rash on infants and as a lubricant in plastic molding.

Acid Anhydrides

Acid and acid anhydride have the same interrelationship between each other as alcohol and ether. When two molecules of an acid lose one molecule of water between them, one molecule of acid anhydride is formed, the two acid groups being joined by an oxygen atom. If the two molecules of acid are of the same kind, a simple acid anhydride results.

Mixed anhydrides require the use of two different acids. The anhydride of acetic acid has the structure

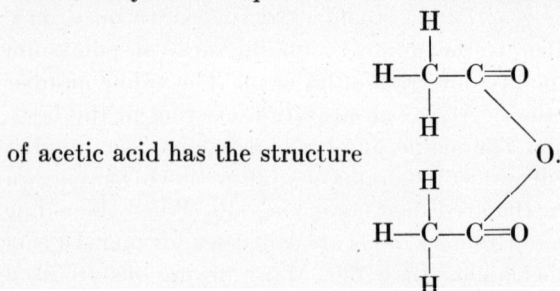

$$
\begin{array}{c}
H \\
| \\
H-C-C=O \\
| \qquad\quad \searrow \\
H \qquad\qquad\quad O. \\
H \qquad\quad \nearrow \\
| \\
H-C-C=O \\
| \\
H
\end{array}
$$

To name an acid anhydride the word *anhydride* is added to the name of the acid or acids from which it is made. Thus the name of the compound $(CH_3CO)_2O$ is ethanoic (acetic) anhydride and the name of $(HCO)O(CH_3CO)$ is methanoic ethanoic anhydride.

Preparation. 1. Simple acid anhydrides of lower acids are produced by the reaction of sulfur monochloride (S_2Cl_2) with the sodium salt of an acid.

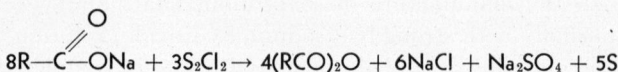

$$
8R-\overset{\overset{O}{\|}}{C}-ONa + 3S_2Cl_2 \rightarrow 4(RCO)_2O + 6NaCl + Na_2SO_4 + 5S
$$

2. In the laboratory either simple or mixed acid anhydride may be prepared by the action of an acyl halide on a dry sodium salt.

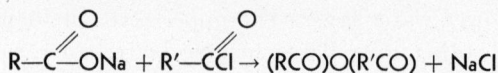

$$
R-\overset{\overset{O}{\|}}{C}-ONa + R'-\overset{\overset{O}{\|}}{C}Cl \rightarrow (RCO)O(R'CO) + NaCl
$$

Reactions. Like other acyl compounds, acid anhydrides undergo reactions of hydrolysis, alcoholysis, and ammonolysis, producing acid, ester, and amide respectively.

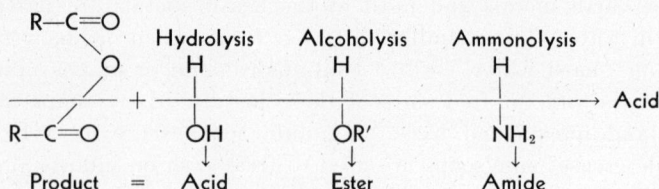

$$
\begin{array}{l}
R-C=O \\
\qquad \searrow \\
\qquad\quad O \\
\qquad \nearrow \\
R-C=O
\end{array}
\quad + \quad
\begin{array}{ccc}
\text{Hydrolysis} & \text{Alcoholysis} & \text{Ammonolysis} \\
H & H & H \\
| & | & | \\
\underline{\qquad\qquad\qquad\qquad\qquad\qquad\qquad} & & \longrightarrow \text{ Acid} \\
OH & OR' & NH_2 \\
\downarrow & \downarrow & \downarrow
\end{array}
$$

Product   =   Acid   Ester   Amide

Acyl Halides

In acyl halides the hydroxyl of the acid is replaced by a halogen. It will be recalled that alkyl halide results from the substitution of a halogen for the —OH of an alcohol.

The name acyl halide is a general name based upon the acyl group,

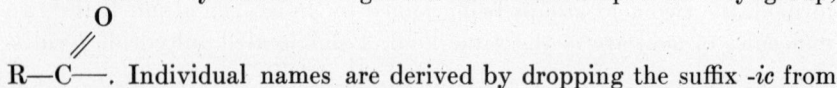

$$
R-\overset{\overset{O}{\|}}{C}-.
$$
Individual names are derived by dropping the suffix *-ic* from

the name of the acid and adding *-yl* plus the separate name of the halide. Thus the compound CH_3COCl is ethanoyl chloride (acetyl chloride).

Acyl halides are very active compounds. The kind of halide in the compound determines the activity. As acylating reagents these derivatives are the most active known and their order of activity is

$$RCOF > RCOCl > RCOBr > RCOI.$$

In this respect the order of activity is directly reversed from that of the alkyl halides. This reversal of activity as observed when Grignard reagents add to acyl compounds, is related to the space-filling ability of the respective halogens as well as to their strongly polar character when bonded to the carbonyl group.

Preparation. Methods of preparing acyl halides from acids are similar to those of producing alkyl halides from alcohols. Phosphorus trihalide, phosphorus pentahalide, and thionyl chloride in reaction with acids will all produce acyl halides. Since acyl halides are separated from impurities in the reacting vessel by distillation, the choice of method should depend upon boiling points and volatility of all compounds used and produced. Thionyl chloride is an especially effective reactant since the two byproducts of the reaction are gases.

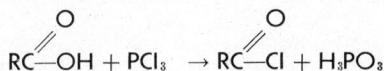

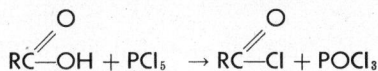

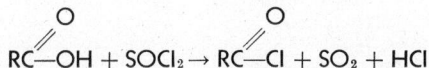

$$RC\overset{O}{\overset{\|}{-}}OH + PCl_3 \rightarrow RC\overset{O}{\overset{\|}{-}}Cl + H_3PO_3$$

$$RC\overset{O}{\overset{\|}{-}}OH + PCl_5 \rightarrow RC\overset{O}{\overset{\|}{-}}Cl + POCl_3$$

$$RC\overset{O}{\overset{\|}{-}}OH + SOCl_2 \rightarrow RC\overset{O}{\overset{\|}{-}}Cl + SO_2 + HCl$$

Reactions. 1. Like other acyl compounds, acyl halides undergo the reactions of hydrolysis, alcoholysis, and ammonolysis. All these reactions are violent at room temperature. Acids, esters, and amides are produced respectively and all three in common yield hydrogen halide. The acyl halides being the most reactive of the acid derivatives, are used to prepare the other derivatives.

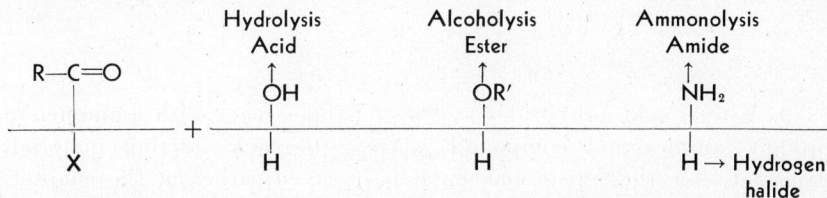

		Hydrolysis Acid	Alcoholysis Ester	Ammonolysis Amide
R—C=O		$\overset{\uparrow}{O}H$	$\overset{\uparrow}{O}R'$	$\overset{\uparrow}{N}H_2$
$\mid$	+	$\mid$	$\mid$	$\mid$
X		H	H	H $\rightarrow$ Hydrogen halide

2. Anhydrous salts of organic acids react with acyl halide to produce acid anhydrides, simple or mixed, in accord with the starting materials.

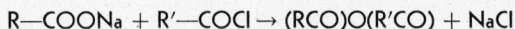

$$R-COONa + R'-COCl \rightarrow (RCO)O(R'CO) + NaCl$$

3. **Reduction.** To convert organic acids to aldehydes the acids are often changed to the acyl halide, dissolved in benzene or xylene and reduced with hydrogen in the presence of palladium and barium sulfate as catalyst.

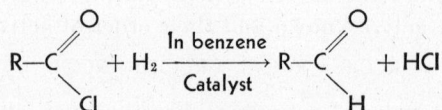

$$R-C\overset{O}{\underset{Cl}{<}} + H_2 \xrightarrow[\text{Catalyst}]{\text{In benzene}} R-C\overset{O}{\underset{H}{<}} + HCl$$

Amides

The amides are compounds in which the group —NH_2 replaces the hydroxyl of an acid. Amides are named by dropping the letter -*e* from the name of the alkyl root and adding the suffix -*amide* to make one word. Thus CH_3CONH_2 is called ethanamide (acetamide).

The structure of an amide is usually accepted in its most common

form R—C—NH_2. Some reactions of the compound suggest the

possibility of a tautomeric form R—C $\overset{OH}{\underset{NH}{<}}$. Amides are practically

neutral compounds. Except for liquid methanamide all the amides are crystalline.

Preparation. 1. Amides may be prepared by the distillation of dry ammonium salts of alkanoic acids.

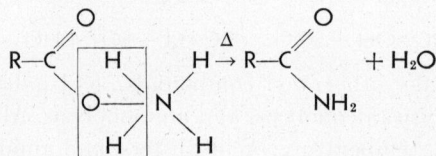

$$R-C\overset{O}{<}\boxed{\begin{matrix} H \\ O\text{---}N \\ H \end{matrix}}\ \overset{\Delta}{H \rightarrow R-C}\overset{O}{\underset{NH_2}{<}} + H_2O$$

2. Higher homologues may be prepared by passing ammonia gas through hot acids.

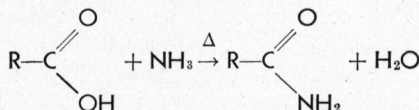

$$R-C\overset{O}{\underset{OH}{<}} + NH_3 \overset{\Delta}{\rightarrow} R-C\overset{O}{\underset{NH_2}{<}} + H_2O$$

3. Esters, acid anhydrides and acyl halides react with ammonia to produce amides and compounds specific to each starting material. Equations for these reactions have been given earlier in the chapter.

Reactions. 1. Hydrolysis occurs with dilute acids or dilute base.

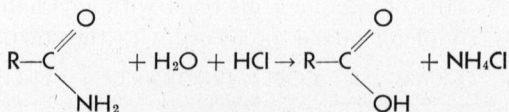

$$R-C\overset{O}{\underset{NH_2}{<}} + H_2O + HCl \rightarrow R-C\overset{O}{\underset{OH}{<}} + NH_4Cl$$

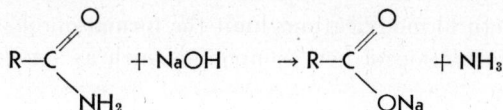

2. With nitrous acid amides produce acid, free nitrogen, and water.

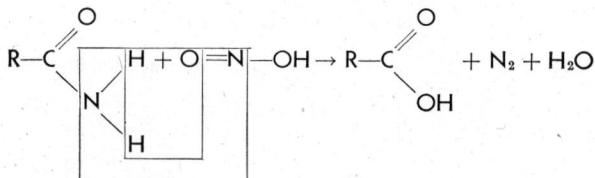

3. With strong dehydrating agents, as P_2O_5, amides are converted into nitriles. This reaction leads to the conclusion that the nitrile itself may be considered as an acid derivative in which both the $=O$ and $-OH$ have been replaced.

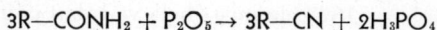

$$3R-CONH_2 + P_2O_5 \rightarrow 3R-CN + 2H_3PO_4$$

4. By decarboxylation of its brominated derivative, amides are changed to primary amines, having one less carbon in the chain. This reaction is called the Hofmann degradation.

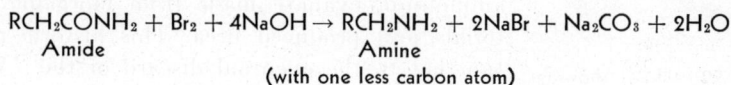

$$RCH_2CONH_2 + Br_2 + 4NaOH \rightarrow RCH_2NH_2 + 2NaBr + Na_2CO_3 + 2H_2O$$
Amide   Amine
(with one less carbon atom)

Acyl Compounds of Alkanedioic Acids

In general the preparation and reactions of compounds follow the preparations and reactions for alkanoic acids.

In contrast with the difficulty of forming anhydrides of monocarboxylic acids, certain dicarboxylic acids, particularly succinic and glutaric acids, readily lose water upon being heated and form internal anhydrides. This behavior is due primarily to the stability of five- and six-membered ring structures. This reaction for glutaric acid is:

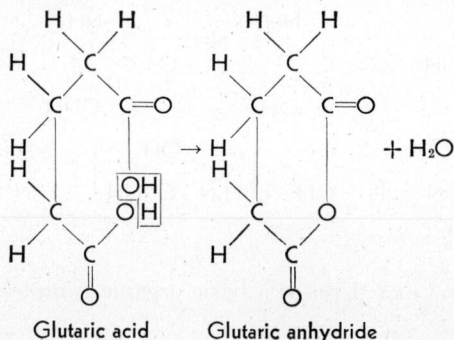

Glutaric acid   Glutaric anhydride

Similar structural modifications limit the formation of some diamides and favor the formation of new compounds such as *-imides*, $=NH$, by deammonization.

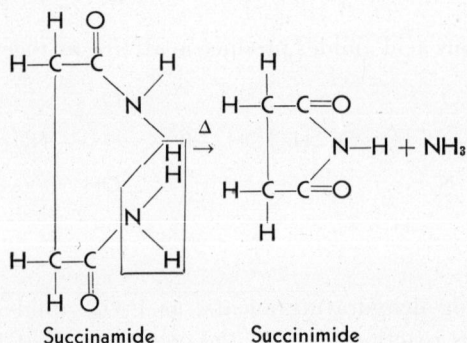

Succinamide   Succinimide

Important Diamides: A. UREA. Urea is an historically important compound in organic chemistry. Previous to 1828 chemists believed that compounds produced by living organisms were different from other chemicals that had not been produced by a vital force. In that year Friedrich Woehler (1800–1882), a young German chemist, heated ammonium cyanate made from inorganic materials and produced urea. This historic reaction led to the eventual discard of the "Vital Force Theory."

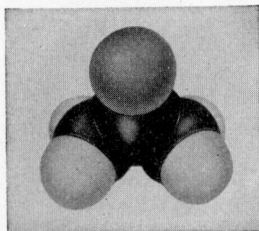

Fig. 22. Urea showing steric proportions.

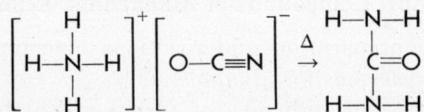

Urea is an important industrial reagent and a fertilizer. There are two processes of nitrogen fixation which yield large tonnages of urea for these uses. The first uses the addition of synthetic ammonia to carbon dioxide; the second utilizes the synthetic product cyanamide.

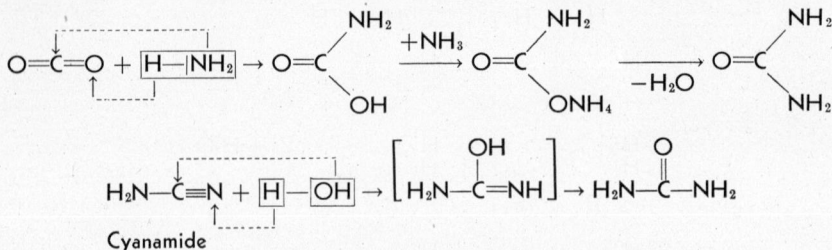

Cyanamide

REACTIONS OF UREA. Urea is a basic organic compound.

1. Urea adds acid to form salts, using only one of its —NH₂ groups.

$$H_2N-\overset{\overset{O}{\|}}{C}-NH_2 + HNO_3 \rightarrow \left[H_2N-\overset{\overset{O}{\|}}{C}-\overset{\overset{H}{|}}{\underset{\underset{H}{|}}{N}}-H \right]^+ NO_3^-$$

2. It hydrolyzes easily in acid or basic medium. The gradual release of ammonia in the soil from urea fertilizer occurs in this manner.

$$H_2N-\overset{\overset{O}{\|}}{C}-NH_2 + 2H_2O \xrightarrow{OH^-} 2NH_3 + CO_2 + H_2O$$

$$H_2N-\overset{\overset{O}{\|}}{C}-NH_2 + 2H_2O + 2H^+ \rightarrow 2NH_4^+ + CO_2 + H_2O$$

3. When urea is oxidized with a hypohalite in basic solution, it forms nitrogen gas and other compounds that are nonvolatile. Measurement of the volume of nitrogen liberated by this reaction is used to determine the urea in urine.

$$(NH_2)_2CO + 3KOX + 2KOH \rightarrow N_2 + K_2CO_3 + 3KX + 3H_2O$$

4. The biuret test for protein is based upon the similarity of structure of proteins and the compound biuret, formed by heating urea. Proteins

Two molecules of urea   Biuret

have many linkages of the type, $-\overset{\overset{\|}{O}}{C}-\overset{\underset{|}{H}}{N}-$. In basic solution biuret and protein turn a bluish purple or pinkish purple color when a drop or two of cupric salt solution is added. This is called the "Biuret Test for Protein."

5. With nitrous acid urea reacts as other amides, liberating nitrogen, carbon dioxide, and water.

6. The reaction of urea with malonic ester in preparing barbituric acid is given on page 106.

B. HEXANEDIAMIDE (ADIPAMIDE). The compound 1,6-hexanediamide is of peculiar interest because it is an intermediate formed and used in the synthesis of nylon. Adipic acid is first converted to the diamide. This is then reduced to the diamine. The diamine condenses with more adipic acid and then polymerizes to form nylon which resembles protein in structure.

Acyl Compounds in General

Acyl compounds are very similar in their preparation and in their chemical behavior. They show a definite gradation of characteristics based upon the modifying influence of the acid formed as a companion product when hydrolysis occurs. The stronger the acid, the more highly polarized will the acyl compound be, and also the more reactive.

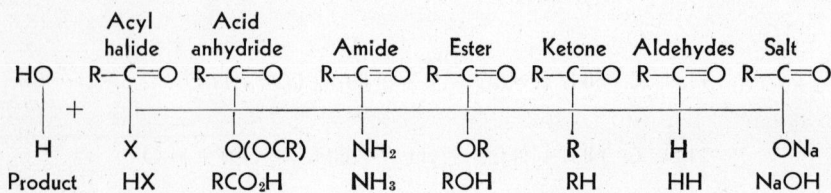

	Acyl halide	Acid anhydride	Amide	Ester	Ketone	Aldehydes	Salt
HO	R—C=O	R—C=O	R—C=O	R—C=O	R—C=O	R—C=O	R—C=O
+	X	O(OCR)	NH$_2$	OR	R	H	ONa
Product	HX	RCO$_2$H	NH$_3$	ROH	RH	HH	NaOH

Fig. 23. Hydrolysis of compounds containing the Acyl $\left(R-C\diagup^{O}_{\diagdown} \right)$ group.

In Figure 23 shown above, ketones and aldehydes are not true acyl compounds and they do not hydrolyze. They are placed in this chart because it is interesting to note that the *theoretical* hydrolysis products of a ketone and an aldehyde would be the definitely nonpolar hydrocarbon, RH, and H$_2$ respectively. Since hydrolysis of salts of weak acids actually produces some —OH ion, the position of salts in the chart shows that acyl compounds are more active when they are more highly polarized, and that their polarity is definitely associated with the acidity of their companion products of hydrolysis.

Study and Review Questions

1. By equations show how ethanoic (acetic) acid can be converted to sodium ethanoate (acetate), and how sodium ethanoate can be converted to ethanoic acid.
2. Write the structural formulas of the esters shown on page 113. (Table of Esters and Odors.)
3. Show by an equation how to prepare ethyl ethanoate.
4. Show the hydrolysis of an ester in neutral, acidic, and basic solutions.
5. Explain what is meant by the alcohol exchange reaction of esters. Under what conditions does it take place?
6. What is produced by the reaction of methyl ethanoate and ethyl Grignard reagent in equal parts?
7. What is the distinction between a wax and a fat? Give one example of each.
8. Give two methods of distinguishing between a fat and an oil.
9. What is a glyceride?

10. Certain oils are hydrogenated to make shortening. What catalyst is used? Explain the change that occurs.

11. Calculate the saponification number of a fat when 3.430 g. of the fat require 0.686 g. of KOH for saponification.

12. If 6.31 g. of unsaturated glyceride adds 7.572 g. of iodine, calculate the iodine number of the glyceride. Classify the glyceride correctly as a nondrying, semidrying, or drying oil.

13. Define the term "soap."

14. What is the specific chemical difference between "soft" and "hard" soap?

15. Are all "soaps" soluble in water? List three distinctly different uses of soaps.

16. Describe briefly how soap is made.

17. List the classes of compounds studied in this chapter that are found in nature. Those that are not.

18. Name the following compounds: (a) $(CH_3CH_2CO)_2O$ and (b) $(CH_3CO)O(CH_3CH_2CO)$.

19. Write an equation for the preparation of each compound given in question 18.

20. Name the following compounds: (a) CH_3CH_2COCl, (b) $CH_3CH_2CH_2COBr$, and (c) $(CH_3)_2CHCOCl$.

21. Discuss the halogen order of activity of acyl halides.

22. By equations, show two methods of preparing propanoyl chloride.

23. Write an equation for the ammonolysis of butanoyl chloride.

24. Write an equation for the reduction reaction of propanoyl chloride.

25. Write equations for three methods of preparing ethanamide, using starting material of your own choice.

26. How does an amide react with nitrous acid? Show by an equation.

27. Compare the reactions for preparing succinic anhydride and succinimide.

28. Why is Woehler's synthesis of urea of historical importance? Show his synthesis by an equation.

29. What is formed by the hydrolysis of urea?

30. What is biuret? What is meant by the "Biuret Test for Protein"?

CHAPTER PROLOGUE

Scientifically and historically the work of scientists in the field of substituted acids constitutes one of the brightest pages in organic chemistry. It is in elapsed time a long story, studded with the names of brilliant, fortunate men who were gifted with momentary and fragmentary glimpses of what lay beyond the ordinary spectacle of routine scientific work. In some cases their achievement has been but the prelude to the life of a scientific giant; in other cases their achievement has been a laurel wreath of final victory. There is the story of young Biot, a pioneer of the polarimeter, looking beyond the fact of a rotated beam of polarized light to the cause of its rotation. There is the picture of young Louis Pasteur, tweezers in hand, separating minute crystals with the aid of a magnifying glass, and repeating his experiment before the eyes of the aging Biot. There are the papers of LeBel and van't Hoff, published only a few weeks apart, telling of their independent codiscovery of the asymmetric carbon atom. There is also the notable work of Emil Fischer on proteins and amino acids coming as a second reward to a giant of science.

In this chapter we will view briefly the discoveries of these men and correlate their specific contributions with those of their less spectacular fellow scientists.

When any part of an acid molecule outside the carboxyl group is modified by the substitution of any element or group for a hydrogen atom of the alkyl group, the resulting compound is called a substituted acid. As defined in Chapter 8, acid derivatives are compounds with modifications of the carboxyl group. This chapter will deal only with certain typical substitution modifications of the alkyl or $R—$ part of the molecule. Thus the presence of a double bond within the alkyl part of the molecule makes the acid an alkenoic acid rather than a substituted acid and the presence of a chlorine atom makes the acid a chloroalkanoic acid.

Different elements and groups are capable of being substituted for hydrogen in the alkyl part of the molecule, such as the halogen elements and hydroxyl, amino, oxy, and thio groups. Only the first three of these substituents will be discussed extensively in this chapter.

In naming substituted acids the correct alkanoic acid name describing the length of the carbon chain is preceded by a description of the substituent and a number showing the location of the modifying element or group. The name as one word is followed by the separate word *acid*. Carbon atoms are numbered from the functional (carboxyl) group as

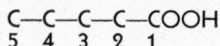

$$\underset{5}{C}—\underset{4}{C}—\underset{3}{C}—\underset{2}{C}—\underset{1}{COOH}$$

carbon atom number one. Thus a substituted acid having the structural formula $CH_3CHClCOOH$ is named 2-chloropropanoic acid.

Theoretically the number of substituted acids from a given alkanoic acid would be limited only by the number of hydrogen atoms in the alkyl part of the molecule. Actually there are usually only one or two such substitutions and the location of the substituent can usually be predicted on a basis of the molecular structure and the influence of the carboxyl group.

Structure and Bonding. The structure and polarity of the carboxyl group as a separate group was discussed in Chapter 7. No attempt was made at that time to explain ionization from structure, since structure of only the carboxyl group was considered. The two oxygen atoms of the carboxyl group were seen to be alternately very negative or slightly less negative as resonance of the bonding electrons occurred.

Due to this resonance the distance between the nuclei of the carbon and oxygen atoms is less than would otherwise be true and the entire group becomes stable. The hydrogen proton may be considered to lie between the oxygen atoms, first held more closely by one and then the other as the electrons resonate. The result is that the entire carboxyl group is electronegative causing the 2-carbon atom, to which it is attached, to be electropositive. The secondary effect is that the hydrogens on the 2-carbon atom are slightly repelled, becoming more labile and thus subject to easier replacement. It may even be expected that there

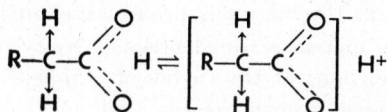

is some tendency for enolization (p. 85) to occur thus further enhancing the opportunity for substitution of the 2-carbon atom. It is on a basis

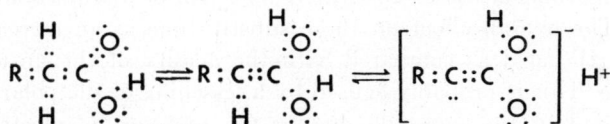

of these structural characteristics that the formation of substituted acids is explained.

Haloalkanoic Acids

From the foregoing discussion it will be clear that the hydrogen on the 2-carbon atom is activated and readily subject to replacement. The manner in which this replacement occurs with chlorine, for example, may be (1) as a direct replacement with the formation of HCl and (2)

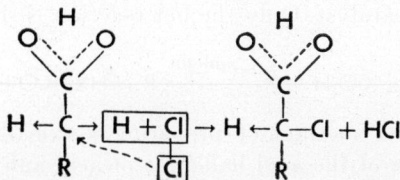

in a lesser degree as an enolization addition with formation of HCl.

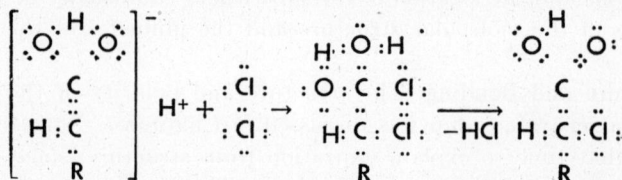

When the reaction cycle is completed the carboxyl group remains the same but the alkyl part of the molecule has been modified by substituting a chlorine on the 2-carbon atom. Chlorine is slightly less electronegative than oxygen. Therefore the intensity of the negative polarity of the carboxyl group in the substituted acid is somewhat modified by the chlorine atom. This modification in negative polarity of the carboxyl group yields two results: (1) The hydrogen of the carboxyl group ionizes more readily and (2) the hydrogen of the 2-carbon atom is more labile. Thus 2-chloroacetic acid is more acidic, that is, it has a higher degree of ionization than the unsubstituted acid. Also substitution of a second chlorine on the 2-carbon atom occurs even more readily. When acetic acid is halogenated in this manner, progressive substitution of chlorine for the hydrogens so increases the electronegativity of the molecule that the proton (hydrogen ion) of the carboxyl is repelled making trichloroacetic acid the strongest known organic acid.

Substitution of elements and groups on carbon atoms other than the 2-carbon occurs indirectly rather than by direct replacement. The manner of preparing some of these substituted acids will be discussed later in this chapter. The overall effect on these substitutions is an increase in the acidity of the acid as compared with the acidity of the unsubstituted homologue. This is probably caused by a lessening of the polarity at the carboxyl end of the molecule due to the presence of electronegative elements or groups in another part of the molecule.

Chlorine and bromine are the halogens used most for such substitutions, chlorine being the more active. Conversely 2-bromoalkanoic acids react more actively at the point of substitution than 2-chloroalkanoic acids.

Preparation. 1. Of 2-Haloalkanoic Acid. a. The direct halogenation method has already been discussed in detail. Sunlight is required for the reaction, or in the absence of sunlight, iodine or phosphorus trichloride may be used as a catalyst. Only the net reaction is shown here.

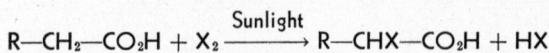

$$R—CH_2—CO_2H + X_2 \xrightarrow{\text{Sunlight}} R—CHX—CO_2H + HX$$

b. A more rapid method of preparing 2-haloalkanoic acids is the direct halogenation of the acyl halide or of acid anhydride, followed by hydrolysis. These two reactions are additions to the enolic forms of the

$$+2Cl_2 \quad -2HCl \qquad +H_2O \longrightarrow 2RCHClCOOH$$

$$+Cl_2 \quad -HCl \qquad +H_2O \longrightarrow RCHClCOOH + HCl$$

anhydride and acyl halide, due to carbonyl groups in both molecules. Enolization is more pronounced because the carboxyl is gone leaving no opportunity for resonance between the oxygens.

c. 2-Chloroalkanoic acids may also be prepared by the action of phosphorus trichloride on 2-hydroxyalkanoic acid.

$$3RCHOHCO_2H + PCl_3 \rightarrow 3RCHClCO_2H + H_3PO_3$$

2. Halogen substitutions other than on the 2-carbon may be prepared by the addition of hydrogen halide to an alkenoic acid. When such additions occur, the halogen is added to the double-bonded carbon atom farthest from the carboxyl group. The carboxyl group is negatively polarized and the negative part of the hydrogen halide, the halogen, adds as far from the carboxyl group as is possible.

$$+ HCl \rightarrow$$

Reactions. Halogen substituents react in much the same way as haloalkanes (Chapter 4). They react (1) with ammonia to produce amino acid salts and ammonium halide. They are converted (2) to hydroxyalkanoic acids by hydrolysis with aqueous sodium hydroxide. With alcoholic potassium hydroxide, (3) alkenoic acids are formed. Esters (4) can be prepared by the addition of sodium alkanoate, and (5) nitriles result from the reaction with potassium cyanide.

(1)

$$+NH_3 \quad -NH_4X \qquad -NH_3$$

(2)
$$R-\underset{X}{\underset{|}{C}}H-C\overset{O}{\underset{OH}{\diagup}} \xrightarrow[\substack{-NaX \\ -H_2O}]{+2NaOH} R-\underset{OH}{\underset{|}{C}}H-C\overset{O}{\underset{ONa}{\diagup}} \xrightarrow[-Na^+]{+H^+} R-\underset{OH}{\underset{|}{C}}H-C\overset{O}{\underset{OH}{\diagup}}$$

(3)
$$R-\underset{H}{\underset{|}{C}}H-\underset{X}{\underset{|}{C}}H-C\overset{O}{\underset{OH}{\diagup}} \xrightarrow[-HX]{KOH \, (alc.)} R-CH=CH-C\overset{O}{\underset{OK}{\diagup}} \xrightarrow[-K^+]{+H^+} R-CH=CH-C\overset{O}{\underset{OH}{\diagup}}$$

(4)
$$R-\underset{X}{\underset{|}{C}}H-C\overset{O}{\underset{OH}{\diagup}} + NaO-\overset{O}{\underset{}{C}}-R' \xrightarrow{-NaX} R-\underset{\underset{\overset{|}{O-C-R'}}{\underset{\overset{\|}{O}}{}}}{\underset{|}{C}}H-\overset{O}{\underset{}{C}}-OH$$

(5)
$$R-\underset{X}{\underset{|}{C}}H-C\overset{O}{\underset{OH}{\diagup}} + KCN \xrightarrow{-KX} R-\underset{CN}{\underset{|}{C}}H-C\overset{O}{\underset{OH}{\diagup}}$$

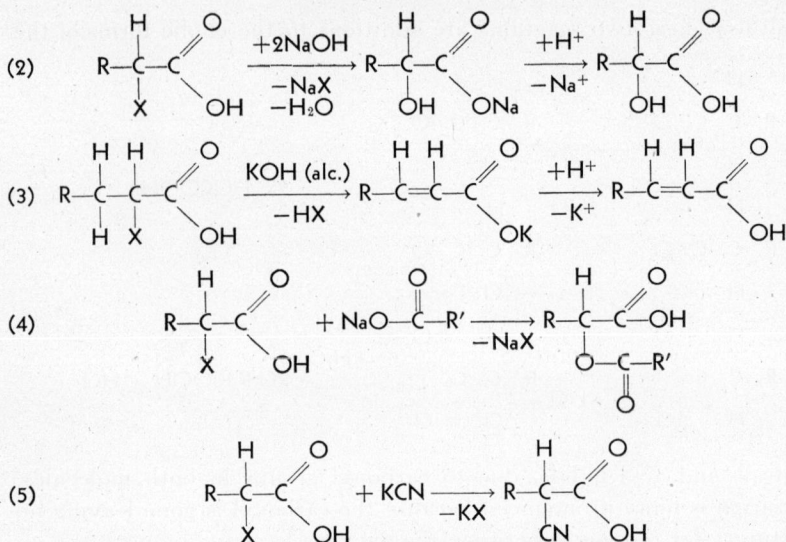

Hydroxyalkanoic Acids

The preparation and chemical behavior of hydroxyalkanoic acids varies with the location of the hydroxy substituent in the molecule. More than one hydroxyl group may be located in a molecule. In this chapter we shall discuss only monohydroxy substitutions on the 2-, 3-, and 4-carbon atoms.

1. 2-Hydroxyalkanoic Acids: PREPARATION. a. By hydrolysis a 2-haloalkanoic acid may be converted to 2-hydroxyalkanoic acid. This reaction is shown above as a reaction of haloalkanoic acids.

b. By hydrolysis of a nitrile formed by the addition of hydrogen cyanide to an aldehyde or a ketone, 2-hydroxyalkanoic acid may be prepared. The hydroxyl group from ketones forms a tertiary alcohol.

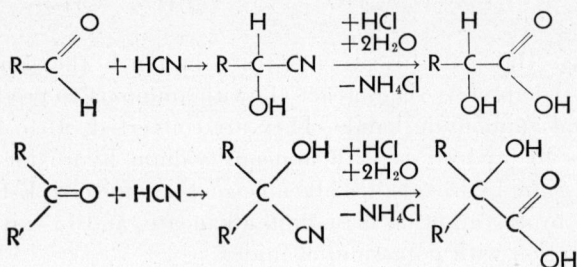

$$R-C\overset{O}{\underset{H}{\diagup}} + HCN \rightarrow R-\underset{OH}{\underset{|}{C}}H-CN \xrightarrow[-NH_4Cl]{\substack{+HCl \\ +2H_2O}} R-\underset{OH}{\underset{|}{C}}H-C\overset{O}{\underset{OH}{\diagup}}$$

$$\underset{R'}{\overset{R}{\diagdown}}C=O + HCN \rightarrow \underset{R'}{\overset{R}{\diagdown}}\underset{CN}{\overset{OH}{\underset{}{C}}} \xrightarrow[-NH_4Cl]{\substack{+HCl \\ +2H_2O}} \underset{R'}{\overset{R}{\diagdown}}\underset{\underset{OH}{}}{\overset{OH}{\underset{}{C}}}C\overset{O}{\diagup}$$

REACTIONS. a. The hydroxyl group acts like an alcohol. With phosphorus trichloride it forms a halide. Esters and ethers may also be formed by the action of proper reagents.

b. Oxidation. If the hydroxyl group is on a primary carbon as in the case of 2-hydroxyethanoic acid, the hydroxyl group may be oxidized with the usual strong oxidizing agent to 2-al-ethanoic acid (glycolic acid).

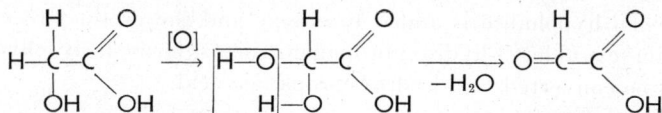

When the hydroxyl group is on a secondary carbon atom, oxidation yields a ketone acid, also called a 2-oxyalkanoic acid.

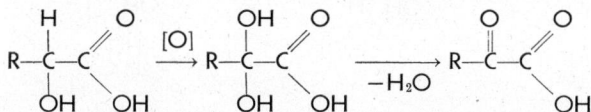

Mild oxidation of 2-hydroxyalkanoic acids containing more than two carbons results in decarboxylation of the molecule and formation of an aldehyde, having one less carbon in the chain.

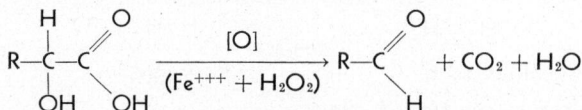

c. Dehydration of 2-hydroxyalkanoic acid causes the formation of cyclic esters called lactides. Also ordinary linear esters may form. This

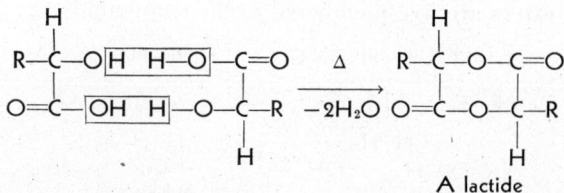

A lactide

case of double esterification is readily accomplished because of the pronounced tendency of organic compounds to form ring structures with six atoms in the ring bonded at angles of approximately 109°.

2. 3-Hydroxyalkanoic Acids: PREPARATION. a. When the aldehyde part of an aldol condensation product is oxidized, 3-hydroxyalkanoic acid is formed.

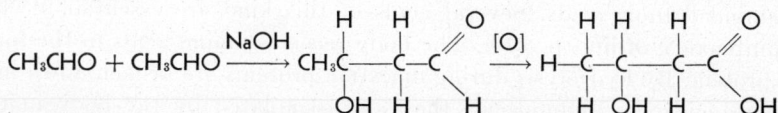

b. When hydrogen chloride is added to 2-alkenoic acid and the product is hydrolized, 3-hydroxyalkanoic acid is formed.

$$R-\overset{\overset{\displaystyle H}{|}}{C}=\overset{\overset{\displaystyle H}{|}}{C}-C\overset{\displaystyle O}{\underset{\displaystyle OH}{}} + HCl \rightarrow R-\overset{\overset{\displaystyle H}{|}}{\underset{\underset{\displaystyle Cl}{|}}{C}}-\overset{\overset{\displaystyle H}{|}}{\underset{\underset{\displaystyle H}{|}}{C}}-C\overset{\displaystyle O}{\underset{\displaystyle OH}{}} \xrightarrow[-HCl]{+H_2O} R-\overset{\overset{\displaystyle H}{|}}{\underset{\underset{\displaystyle OH}{|}}{C}}-\overset{\overset{\displaystyle H}{|}}{\underset{\underset{\displaystyle H}{|}}{C}}-C\overset{\displaystyle O}{\underset{\displaystyle OH}{}}$$

c. When hypohalite is added to ethene and the product reacts with potassium cyanide, 3-hydroxypropanenitrile is formed. By hydrolysis this can be converted to 3-hydroxypropanoic acid.

$$\overset{H}{\underset{H}{}}C=C\overset{H}{\underset{H}{}} + HOCl \rightarrow H-\overset{\overset{\displaystyle H}{|}}{\underset{\underset{\displaystyle OH}{|}}{C}}-\overset{\overset{\displaystyle H}{|}}{\underset{\underset{\displaystyle H}{|}}{C}}-Cl + KCN \rightarrow H-\overset{\overset{\displaystyle H}{|}}{\underset{\underset{\displaystyle OH}{|}}{C}}-\overset{\overset{\displaystyle H}{|}}{\underset{\underset{\displaystyle H}{|}}{C}}-CN$$

$$\xrightarrow[-NH_4Cl]{\substack{+HCl \\ +2H_2O}} H-\overset{\overset{\displaystyle H}{|}}{\underset{\underset{\displaystyle OH}{|}}{C}}-\overset{\overset{\displaystyle H}{|}}{\underset{\underset{\displaystyle H}{|}}{C}}-C\overset{\displaystyle O}{\underset{\displaystyle OH}{}}$$

REACTIONS. When 3-hydroxyalkanoic acids are dehydrated, 2-alkenoic acid is formed.

$$R-\overset{\overset{\displaystyle H}{|}}{\underset{\underset{\displaystyle OH}{|}}{C}}-\overset{\overset{\displaystyle H}{|}}{\underset{\underset{\displaystyle H}{|}}{C}}-C\overset{\displaystyle O}{\underset{\displaystyle OH}{}} \xrightarrow[-H_2O]{H_2SO_4} R-\overset{\overset{\displaystyle H}{|}}{C}=\overset{\overset{\displaystyle H}{|}}{C}-C\overset{\displaystyle O}{\underset{\displaystyle OH}{}}$$

3. 4-Hydroxyalkanoic Acids. When these compounds are heated a molecule of water is split off forming a class of compounds called *lactones*. These inner esters are five-membered cyclic compounds.

$$R-\overset{\overset{\displaystyle H}{|}}{\underset{\underset{\displaystyle O-\boxed{H}}{|}}{C}}-\overset{\overset{\displaystyle H}{|}}{\underset{\underset{\displaystyle H}{|}}{C}}-\overset{\overset{\displaystyle H}{|}}{\underset{\underset{\displaystyle H}{|}}{C}}-C\overset{\displaystyle O}{\underset{\displaystyle \boxed{OH}}{}} \rightarrow R-\overset{\overset{\displaystyle H}{|}}{C}-\overset{\overset{\displaystyle H}{|}}{\underset{\underset{\displaystyle H}{|}}{C}}-\overset{\overset{\displaystyle H}{|}}{\underset{\underset{\displaystyle H}{|}}{C}}-C=O$$

A lactone

Aminoalkanoic Acids

Aminoalkanoic acids have an amine group ($-NH_2$) substituted in the alkyl part of the acid. Such a substitution may take place at any desired point in the molecule where halogen may be substituted. By common usage however the name *amino acid* is associated with the 2-aminoalkanoic acids. Several acids of this kind are essential for the maintenance of life (p. 295). The body receives amino acids in the form of protein. By hydrolysis during digestion proteins are broken down into amino acids. As amino acids they are assimilated by the body in the formation of new tissue. In this brief discussion chiefly 2-aminoalkanoic acids will be considered.

Preparation. Some 2-aminoalkanoic acids are best prepared from simpler compounds by synthesis, others by breaking down complex proteins and separating the specific acid desired.

1. An example of synthesis from simpler compounds for 2-amino-alkanoic acid is given in the following sequence:

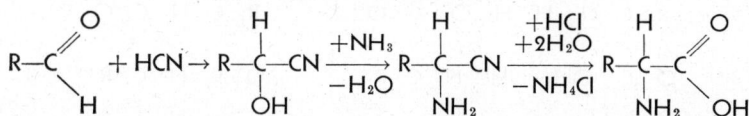

$$R-\underset{H}{\overset{O}{C}} + HCN \rightarrow R-\underset{OH}{\overset{H}{C}}-CN \xrightarrow[-H_2O]{+NH_3} R-\underset{NH_2}{\overset{H}{C}}-CN \xrightarrow[-NH_4Cl]{+HCl\ +2H_2O} R-\underset{NH_2}{\overset{H}{C}}-\overset{O}{\underset{OH}{C}}$$

2. Another synthesis illustrates the circuitous reactions required for separating the pure final product.

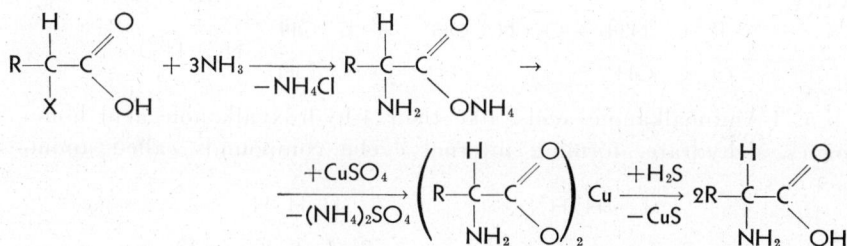

$$R-\underset{X}{\overset{H}{C}}-\overset{O}{\underset{OH}{C}} + 3NH_3 \xrightarrow{-NH_4Cl} R-\underset{NH_2}{\overset{H}{C}}-\overset{O}{\underset{ONH_4}{C}} \rightarrow$$

$$\xrightarrow[-(NH_4)_2SO_4]{+CuSO_4} \left(R-\underset{NH_2}{\overset{H}{C}}-\overset{O}{\underset{O}{C}} \right)_2 Cu \xrightarrow[-CuS]{+H_2S} 2R-\underset{NH_2}{\overset{H}{C}}-\overset{O}{\underset{OH}{C}}$$

Reactions. 1. 2-Aminoalkanoic acids react with bases to form regular 2-aminoalkanoate salts and with acid by addition to the amino group.

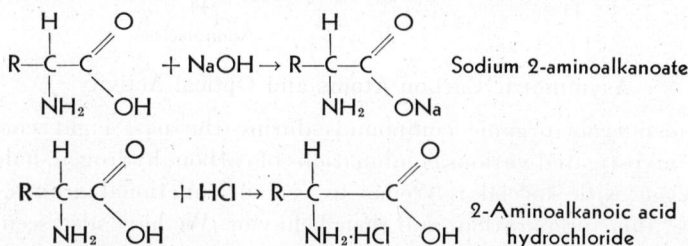

$$R-\underset{NH_2}{\overset{H}{C}}-\overset{O}{\underset{OH}{C}} + NaOH \rightarrow R-\underset{NH_2}{\overset{H}{C}}-\overset{O}{\underset{ONa}{C}} \qquad \text{Sodium 2-aminoalkanoate}$$

$$R-\underset{NH_2}{\overset{H}{C}}-\overset{O}{\underset{OH}{C}} + HCl \rightarrow R-\underset{NH_2\cdot HCl}{\overset{H}{C}}-\overset{O}{\underset{OH}{C}} \qquad \begin{array}{l}\text{2-Aminoalkanoic acid}\\ \text{hydrochloride}\end{array}$$

2. Proteins are complex molecules whose formation has resulted from the reaction of amino groups with the acid groups of other 2-amino-alkanoic acids. This type of union is called the "peptide linkage." This peptide linkage continues from acid to acid until a molecular unit of

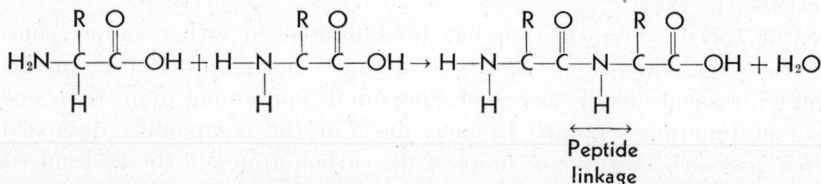

$$H_2N-\underset{H}{\overset{R}{C}}-\overset{O}{C}-OH + H-\underset{H}{\overset{R}{N}}-\underset{}{\overset{O}{C}}-\overset{O}{C}-OH \rightarrow H-\underset{H}{\overset{R}{N}}-\overset{O}{C}-\overset{O}{C}-\underset{H}{\overset{R}{N}}-\overset{O}{C}-\overset{O}{C}-OH + H_2O$$

$$\underset{\text{linkage}}{\underset{\text{Peptide}}{\longleftrightarrow}}$$

protein having a molecular weight minimum of about 35,500 is produced.

3. With the loss of two moles of water, two molecules of 2-amino-alkanoic acid may form a cyclic peptide called a "diketopiperazine." The similarity of this reaction to that in which a lactide (p. 131) forms from 2-hydroxyalkanoic acids is readily apparent.

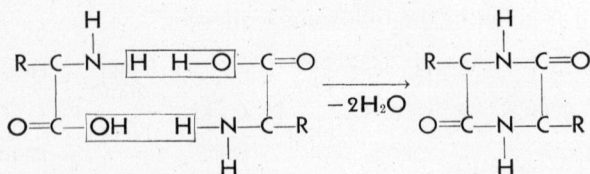

$$
\begin{array}{ccc}
\text{H} & & \text{H} \\
| & & | \\
\text{R—C—N—}\boxed{\text{H} \quad \text{H—O}}\text{—C=O} & & \text{R—C—N—C=O} \\
| \qquad\qquad\qquad | & \xrightarrow{-2\text{H}_2\text{O}} & | \qquad\quad | \\
\text{O=C—}\boxed{\text{OH}} \quad \boxed{\text{H}}\text{—N—C—R} & & \text{O=C—N—C—R} \\
& & \quad | \\
\text{H} & & \text{H}
\end{array}
$$

4. Aminoalkanoic acids in general react with nitrous acid forming the hydroxyalkanoic acid homologue with the liberation of nitrogen and water.

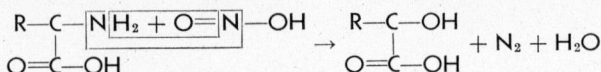

$$
\begin{array}{ccc}
\text{R—C—}\boxed{\text{NH}_2 + \text{O}=\text{N}}\text{—OH} & & \text{R—C—OH} \\
| & \rightarrow & | \qquad\qquad + \text{N}_2 + \text{H}_2\text{O} \\
\text{O=C—OH} & & \text{O=C—OH}
\end{array}
$$

5. 4-Aminoalkanoic acids, like their 4-hydroxyalkanoic acid homologues, dehydrate, forming internal cyclic compounds called aminolactones.

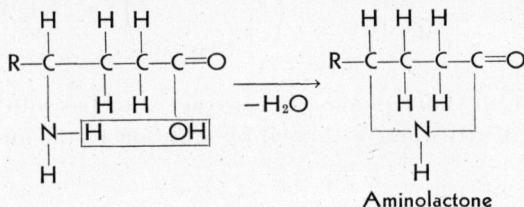

$$
\begin{array}{ccc}
\text{H} \quad \text{H} \;\; \text{H} & & \text{H} \;\; \text{H} \;\; \text{H} \\
| \quad\; | \;\;\; | & & | \;\;\; | \;\;\; | \\
\text{R—C——C—C—C=O} & \xrightarrow{-\text{H}_2\text{O}} & \text{R—C—C—C—C=O} \\
| \quad\;\; | \;\; | \;\; | & & | \;\;\; | \;\; | \\
\text{N—}\boxed{\text{H} \qquad \text{OH}} & & \text{N} \\
| & & | \\
\text{H} & & \text{H}
\end{array}
$$

Aminolactone

Asymmetric Carbon Atoms and Optical Activity

In considering organic compounds during the past eight chapters, we have investigated various combinations of carbon, hydrogen, halogens, and oxygen with special reference to certain functional groups, their structure, their preparation, and their behavior. We have also seen three kinds of isomerism: (a) chain isomerism as exemplified in such hydrocarbons as butane and 2-methylpropane (p. 18); (b) position isomerism as exemplified by the cis- and trans- compounds of 2-butene (p. 39); and (c) metameric isomerism in which different functional compounds have the same molecular formula as in the case of ethanol and methoxymethane (p. 77).

Our specific investigation has been limited to rather simple combinations containing only one type of functional group. Earlier in this chapter we considered classes of compounds containing more than one kind of functional group. In fact, most of the compounds discussed herein are such that one or more of the carbon atoms in the molecule is bonded to four distinctly different elements or groups. Thus in 2-hydroxy-propanoic acid the central carbon atom is joined (a) to a carboxyl group,

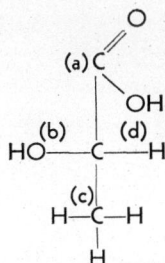

(b) to a hydroxyl group, (c) to a methyl group, and (d) to a hydrogen atom. Such a carbon atom is called an *asymmetric carbon atom*, that is, one without symmetrical structure in terms of its union with other groups. For such a molecule there are two possible spacial arrangements as shown in Figure 24 below. Also the arrangements are the mirror images

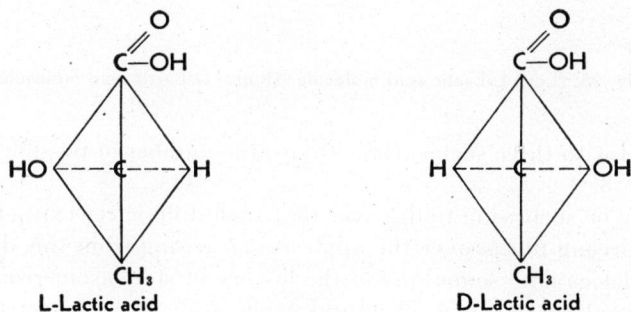

L-Lactic acid   D-Lactic acid

Fig. 24.

of each other and they are isomers. This example introduces a new type of isomerism, namely that involving position in three dimensional space, known as *stereoisomerism*. In molecular weight and molecular formula the two compounds are identical. The molecules are not, however, identical in structure. If you should walk up to a full length mirror and reach out your right hand, your image would reach out with its left hand to shake hands with you; it is essentially the same with a compound and its steric isomer.

For more complex molecules there may be more than one asymmetric carbon atom in the chain. When the number of asymmetric carbon atoms in a molecule increases the number of possible stereoisomers increases in geometric proportion. If the number of asymmetric carbon atoms in a molecule is represented by a, the total number of *possible* stereoisomers N would be

$$N = 2^a$$

Thus in the example of 2-hydroxypropanoic (Lactic) acid there is one asymmetric carbon and the number of steric isomers, $N = 2^1 = 2$. In a

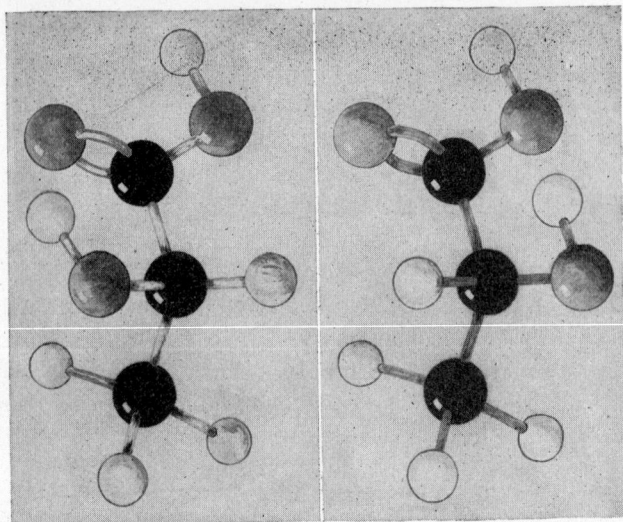

Fig. 25. (Left) L-Lactic acid molecule. (Right) D-Lactic acid molecule.

compound with three such carbon atoms, the number of possible isomers, $N = 2^3 = 8$.

It may be surprising to discover that such differences exist. It is even more surprising to discover the applications arising from this difference. But first let us note something of the history of stereoisomerism.

The identification and structural explanation of asymmetric carbon atoms by van't Hoff and Le Bel in 1874 culminated more than a half century of research by many of the best scientists of Europe. In 1808 Malus discovered the property of a crystal called Iceland spar, whereby a beam of light in passing through the crystal is polarized, that is, made to vibrate in a single plane. Biot in 1815 discovered that certain organic substances, either as liquids or as solutions, were capable of rotating the plane of this polarized light about the path of the light ray. He correctly deduced that rotation of polarized light was caused by something within the molecules themselves. During the next 30 years many organic compounds were found to have this effect, rotating the plane of polarized light either to the right (dextrorotatory, d-) or to the left (levorotatory, l-). Among these compounds tartaric acid (2,3-dihydroxybutanedioic acid) was found to be dextrorotatory. Another acid having the same molecular weight and similar properties, called "racemic acid," was not "optically active" in this manner. Louis Pasteur on examining a salt of "racemic acid" discovered that two shapes of crystals were formed when a solution of the salt evaporated slowly, the one the mirror image of the other. Thus racemic acid proved to be a mixture. After separating the crystals with the aid of a magnifying glass and by using a pair of tweezers, Pasteur found that one kind was dextrorotatory and the other was levorotatory.

When equal amounts of each were dissolved in the same solution, the solution showed no optical activity. Another quarter of a century elapsed before van't Hoff and Le Bel, working independently of each other, explained the mystery of the asymmetric carbon atom. We now know that optical activity depends upon the spacial relationship of atoms or groups within the molecule when four different elements or groups are bonded to the same carbon atom.

In representing optical activity of a compound, the actual direction of rotation to the right is shown by the small letter *d-* in parentheses (*d-*) or by the plus sign (+); the small letter *l-* in parentheses (*l-*) or the minus sign (−) is used to show actual rotation to the left.

In representing the structure of these optical active compounds, there is a conventional agreement for writing graphic formulas because exact coordination of each structure with rotation has not yet been established. In writing formulas therefore the capital letter *D-* (Dextro-) shows spacial arrangement of the hydroxyl group on the right hand side of the asymmetric carbon atom farthest removed from the 1-carbon atom of the molecule. Capital letter *L-* (Levo-) shows the hydroxyl at the left as in the examples of lactic acid shown below.

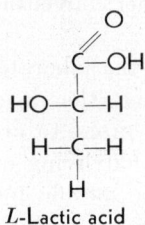

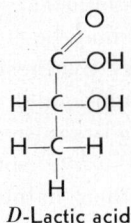

L-Lactic acid   *D*-Lactic acid

Racemic compounds for certain optically active substances are also known; they depend upon internal compensation of optical activity by structural arrangement. Thus for tartaric acid containing two asymmetric carbon atoms, there are the *d-* form, the *l-* form, the meso form (racemic within its own structure), and the possibility for equal amounts of *d-* acid and *l-* acid, producing a racemic mixture.

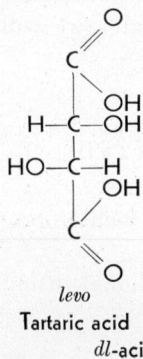

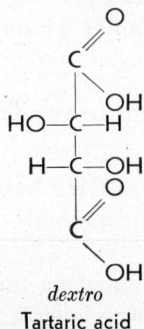

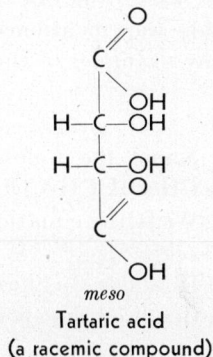

levo   dextro   meso
Tartaric acid   Tartaric acid   Tartaric acid
dl-acid (a racemic mixture)   (a racemic compound)

Racemic mixtures of optically active compounds, being mixtures, do not have a graphic formula and are usually designated by dl-, as dl-tartaric acid.

Separation of Racemic Mixtures. Racemic mixtures may be resolved to their d- and l- compounds by various methods.

a. A few have been resolved by the tedious process followed by Pasteur—by hand separation of crystals.

b. Certain molds, varieties of *Penicillium*, capable of growing in racemic mixtures of some compounds, e.g. tartaric acid, use only one part of the mixture as their food, leaving the optical isomer in the solution. Yes, molds have food idiosyncrasies, some preferring l-isomers and others attacking only d-isomers.

c. Racemic mixtures may also be separated chemically by reaction with another optically active compound. For example, a racemic acid mixture may react with a d-base. Two salts would be formed, one a dd-salt and the other a dl-salt. Salts such as these will have different physical and chemical properties including their solubilities. Accordingly they may be separated by fractional crystallization and the respective acids obtained by chemical treatment of the salts. If difference in solubilities is not sufficient to permit separation, other conventional methods of extraction may be employed.

Laboratory Synthesis vs. Vital Activity. In the laboratory synthesis of organic compounds that are optically active, the racemic mixture results; living organisms on the other hand are prone to produce one of the optically active isomers. The corollary is that living organisms are selective in their habits, not only producing a specific form, but also requiring a specific optically active form of many materials. Man, in his efforts to improve his welfare by producing more effective drugs has discovered that the certain optically active forms are far more efficacious than others. This marked difference in physiological activity has led to the search for the compound that will best serve man's needs. As an example of this applied function of optical activity we may cite the fact that l-Adrenaline is about 12 times as effective as d-Adrenaline when used as a drug. In the preceding section the selective habits of certain molds was mentioned. In Chapter 11 on Carbohydrates we will meet many examples of the principles of stereoisomerism.

Study and Review Questions

1. Name the following compounds: (a) $CH_3CHClCOOH$, (b) $CH_3CH(OH)CH_2COOH$, and (c) $CH_2(NH_2)CH_3COOH$.

2. Write the equations for four methods of preparing 2-chloropropanoic acid.

3. How can 3-chlorobutanoic acid be made from ethanal, using additional chemicals of your own choice?

4. How can 2-propenoic acid be prepared from propanoic acid? Give equations.
5. By equations, show how to prepare 2-methoxypropanoic acid from propanoic acid.
6. Write the equations for preparing 2-methylpropanedioic acid from propanoic acid.
7. Show a method of preparing 2-hydroxypropanoic acid from ethanal.
8. Show two other methods of preparing 2-hydroxypropanoic acid.
9. Write the equations and name the final products when the following are heated and dehydrated: (a) 2-hydroxybutanoic acid, (b) 3-hydroxybutanoic acid, and (c) 4-hydroxybutanoic acid.
10. Show by equations how to convert 2-hydroxybutanoic acid to 3-hydroxybutanoic acid.
11. What is meant by a "peptide linkage"?
12. What simple reaction is required in converting 2-aminopropanoic acid to 2-hydroxypropanoic acid?
13. In what respect is a "diketopiperazine" reaction like the reaction for forming a lactide?
14. Explain the requirements for an asymmetric carbon atom.
15. How many stereoisomers are possible for a compound having three asymmetric carbon atoms?
16. Explain the convention of D- and L- compounds, using tartaric acid as an example.
17. What is a "racemic mixture"? How does it differ from a racemic compound?
18. Explain briefly three methods of resolving racemic mixtures to their optically active constituents.

CHAPTER PROLOGUE

Nitrogen at ordinary temperatures is a very inactive element. At high temperatures it combines with many elements, including carbon, to form several binary compounds. It also unites with carbon and other elements in various combinations to produce some of the most complex and interesting compounds in organic chemistry. A brief survey of a few simple carbon-nitrogen-hydrogen-oxygen compounds reveals certain similarities in structure with common inorganic chemicals. Cyanogen, for example, is very similar in structure in its field to hydrogen peroxide and hydrazine among water and ammonia derivatives. Cyanamide bears a structural relationship to hydrazine and cyanogen similar to the relationship of hydroxylamine to hydrazine and hydrogen peroxide.

H—OH Water	H—NH$_2$ Ammonia	H—CN Hydrogen cyanide	R—H Alkane
HO—OH Hydrogen peroxide	H$_2$N—NH$_2$ Hydrazine	NC—CN Cyanogen	R—R Alkane
	HO—NH$_2$ Hydroxylamine	HO—CN Cyanic acid	R—OH Alcohol
		H$_2$H—CN Cyanamide	R—NH$_2$ Amine
			R—CN Alkane nitrile

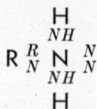

Two of these nitrogen-carbon compounds, nitriles and amines, will be examined in this chapter. Cyclic compounds of carbon and nitrogen will be discussed in Chapter 15.

Names of Compounds. The naming of compounds containing carbon, nitrogen, hydrogen, and oxygen becomes more complex with the complexity of the compounds. Nitrogen may be attached to a carbon directly or through an intermediate atom such as oxygen. Certain combinations of these elements have been accorded a standardized acceptance in the naming of a compound as a prefix or as a suffix or as a separate added word. A list of the more common nitrogen-containing functional groups and their names is given in Table 8. The absence of a hyphen indicates a separate word.

Structure and Bonding. The nitrogen atom unites with elements first by covalent bonding. There are five valence electrons in the outer shell of nitrogen. In combination three of these electrons are shared by covalent bonding and the two remaining electrons may unite with an element by coordinate bonding. When three electrons are shared with hydrogens, the last two electrons are responsible for the formation of many substances known as ammonium compounds.

$$R\ {}^R_C C\ {}^{C\ N}_{\substack{C\ N}}\ N\ N^1 \qquad R\ {}^R_N N\ {}^{NH}_{\substack{NH}}\ N \qquad \left[H\ {}^H_N N\ {}^{NH}_{\substack{NH}}\ H \right]^+ \left[{}_H^O O\ {}^{OO}_{\substack{OO}}\ {}^O_H H \right]^-$$

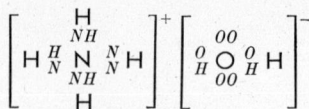

H	H	
Alkanenitrile	Amine	Ammonium hydroxide

[1] The use of small letters in italics for electrons shows the source of the electron.

Table 8

NITROGEN GROUPS AND THEIR NAMES

Formula	Name
—NH$_2$	amine (amino-) (to carbon not carboxyl)
	-amide (amido-) (replacing —OH in carboxyl)
=NH	imine (imino-) (to carbon not carboxyl)
	-imide (imido-) (replacing =O in carboxyl)
=NOH	-oxime
—N=O	nitroso-
—O—N=O	nitrite
—N(=O)(O)	nitro- (replacing —H)
—O—N(=O)(O)	nitrate
—C≡N	-nitrile
—N≡C	-isonitrile
—NH—NH$_2$	-hydrazine
=NH—NH$_2$	-hydrazone
—N=N—	azo-

When organic groups replace the hydrogens bonded to nitrogen in ammonium compounds, quaternary ammonium salts are formed.

Nitriles

The simplest nitrile is methanenitrile, HCN. This compound has many chemical properties typical of inorganic hydro acids and so it is also known commonly as hydrocyanic acid or hydrogen cyanide. All nitriles are poisonous. At room temperatures methanenitrile is a gas and other nitriles in the range C2 to C14 are liquids. They are named by adding the word nitrile to the full name of the alkane represented by the number of carbon atoms, including —CN, in the longest chain. The name is written as one word. The compound CH_3CH_2CN is thus called propanenitrile.

Preparation. 1. Nitriles may be prepared by the reaction of halo-alkanes with potassium cyanide (p. 61).

$$R—\boxed{X + K}—CN \rightarrow RCN + KX$$

2. Nitriles may also be prepared by the addition reaction of aldehydes or ketones with HCN. Both reactions form 2-hydroxynitriles (p. 88).

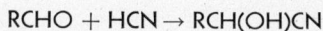

$$RCHO + HCN \rightarrow RCH(OH)CN$$

3. Strong dehydrating agents like phosphorus pentoxide when distilled with an alkanamide will also form nitriles.

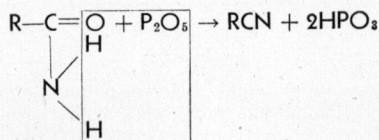

$$R-C=\boxed{O + P_2O_5} \rightarrow RCN + 2HPO_3$$

4. Thionyl chloride dehydrates an alkanamide, yielding a nitrile and two gases, SO_2 and HCl.

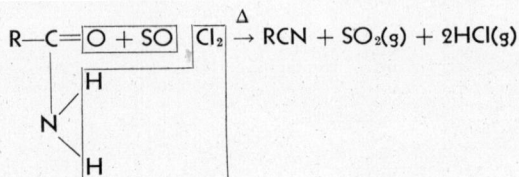

$$R-C=\boxed{O + SO}\ \boxed{Cl_2} \xrightarrow{\Delta} RCN + SO_2(g) + 2HCl(g)$$

Reactions: 1. HYDROLYSIS. Hydrolysis occurs with nitriles in either acid or basic solution. In basic solution the salt of an organic acid is formed.

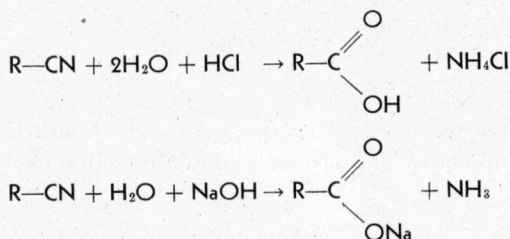

$$R-CN + 2H_2O + HCl \rightarrow R-C\begin{matrix}O\\\\OH\end{matrix} + NH_4Cl$$

$$R-CN + H_2O + NaOH \rightarrow R-C\begin{matrix}O\\\\ONa\end{matrix} + NH_3$$

2. REDUCTION. Nitriles are reduced to amines, chiefly primary, by hydrogen from the reaction of sodium with ethanol.

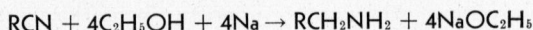

$$RCN + 4C_2H_5OH + 4Na \rightarrow RCH_2NH_2 + 4NaOC_2H_5$$

Hydrogen and alkanenitrile passed over hot nickel catalyst (200°C) yields a primary amine.

$$R-CN + 2H_2 \xrightarrow[200°C]{Ni} RCH_2NH_2$$

Unsaturated Nitriles. Unsaturated nitriles, such as propenenitrile (acrylonitrile) used in the manufacture of Buna N rubber, can be prepared from ethene.

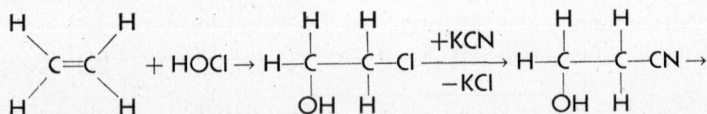

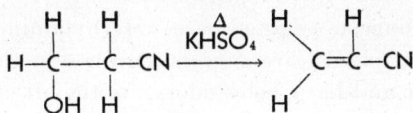

Amines

Amines are basic organic compounds. There are three kinds of amines, primary, secondary, and tertiary, the name varying with the kind and number of alkyl groups replacing hydrogen atoms in ammonia.[2]

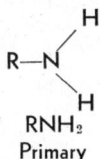

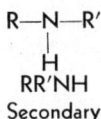

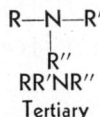

Quaternary ammonium compounds will be discussed later in this chapter.

In the I.U.C. system of nomenclature many of the names used for compounds during the past century have been officially retained with modifications only where needed to avoid confusion. The size of an alkyl group is designated by names such as methyl-, ethyl-, etc. Each alkyl substitution must be shown in the name which is written as one word. Thus, CH_3NHCH_3 is dimethylamine and $CH_3CH_2N(CH_3)$-$CH_2CH_2CH_3$ is methylethylpropylamine.

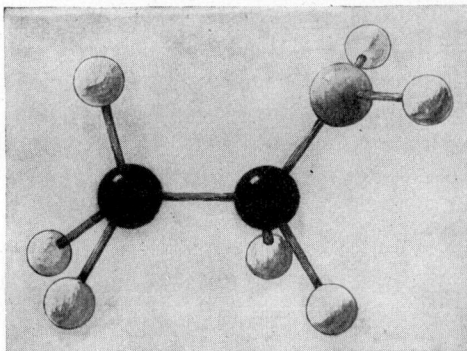

Fig. 26. The ethyl amine molecule.

Amines have basic properties. Like their prototype, ammonia, all amines have a strong tendency to accept a proton to join with the unshared electron pair of the nitrogen atom. The newer concept of acid and base as defined by Brönsted and by Lewis (see an inorganic text) simplifies our understanding of the basic properties of the amines. Dissociation of these addition products is slight in primary amines and increases through secondary and tertiary compounds reaching maximum dissociation in the quaternary compound, tetramethylammonium

[2] It will be recalled from Chapter 5 on Alcohols that primary, secondary, and tertiary *carbon* atoms are those joined respectively to one, two, or three other carbon atoms. In classifying amines we refer to the *nitrogen* atom being joined to one, two, or three carbon atoms respectively.

hydroxide, the strongest organic base. Methylamine is a gas. Other primary amines C2 to C12 are liquids. Amines in the range C1 to C6 are soluble in water and have fishy odors. As the alkyl groups in amines increase in size, solubility in water decreases.

Structure and Bonding. The structure of amines is dependent upon nitrogen. Nitrogen is tricovalent, allowing as many as three alkyl groups to be bonded covalently to the nitrogen. It is due to the presence in each structure of two electrons capable of coordinate bonding that amines

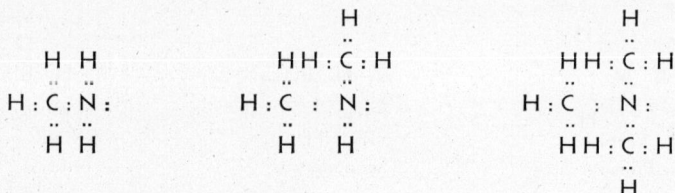

$$
\begin{array}{ccc}
 & \overset{H}{\underset{\cdot\cdot}{}} & \overset{H}{\underset{\cdot\cdot}{}} \\
\text{H H} & \text{H H : C : H} & \text{H H : C : H} \\
\text{H : C : N :} & \text{H : C : N :} & \text{H : C : N :} \\
\text{H H} & \text{H H} & \text{H H : C : H} \\
 & & \text{H}
\end{array}
$$

have their basic property and that they form salts with acids by addition. Thus a polar compound such as HCl will readily add to an amino group forming a salt in which the $RNH_3{}^+$ is cationic to the chloride ion.

Preparation. 1. Commercially, methyl amines are made by passing hot ammonia and methanol vapors over alumina gel heated to 450°C. The mixture of primary, secondary, and tertiary amines which forms is then separated by fractional distillation. The yield is not very high but fresh ammonia and methanol may be added to residues, which are again passed over the catalyst in a continuous operation.

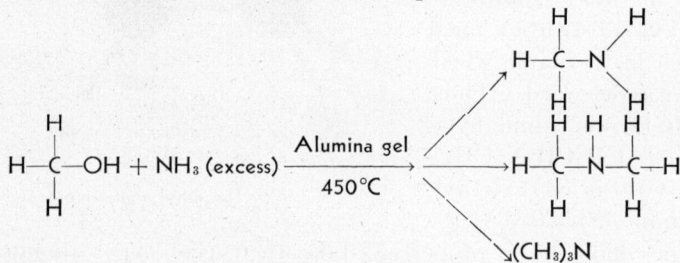

2. When a nitrile is reduced in the presence of platinum, colloidal palladium, or heated nickel, a primary amine is produced.

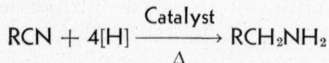

$$RCN + 4[H] \xrightarrow[\Delta]{\text{Catalyst}} RCH_2NH_2$$

3. The Hofmann rearrangement of an amide (p. 121) results in the formation of a primary amine having *one less carbon atom* than the parent amide in the chain. Some authors refer to this reaction as the Hofmann degradation because carbon is removed from the chain. It should not be confused with another reaction (p. 148) also called the Hofmann deg-

radation, used in the characterization of amines by exhaustive methylation. The rearrangement reaction is shown here in three steps.

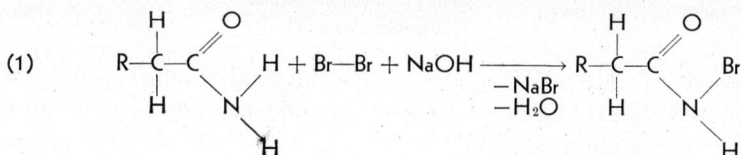

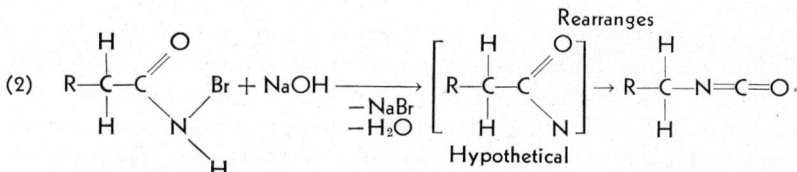

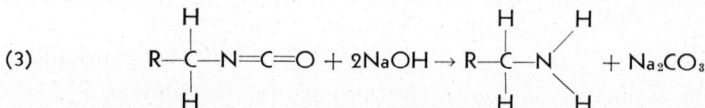

(1)

(2)

Rearranges

Hypothetical

(3)

$$RCH_2CONH_2 + Br_2 + 4NaOH \rightarrow RCH_2NH_2 + Na_2CO_3 + 2NaBr + 2H_2O$$

4. The Hofmann Reaction. The Hofmann reaction for preparing amines is often confused with the rearrangement reaction and degradation reaction mentioned in the preceding paragraph. The Hofmann reaction explained in this paragraph was discovered more than a hundred years ago, 30 years before the discovery of the Hofmann rearrangement.

When a haloalkane reacts with ammonia, an alkylammonium halide is formed by addition. Alkylammonium halide reacts with ammonia

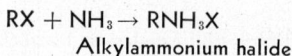

$$RX + NH_3 \rightarrow RNH_3X$$
Alkylammonium halide

forming a primary amine. This primary amine reacts with haloalkane and forms dialkylammonium halide. The dialkylammonium halide

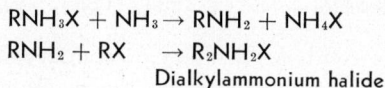

$$RNH_3X + NH_3 \rightarrow RNH_2 + NH_4X$$
$$RNH_2 + RX \rightarrow R_2NH_2X$$
Dialkylammonium halide

reacts with more ammonia, producing ammonium halide and dialkyl amine. This secondary amine reacts with more haloalkane forming trialkylammonium halide. The trialkylammonium halide reacts with more

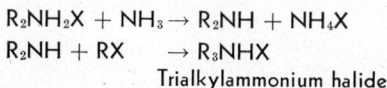

$$R_2NH_2X + NH_3 \rightarrow R_2NH + NH_4X$$
$$R_2NH + RX \rightarrow R_3NHX$$
Trialkylammonium halide

ammonia, forming ammonium halide and trialkylamine. This tertiary amine adds haloalkane to form the quaternary salt, tetraalkylammonium halide.

$$R_3NHX + NH_3 \rightarrow R_3N + NH_4X$$
$$R_3N + RX \rightarrow R_4NX$$

Tetraalkylammonium halide
(a quaternary salt)

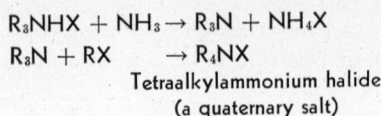

As a result of this continuous reaction the final product will be a mixture of three kinds of amines, primary, secondary, and tertiary, plus a quaternary ammonium salt. The mixture can be separated by fractional distillation, except in the case of the C1 amines whose boiling points are too close together for effective separation.

In this reaction the presence of an excess of ammonia favors the production of primary and secondary amines, whereas an excess of halo-alkane favors the production of tertiary amines and quaternary ammonium salts. One single amine is never the sole product of this reaction. Indeed, no simple method has yet been devised for the production of pure secondary or tertiary amines.

Reactions. 1. The tricovalent nature of nitrogen and its consequent basic property (ability to attract a proton) permit the amines to form salts by the addition of acids. These salts are ionic in character as is shown by the reaction below, using electronic formulas. The conventional

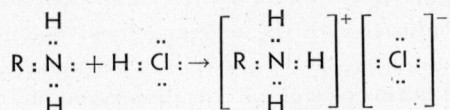

$$R:\overset{H}{\underset{H}{N}}: + H:\overset{..}{\underset{..}{Cl}}: \rightarrow \left[R:\overset{H}{\underset{H}{N}}:H \right]^{+} \left[:\overset{..}{\underset{..}{Cl}}: \right]^{-}$$

structural formula of such salts shows a dot between the amine and the hydrogen halide, and the word "hydrohalide" must follow the name of the amine, thus:

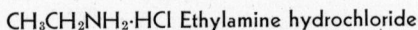

$$CH_3CH_2NH_2{\cdot}HCl \text{ Ethylamine hydrochloride}$$

These salts react with strong bases like sodium hydroxide to liberate the amine.

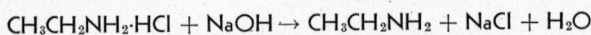

$$CH_3CH_2NH_2{\cdot}HCl + NaOH \rightarrow CH_3CH_2NH_2 + NaCl + H_2O$$

2. With Nitrous Acid. a. Primary, secondary, and tertiary amines react differently with nitrous acid, and the reaction serves as a test for the type of amine. Primary amines produce an alcohol, nitrogen, and water.

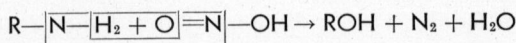

$$R{-}\boxed{N{-}H_2 + O}{=}N{-}OH \rightarrow ROH + N_2 + H_2O$$

b. Secondary amines form a yellow nitrosamine which is soluble in water.

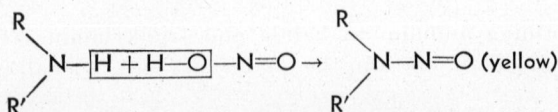

$$\overset{R}{\underset{R'}{>}}N{-}\boxed{H + H{-}O}{-}N{=}O \rightarrow \overset{R}{\underset{R'}{>}}N{-}N{=}O \text{ (yellow)}$$

c. Tertiary amines add nitrous acid to form a salt, trialkylammonium nitrite. There are no visible signs of reaction for the new compound is ionic in character.

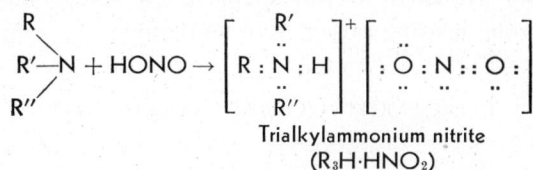

$$
\begin{array}{c} R \\ \diagdown \\ R'\!-\!N + HONO \rightarrow \\ \diagup \\ R'' \end{array}
\left[\begin{array}{c} R' \\ \ddots \\ R : N : H \\ \ddots \\ R'' \end{array} \right]^{+}
\left[\begin{array}{c} \ddots \quad\quad \\ : O : N :: O : \\ \ddots \quad\quad \ddots \end{array} \right]
$$

Trialkylammonium nitrite
($R_3H \cdot HNO_2$)

Significance of the Amines. The amines find important application as compounds employed in synthesis. In the preceding chapters of this text a surprisingly large number of reactions involve the amino group. Reference to the amino acids in Chapter 9 and to the applied biochemical material in Chapters 19 and 20 illustrates the imposing significance of amines in physiological chemistry and biology. In industry amines and their derivatives find use as solvents, emulsifying agents, cleaning compounds, accelerators in rubber vulcanization, disinfectants, fungicides, and lubricants.

Quaternary Alkylammonium Compounds

The increased use of quaternary ammonium compounds during the past decade has finally brought prominence to compounds whose existence has been known for more than a century. Quaternary ammonium salt in even small concentrations, 100 to 200 parts per million, has definite bactericidal properties. Because of its ability to lessen the surface tension of water it also has cleansing properties similar to other types of detergents.

Quaternary alkylammonium compounds may be prepared by heating tertiary amines with a haloalkane. Tetraalkylammonium halide, a solid, is formed. This is readily soluble in water and may be converted in solution by silver hydroxide to tetraalkylammonium hydroxide. When silver halide is filtered off, the solution can be evaporated to a crystalline

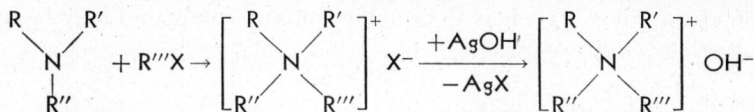

$$
\begin{array}{c} R \quad\quad R' \\ \diagdown \diagup \\ N \\ | \\ R'' \end{array} + R'''X \rightarrow
\left[\begin{array}{c} R \quad\quad R' \\ \diagdown \diagup \\ N \\ \diagup \diagdown \\ R'' \quad\quad R''' \end{array} \right]^{+} X^{-}
\xrightarrow[-AgX]{+AgOH}
\left[\begin{array}{c} R \quad\quad R' \\ \diagdown \diagup \\ N \\ \diagup \diagdown \\ R'' \quad\quad R''' \end{array} \right]^{+} OH^{-}
$$

solid. The alkyl groups need not be all alike. In fact disinfectants and detergents of this type vary in accord with the alkyl groups attached to the nitrogen. Tetramethylammonium hydroxide is the strongest known organic base. In solution its strength is similar to that of sodium or potassium hydroxide.

When tetramethylammonium hydroxide is heated strongly, it decomposes to form methanol and a tertiary amine. If alkyl groups other

$$(CH_3)_4NOH \xrightarrow{\Delta} (CH_3)_3N + CH_3OH$$

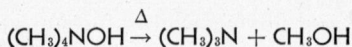

than methyl are attached to the nitrogen, an alkene and water are produced by strong heating rather than methanol.

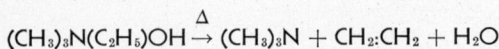

$$(CH_3)_3N(C_2H_5)OH \xrightarrow{\Delta} (CH_3)_3N + CH_2{:}CH_2 + H_2O$$

In the characterization of amines the type of alkene formed by heating is very important. If an original amine is successively methylated with methyl halide until all hydrogens of the original amine are replaced, the resulting tetraalkylammonium salt may then be converted with silver hydroxide to tetraalkylammonium hydroxide. The alkene formed by

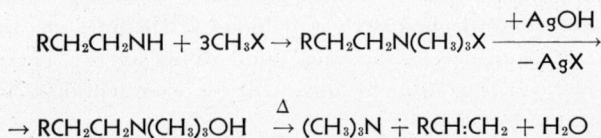

$$RCH_2CH_2NH + 3CH_3X \rightarrow RCH_2CH_2N(CH_3)_3X \xrightarrow[-AgX]{+AgOH}$$

$$\rightarrow RCH_2CH_2N(CH_3)_3OH \xrightarrow{\Delta} (CH_3)_3N + RCH{:}CH_2 + H_2O$$

heating this hydroxide may then be characterized. This process of exhaustive methylation of an amine and the breaking down of the product by heating is called the Hofmann degradation.

The term "Hofmann degradation" has therefore a reference to two types of reactions (p. 144) and students must avoid confusing them.

Changing Length of Carbon Chains

In considering the preparation and the reactions of acyclic carbon compounds thus far some few methods of increasing or decreasing the length of carbon chains have been emphasized. These reactions are assembled here again to emphasize the importance they bear to synthesis. Only one method is given here. Students should feel free to use any combination of correct reactions to increase or decrease the length of a carbon chain. Since a cycle is to be achieved, any compound may be used as a starting point.

Increasing the length of a chain.

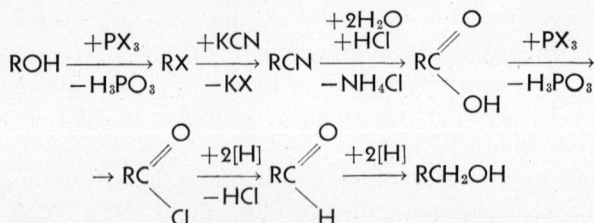

$$ROH \xrightarrow[-H_3PO_3]{+PX_3} RX \xrightarrow[-KX]{+KCN} RCN \xrightarrow[-NH_4Cl]{+2H_2O \atop +HCl} RC{\overset{O}{\underset{OH}{\diagup\!\diagdown}}} \xrightarrow[-H_3PO_3]{+PX_3}$$

$$\rightarrow RC{\overset{O}{\underset{Cl}{\diagup\!\diagdown}}} \xrightarrow[-HCl]{+2[H]} RC{\overset{O}{\underset{H}{\diagup\!\diagdown}}} \xrightarrow{+2[H]} RCH_2OH$$

Decreasing the length of a chain.

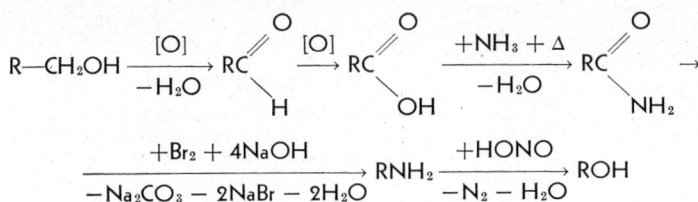

$$R\!-\!CH_2OH \xrightarrow[-H_2O]{[O]} RC\!\!\!\diagdown\!\!\!\overset{\displaystyle O}{\diagup}_H \xrightarrow{[O]} RC\!\!\!\diagdown\!\!\!\overset{\displaystyle O}{\diagup}_{OH} \xrightarrow[-H_2O]{+NH_3+\Delta} RC\!\!\!\diagdown\!\!\!\overset{\displaystyle O}{\diagup}_{NH_2} \rightarrow$$

$$\xrightarrow[-Na_2CO_3-2NaBr-2H_2O]{+Br_2+4NaOH} RNH_2 \xrightarrow[-N_2-H_2O]{+HONO} ROH$$

Study and Review Questions

1. Name the following compounds: (a) CH_3CN, (b) $CH_3CH_2NH_2$, (c) $(CH_3)_2NH$, (d) $(CH_3)_3N$, (e) CH_3CHNOH, (f) CH_3CONH_2, (g) $CH_3CH_2NHNH_2$, (h) $CH_3NHCH_2CH_3$, (i) $(CH_3)_2CHCONH_2$, and (j) $(CH_3)_2NH_2Cl$.

2. Write the equations for two methods of preparing propanenitrile.

3. Show two methods of reducing propanenitrile to a primary amine. Name the product.

4. Propenenitrile is used in the manufacture of Buna N rubber. Show in steps how it can be prepared from ethene.

5. Write the equations for the commercial preparation of amines. How are amines separated when prepared in this manner?

6. The Hofmann degradation reaction for preparing amines starts with an amide. Show this reaction with equations in steps starting with propanamide. What kind of amines are formed?

7. The Hofmann reaction for preparing amines starts with a haloalkane and ammonia. Show this reaction with equations in steps. What kind of amines are formed?

8. Show by an equation how a primary amine forms salts with hydrochloric acid. Name your product.

9. Contrast in a small chart the reaction of nitrous acid with (a) ethanamide, (b) ethyl amine, (c) dimethyl amine, and (d) trimethyl amine.

10. By a series of equations, show how tetramethylammonium hydroxide may be prepared from methyl amine and chloromethane.

11. How does heating affect tetramethylammonium hydroxide?

12. By a series of equations, showing all reagents and catalysts when needed, devise a method of preparing 2-propanol from ethanol. (The student is free to use any correct combination of reactions.)

13. By a series of equations, showing all reagents and catalysts when needed, devise a method of preparing ethanol from 2-propanol.

CHAPTER PROLOGUE

The most extensive synthetic chemical reaction taking place in the world today is the making of carbohydrates. Indeed this reaction has held its position of priority since life began. Before man appeared upon earth, living plants under the catalytic influence of chlorophyll combined carbon dioxide, water, and a few calories of radiant energy from the sun into carbohydrate, the principal component of plant tissue. Conversely the most extensive nonsynthetic chemical reaction in the world is the ultimate reconversion of carbohydrates to carbon dioxide and water. The Great Cycle of Carbon in Nature (p. 10) follows this never-ending sequence of reactions.

For many decades chemists wondered how the plant world accomplished this amazing synthesis. Some assumed that simple molecules of formaldehyde, CH_2O, were first formed and then polymerized to form the simple sugars, $(CH_2O)_x$. Other explanations have been born, flourished for a time, and then lost favor. Nature still holds her secret, but research workers are probing ever closer to the truth. Strangely enough, it may be that radioactive isotopes, products of our atomic energy research, will be the key that will open the door and disclose the methods nature uses. Recent experiments at the University of California Radiation Laboratories, using carbon dioxide made from radioactive carbon, have found that lowly algae, among the simplest of plant organisms, accomplish the synthesis of complex carbohydrates in an amazingly brief time. This work has shown that a sugar is the primary product of the photosynthetic process.

Here again is a new horizon in organic chemistry, the investigation of reaction processes by the use of "tagged" (radioactive) atoms. How much will come of this work we do not know, but we do know that man's search for truth will go on and on.

Definition of Carbohydrate. The name carbohydrate is derived from the French "hydrate de carbon," hydrate of carbon, a term used by early investigators who thought that these compounds of carbon, hydrogen and oxygen, must always have the two latter elements present in the same proportion as water. The general formula for carbohydrates is often given as $C_x(H_2O)_y$. This general formula fits most carbohydrates but not all. A more concise definition of a carbohydrate is based upon the structure of the compound or upon the structures of the products resulting from hydrolysis of the compound. The structure of a carbohydrate requires that it be a polyhydroxy aldehyde or a polyhydroxy ketone, or that it be a polyhydroxy ether or hemiacetal whose hydrolysis reaction produces either polyhydroxy aldehydes or polyhydroxy ketones or mixtures of these compounds.

General Classification of Carbohydrates. Polyhydroxy aldehydes and polyhydroxy ketones are called *sugars* because they are sweet in varying degrees. They are also called *monosaccharides*. *Disaccharides* are carbohydrates made up of two similar or dissimilar monosaccharides united by an ether linkage. When three monosaccharides are united by two ether linkages the compound is a *trisaccharide*. When many monosaccharides are joined by ether linkages into one molecule, such compounds are known as *polysaccharides*. The specific names of all carbohydrates end with the suffix -ose.

This general classification of carbohydrates will serve as an outline for the development of this chapter.

   I. Monosaccharides (a simple sugar unit)
  II. Disaccharides (two sugar units connected by one ether linkage with elimination of 1 molecule of H_2O)
 III. Trisaccharides (three sugar units connected by two ether linkages with elimination of 2 molecules of H_2O)
 IV. Polysaccharides (many (n) sugar units connected by (n-1) ether linkages with elimination of (n-1) molecules of H_2O)

Monosaccharides

Structures and Names. The elements in monosaccharides are united by covalent bonding. In structure there are two kinds of monosaccharides, *aldoses* having an aldehyde group as carbon atom 1 and *ketoses* having a ketone group as carbon atom 2. The number of carbon atoms in a

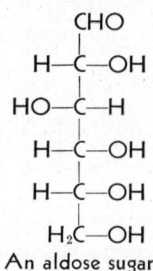

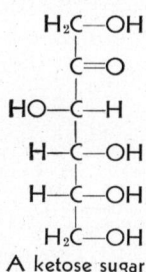

<pre>
        CHO   H₂C—OH
         | |
    H—C—OH   C=O
         | |
   HO—C—H   HO—C—H
         | |
    H—C—OH   H—C—OH
         | |
    H—C—OH   H—C—OH
         | |
      H₂C—OH   H₂C—OH
   An aldose sugar   A ketose sugar
</pre>

monosaccharide may vary from 3 to 10,[1] giving rise to another general type name specifying the number of carbon atoms as triose, tetrose, pentose, hexose, etc. By combining the structure and number types of names as in aldopentose or ketohexose, a monosaccharide can be more accurately identified.

Completely accurate identification of monosaccharides, however, is also dependent upon the exact configuration of hydroxy groups and hydrogen within the molecule. Each carbon atom, except for carbonyl groups, has at least one hydroxy group and one hydrogen atom bonded to it. All monosaccharides, therefore, except ketotriose must have one or more asymmetric carbon atoms in the molecule. The spacial configuration of hydroxyl groups on asymmetric carbon atoms determines the exact

---

[1] This number of carbon atoms conforms (a) to the accepted definition that monosaccharides are aldo- or keto-polyalcohols, the simplest number of —OH groups being 2, and (b) to the fact that, beyond 10 carbon atoms, the compounds known can be hydrolized to give two or more simple sugar units.

Aldotriose

L-
CHO / HO—C—H / H_2C—OH

D-
CHO / H—C—OH / H_2C—OH

Aldotetrose

Erythrose — L-: CHO / HO—C—H / HO—C—H / H_2C—OH ; D-: CHO / H—C—OH / H—C—OH / H_2C—OH

Threose — L-: CHO / H—C—OH / HO—C—H / H_2C—OH ; D-: CHO / HO—C—H / H—C—OH / H_2C—OH

Aldopentose

Ribose — L-: CHO / HO—C—H / HO—C—H / HO—C—H / H_2C—OH ; D-: CHO / H—C—OH / H—C—OH / H—C—OH / H_2C—OH

Arabinose — L-: CHO / H—C—OH / HO—C—H / HO—C—H / H_2C—OH ; D-: CHO / HO—C—H / H—C—OH / H—C—OH / H_2C—OH

Xylose — L-: CHO / HO—C—H / H—C—OH / HO—C—H / H_2C—OH ; D-: CHO / H—C—OH / HO—C—H / H—C—OH / H_2C—OH

Lyxose — L-: CHO / H—C—OH / H—C—OH / HO—C—H / H_2C—OH ; D-: CHO / HO—C—H / HO—C—H / H—C—OH / H_2C—OH

Aldohexose

Allose — L-: CHO / HO—C—H / HO—C—H / HO—C—H / HO—C—H / H_2C—OH ; D-: CHO / H—C—OH / H—C—OH / H—C—OH / H—C—OH / H_2C—OH

Altrose — L-: CHO / H—C—OH / HO—C—H / HO—C—H / HO—C—H / H_2C—OH ; D-: CHO / HO—C—H / H—C—OH / H—C—OH / H—C—OH / H_2C—OH

Glucose — L-: CHO / HO—C—H / H—C—OH / HO—C—H / HO—C—H / H_2C—OH ; D-: CHO / H—C—OH / HO—C—H / H—C—OH / H—C—OH / H_2C—OH

Mannose — L-: CHO / H—C—OH / H—C—OH / HO—C—H / HO—C—H / H_2C—OH ; D-: CHO / HO—C—H / HO—C—H / H—C—OH / H—C—OH / H_2C—OH

Gulose — L-: CHO / HO—C—H / HO—C—H / H—C—OH / HO—C—H / H_2C—OH ; D-: CHO / H—C—OH / H—C—OH / HO—C—H / H—C—OH / H_2C—OH

Idose — L-: CHO / H—C—OH / HO—C—H / H—C—OH / HO—C—H / H_2C—OH ; D-: CHO / HO—C—H / H—C—OH / HO—C—H / H—C—OH / H_2C—OH

Galactose — L-: CHO / HO—C—H / H—C—OH / H—C—OH / HO—C—H / H_2C—OH ; D-: CHO / H—C—OH / HO—C—H / HO—C—H / H—C—OH / H_2C—OH

Talose — L-: CHO / H—C—OH / H—C—OH / H—C—OH / HO—C—H / H_2C—OH ; D-: CHO / HO—C—H / HO—C—H / HO—C—H / H—C—OH / H_2C—OH

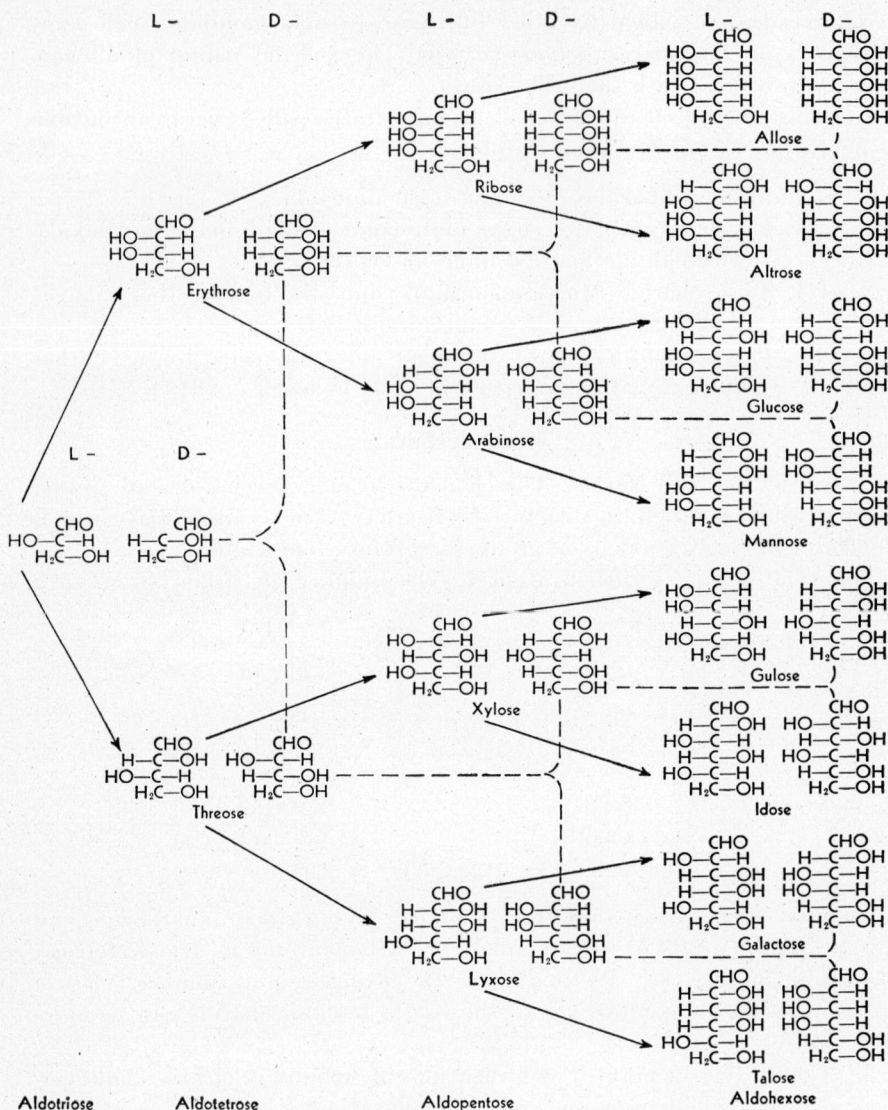

Fig. 27. Simple aldose sugars.

difference in monosaccharides. From van't Hoff's rule (p. 135) the number of possible stereoisomers, N, formed by the number of dissimilar asymmetric carbon atoms, a, is $N = 2^a$. In an aldohexose with 4 asymmetric carbon atoms there are 2^4 or 16 isomers (above). When all the possible different arrangements of groups and atoms are written in graphic formulas, half of the compounds formed are found to be mirror images of the remainder. Each mirror-image pair has been given a specific name. By an accepted convention[2] among chemists one of each name pair is called the structural Dextro- (capital D-) compound when there is an hydroxyl group at the *right* of the asymmetric carbon atom farthest removed from the carbonyl group. When the hydroxy group is at the left of this asymmetric carbon atom in a mirror-image graphic formula, the compound is Levo- (capital L-) structurally. The names, graphic formulas and structural designations of all C_3-C_6 aldoses are shown on p. 152.

The specific name of each monosaccharide refers to a definite molecular configuration. Thus, D-glucose refers to a molecule of specific structure. This system of naming sugars has been used for many decades and has been adopted by the I.U.C. system of nomenclature. Students are not expected to learn all the different names and formulas of simple sugars. Although practically all of them have been synthesized and identified, only a few are found in nature. Students are expected to know about only the more common sugars whose abundance in nature has made them commercially important. In the following brief classification of monosaccharides these more important specific sugars are listed under the appropriate headings.

Monosaccharides:

A. Trioses ($C_3H_6O_3$)
B. Tetroses ($C_4H_8O_4$)
C. Pentoses ($C_5H_{10}O_5$)
   Example: arabinose
D. Hexoses ($C_6H_{12}O_6$)
   Examples:
   (a) Aldoses: Glucose, mannose, galactose
   (b) Ketoses: fructose
E. Heptoses, Octoses, Nonoses, and Decoses.

Class name based upon the number of carbon atoms in the simple sugar.
Subclass name based upon presence of keto- or aldo-structure.
Individual names adopted by I.U.C.

Physical Properties. The monosaccharides are white crystalline solids. They are soluble in water forming true solutions diffusible through animal membranes. Nearly all sugars are optically active since nearly all have at least one asymmetric carbon atom. Each sugar has a specific rotation (α) at a definite temperature for a specific wave length of light. Rotation of plane polarized light to the right is designated by d- or the plus sign ($+$) and to the left by l- or a minus sign ($-$).[3] The direction of specific rotation is entirely independent of the conventional structural designations D- and L-.

[2] Exact three dimensional configuration has been impossible to determine. The recently developed electron-microscope will probably provide the answer.

[3] The designation d- and l- respectively for ($+$) and ($-$) will be used sparingly in this text.

Chemical Properties. Monosaccharides are polyfunctional molecules. They have one carbonyl group and several alcohol groups in accord with their size. They are capable of reacting in many ways, depending on structural modifications in size, arrangement of groups and molecular configuration. The reactions of several individual monosaccharides are of little importance because they are not found in nature. On the other hand, D-glucose is not only found in abundant supply, but also is commercially a very important substance. It has been said that glucose has been more completely studied than any other organic compound. By examining the chemical properties of glucose, therefore, we can best summarize the chemical behavior of aldoses.

In its reactions glucose may be considered in accord with the functional groups present (A) as an aldehyde, (B) as an alcohol, (C) as a full molecule, and (D) as a glucoside unit (ring structure) in synthesis.

A. GLUCOSE AS AN ALDEHYDE: 1. GLUCOSE IS A TYPICAL ALDEHYDE adding a molecule of HCN to give one more carbon atom in an open chain. The resulting nitrile may be hydrolyzed and then converted to an aldoheptose by reduction. This reaction is called the Kiliani synthesis

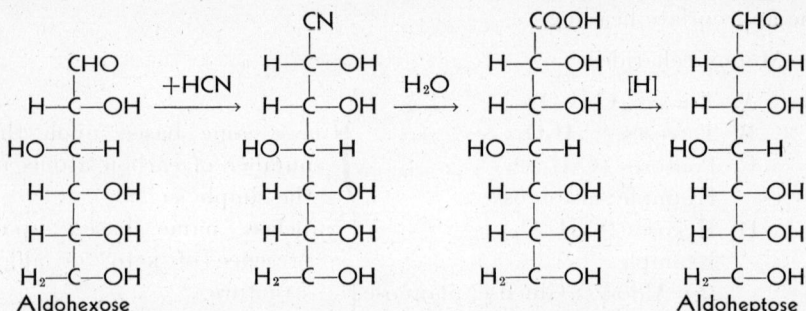

Aldohexose   Aldoheptose

and it has been used in preparing higher homologues from simpler sugars. It is possible by this reaction to start with any simple sugar and from it to synthesize those that are more complex.

The reaction of aldehydes with hydroxylamine has been used in decreasing the number of carbon atoms in a chain. Hydroxylamine reacts with the aldehyde part of glucose, forming glucose oxime. This is reduced through several intermediate steps (shown in larger texts) to the aldopentose, D-arabinose.

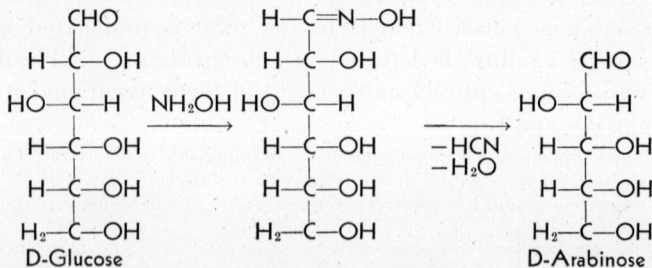

D-Glucose   D-Arabinose

2. OXIDATION OF THE ALDEHYDE GROUP. Glucose can be oxidized by both Fehling's solution and Tollen's solution to produce gluconic acid, together with a variety of secondary products, in which the carbonyl group is converted to a carboxyl group (p. 91). The ability of the aldehyde group to oxidize readily makes glucose a reducing agent, and as such it may be used to reduce silver ion in the formation of silvered mirrors. Sugars capable of acting as reducing agents with Fehling's solution are called *reducing sugars*.

Bromine water also oxidizes glucose to gluconic acid.

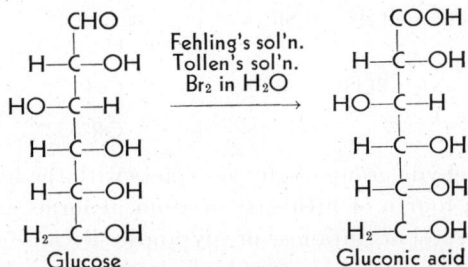

$$
\begin{array}{ccc}
\text{CHO} & & \text{COOH} \\
\text{H—C—OH} & \begin{array}{c}\text{Fehling's sol'n.}\\\text{Tollen's sol'n.}\\\text{Br}_2\text{ in H}_2\text{O}\end{array} & \text{H—C—OH} \\
\text{HO—C—H} & \xrightarrow{\hspace{2cm}} & \text{HO—C—H} \\
\text{H—C—OH} & & \text{H—C—OH} \\
\text{H—C—OH} & & \text{H—C—OH} \\
\text{H}_2\text{—C—OH} & & \text{H}_2\text{—C—OH} \\
\text{Glucose} & & \text{Gluconic acid}
\end{array}
$$

Nitric acid not only oxidizes the aldehyde end of the molecule but also the primary alcohol to form the tetrahydroxyhexanedioic acid called saccharic acid.

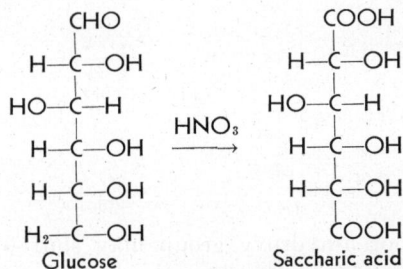

$$
\begin{array}{ccc}
\text{CHO} & & \text{COOH} \\
\text{H—C—OH} & & \text{H—C—OH} \\
\text{HO—C—H} & \text{HNO}_3 & \text{HO—C—H} \\
\text{H—C—OH} & \xrightarrow{\hspace{1.5cm}} & \text{H—C—OH} \\
\text{H—C—OH} & & \text{H—C—OH} \\
\text{H}_2\text{—C—OH} & & \text{COOH} \\
\text{Glucose} & & \text{Saccharic acid}
\end{array}
$$

3. REDUCTION. The aldehyde group of glucose is reduced to the alcohol by sodium amalgam forming sorbitol, used in the preparation of Vitamin C.

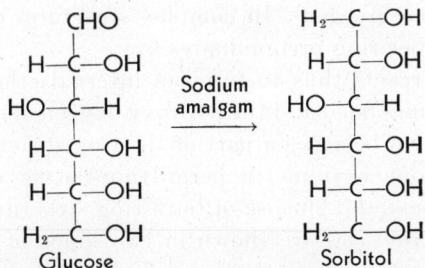

$$
\begin{array}{ccc}
\text{CHO} & & \text{H}_2\text{—C—OH} \\
\text{H—C—OH} & & \text{H—C—OH} \\
\text{HO—C—H} & \begin{array}{c}\text{Sodium}\\\text{amalgam}\end{array} & \text{HO—C—H} \\
\text{H—C—OH} & \xrightarrow{\hspace{1.5cm}} & \text{H—C—OH} \\
\text{H—C—OH} & & \text{H—C—OH} \\
\text{H}_2\text{—C—OH} & & \text{H}_2\text{—C—OH} \\
\text{Glucose} & & \text{Sorbitol}
\end{array}
$$

Glucose may be completely reduced to hexane by hydrogen iodide in the presence of red phosphorus.

$$
\begin{array}{ccc}
\text{CHO} & & \text{CH}_3 \\
\text{H—C—OH} & & \text{CH}_2 \\
\text{HO—C—H} & \xrightarrow[\text{Red phosphorus}]{\text{HI}} & \text{CH}_2 \\
\text{H—C—OH} & & \text{CH}_2 \\
\text{H—C—OH} & & \text{CH}_2 \\
\text{H}_2\text{—C—OH} & & \text{CH}_3 \\
\text{Glucose} & & \text{Hexane}
\end{array}
$$

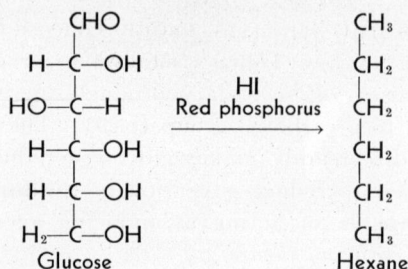

4. HEMIACETAL REACTION OF ALDEHYDE. Aldehydes react with alcohols to form hemiacetals (p. 89).

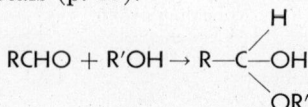

$$\text{RCHO} + \text{R'OH} \rightarrow \text{R—}\overset{\displaystyle \text{H}}{\underset{\displaystyle \text{OR'}}{\text{C}}}\text{—OH}$$

When the aldehyde group of glucose reacts with the hydroxy (alcohol) group on its own fourth or fifth carbon atom, it forms an internal hemiacetal, called a -lactol or -furanose or -pyranose. Because of the proximity of the carbonyl group to these two hydroxy groups, a hydrogen proton

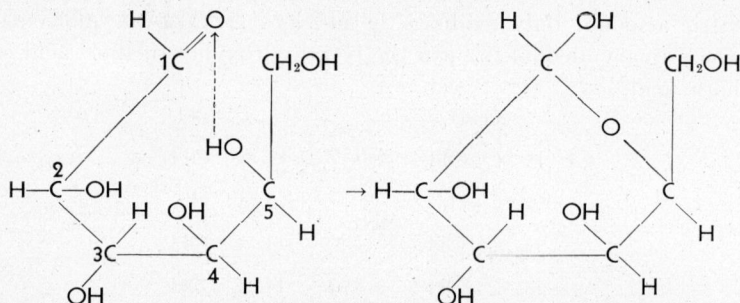

from either the 4- or 5-hydroxy group may shift to the double bond of the enolized carbonyl group. Glucose thus can exist in at least six tautomeric forms, as an open-chain aldehyde, as the enolized open-chain compound, as the cis- and trans- isomers of a five-membered ring (containing one oxygen), and as the cis- and trans- isomers of a six-membered ring (containing one oxygen). In complex structures containing glucose units the six-member ring predominates.

When glucose reacts thus to form an internal ether, the position of the hydrogen becomes fixed and the hydroxy oxygen of either the 4-carbon or the 5-carbon atom becomes a part of the ring structure. The 1-carbon atom has thus become asymmetric permitting the cis- and trans- isomers known as α-glucose and β-glucose of both ring structures.

These cyclic structures are shown in two kinds of graphic formulas. Both kinds of formulas show the same compounds in two methods of representation. The lactol formula is the conventional graphic formula in two dimensions, with the ring forming in the 4-carbon (gamma γ-)

or in the 5-carbon (delta δ-) position. When the hydroxy group on the
1-carbon atom is shown at the right, α-glucose is shown. When the
hydroxy group is at the left, the compound is β-glucose.

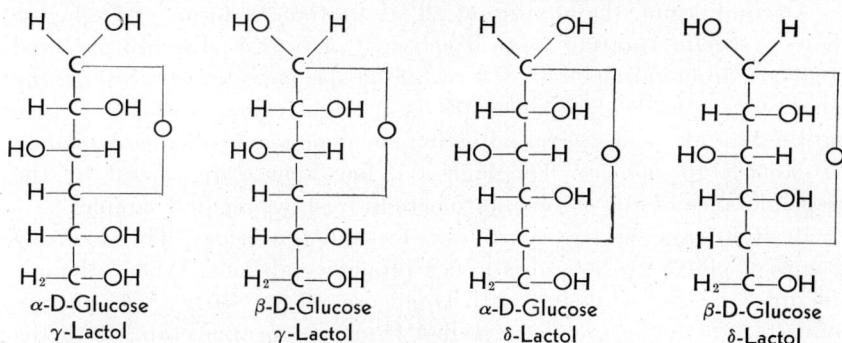

| α-D-Glucose | β-D-Glucose | α-D-Glucose | β-D-Glucose |
| γ-Lactol | γ-Lactol | δ-Lactol | δ-Lactol |

The second method of showing the cyclic structure of glucose is in
perspective which shows three dimensional relationship, with heavy
lines representing the foreground and the junction of two lines representing a carbon atom. Groups below the ring are those at the right in a
graphic formula. The carbonyl carbon, now asymmetric, is shown at

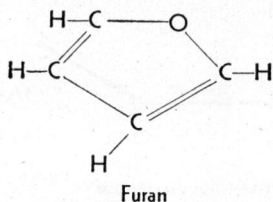

Furan

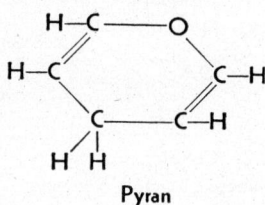

Pyran

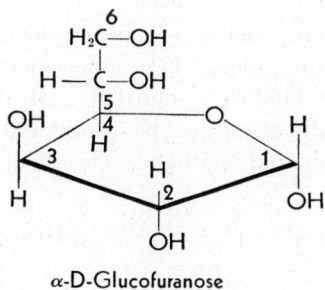

α-D-Glucofuranose

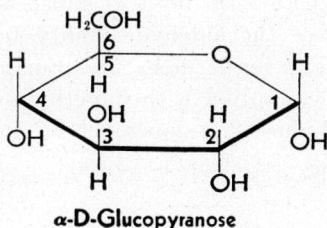

α-D-Glucopyranose

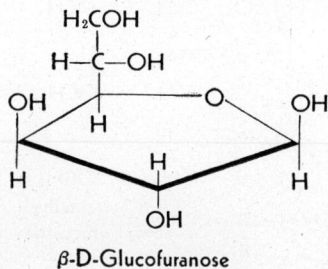

β-D-Glucofuranose

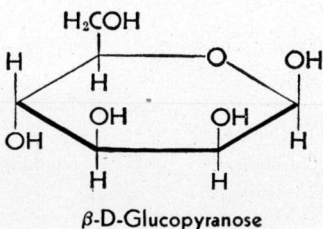

β-D-Glucopyranose

the extreme right. These formulas take their name from furan and pyran (two heterocyclic compounds, p. 241), and are called the -furanose and the -pyranose forms of glucose.

At equilibrium, the mixture of all six tautomeric forms of D-glucose shows a specific rotation (optical activity) of $+52.5°$. Freshly prepared D-glucose from various sources show initial specific rotations through the range of $+19°$ to $+113°$. These specific rotation values gradually change until $+52.5°$ is reached. This adjusting phenomenon is called *mutarotation* (L. *mutare*, to change). The high and low values are caused by the preponderance of one or two tautomers in freshly prepared samples.

B. REACTION AS AN ALCOHOL: 1. ESTERIFICATION. The hydroxy groups of glucose form esters under proper conditions. When ethanoic anhydride reacts completely with D-glucose, all five hydroxy groups are converted to the acetate ester, called D-glucose pentaacetate. Both the α- and β- stereoisomers are formed. Only the α-D-glucose pentaacetate is shown below in both δ-lactol and pyranose forms.

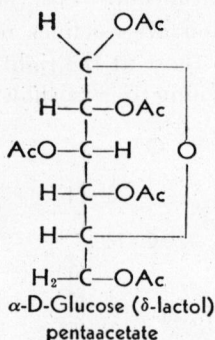

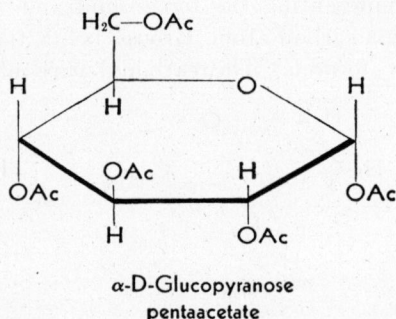

α-D-Glucose (δ-lactol)
pentaacetate

α-D-Glucopyranose
pentaacetate

2. ETHERIFICATION. Glucose reacts with methanol and hydrogen chloride to form both α- and β-methyl glucosides. These glucosides do not have the aldehyde ability to reduce Fehling's solution. β-Methyl glucosides occur in nature. Further etherification can be completed in alkaline solution with dimethyl sulfate and iodomethane. The reaction of α-D-glucose is shown below.

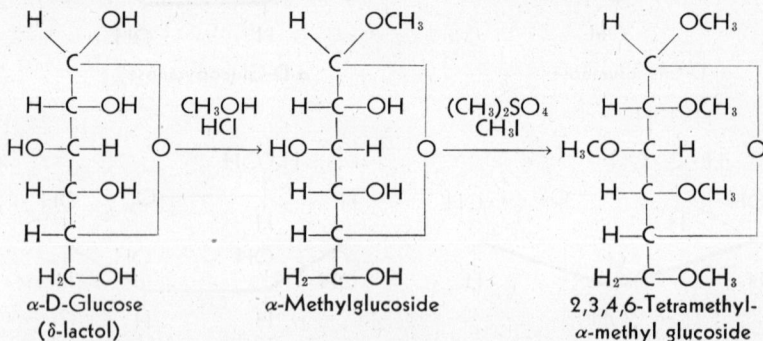

α-D-Glucose
(δ-lactol)

α-Methylglucoside

2,3,4,6-Tetramethyl-
α-methyl glucoside

C. As a Molecule: 1. fermentation. Glucose can be fermented by the enzyme *zymase* in the presence of phosphates. The fermentation end products are ethanol and carbon dioxide. Large amounts of ethanol are produced each year by the fermentation of molasses.

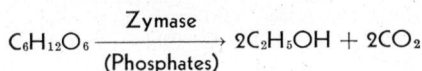

$$C_6H_{12}O_6 \xrightarrow[\text{(Phosphates)}]{\text{Zymase}} 2C_2H_5OH + 2CO_2$$

2. with phenylhydrazine. The reaction of monosaccharides with phenylhydrazine is one of the outstanding reactions of organic chemistry. Emil Fischer received the first Nobel prize award in 1902 for this thorough investigation of sugars. The phenylhydrazine reaction was used in the ultimate identification of hexoses. Phenylhydrazine reacts with aldehydes as shown on p. 90 forming a hydrazone. A second molecule of phenylhydrazine in some manner not yet explained oxidizes the 2-hydroxy

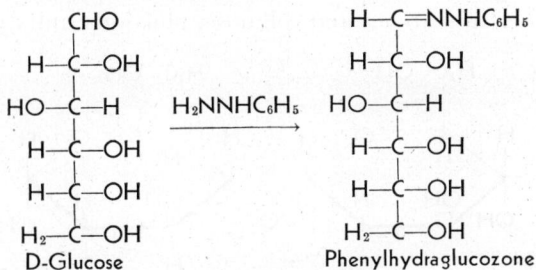

D-Glucose   Phenylhydraglucozone

group to a carbonyl group, being reduced to aniline and ammonia. Then a third molecule of phenylhydrazine reacts with the new carbonyl group forming the osazone. The osazones are yellow crystalline compounds,

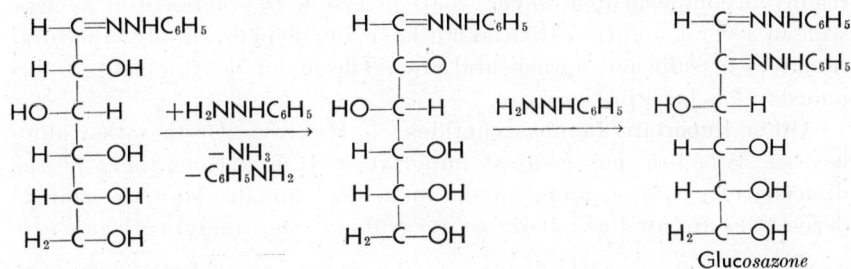

Glucosazone

insoluble in water. All hexoses form osazones. Each hexose, however, requires a different time interval for the inception of crystal formation. By timing the reaction to the first showing of crystals, various sugars may be identified.

The osazone reaction may be used in converting an aldose to a ketose. Osazone, prepared as shown above, is hydrolyzed to an osone (aldoketose) which is then preferentially reduced to a ketose.

$$
\begin{array}{ccc}
\text{H—C}{=}\text{NNHC}_6\text{H}_5 & \text{H—C}{=}\text{O} & \text{H}_2\text{—C—OH} \\
\text{C}{=}\text{NNHC}_6\text{H}_5 & \text{C}{=}\text{O} & \text{C}{=}\text{O} \\
\text{HO—C—H} & \text{HO—C—H} & \text{HO—C—H} \\
\text{H—C—OH} & \text{H—C—OH} & \text{H—C—OH} \\
\text{H—C—OH} & \text{H—C—OH} & \text{H—C—OH} \\
\text{H}_2\text{—C—OH} & \text{H}_2\text{—C—OH} & \text{H}_2\text{—C—OH} \\
\text{Osazone} & \text{Osone} & \text{Ketose (fructose)}
\end{array}
$$

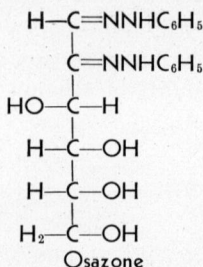

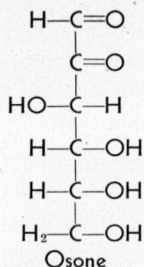

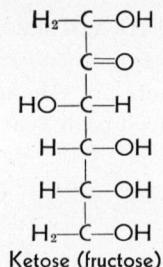

D. As GLUCOSIDE UNITS. Glucose in the form of the glucoside unit exists in many complex compounds. When these compounds are hydrolyzed, glucose is one of the products. Disaccharides, trisaccharides, and polysaccharides are composed of monosaccharides united through ether linkages. Thus two molecules unite to form one molecule of maltose and, conversely, a molecule of maltose may be hydrolized to form two molecules of glucose. In starches and cellulose, glucoside units may make up

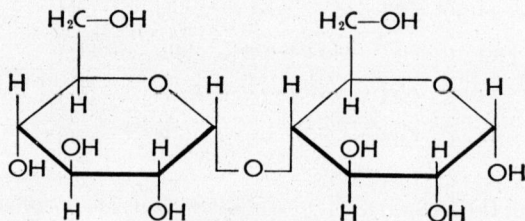

Maltose (1-α-D-glucosyl-4-α-D-glucopyranoside) showing ether linkage.

the entire molecule. In di- and trisaccharides the ether linkage is readily hydrolyzed by boiling in acid solution. Hydrolysis of cellulose is not so readily accomplished, however, the conversion is practiced on a large scale in several countries that do not have an adequate supply of natural resources or sufficient agricultural area. The major development of this industry has been in Germany.

Other Important Monosaccharides: 1. HEXOSES. Of the other aldohexoses, galactose has greatest importance. It is a constituent of the disaccharide, lactose, found in the milk of mammals. Mannose and its derivatives are used as laboratory reagents in some analytical work.

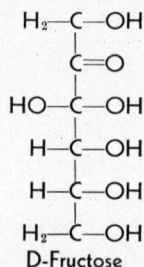

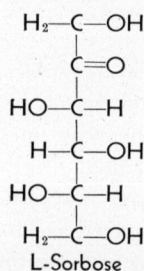

$$
\begin{array}{cc}
\text{H}_2\text{—C—OH} & \text{H}_2\text{—C—OH} \\
\text{C}{=}\text{O} & \text{C}{=}\text{O} \\
\text{HO—C—OH} & \text{HO—C—H} \\
\text{H—C—OH} & \text{H—C—OH} \\
\text{H—C—OH} & \text{HO—C—H} \\
\text{H}_2\text{—C—OH} & \text{H}_2\text{—C—OH} \\
\text{D-Fructose} & \text{L-Sorbose}
\end{array}
$$

Two ketohexoses share commercial importance, D-fructose and L-sorbose. D-Fructose, as a furanose ring united by ether linkage to glucose, forms sucrose, our table sugar. D-Fructose (−) is commonly known as levulose because it is levorotatory. It occurs in honey and ripe fruits. L-Sorbose (−) can be converted to ascorbic acid (vitamin C) (p. 271). Ascorbic acid is found in many vegetables and particularly in citrus fruits. Lack of vitamin C leads to the deficiency disease scurvy.

2. PENTOSES. Pentoses are obtainable from pentosans, complex carbohydrates that occur in nature, especially among the waste products of agriculture. The hulls of oats, wheat, rice, and the cobs of corn are rich in pentosans. Pentosans can be hydrolyzed by heating in strong mineral acid to aldopentose. The aldopentoses (regardless of structure) form furfuraldehyde when dehydrated.

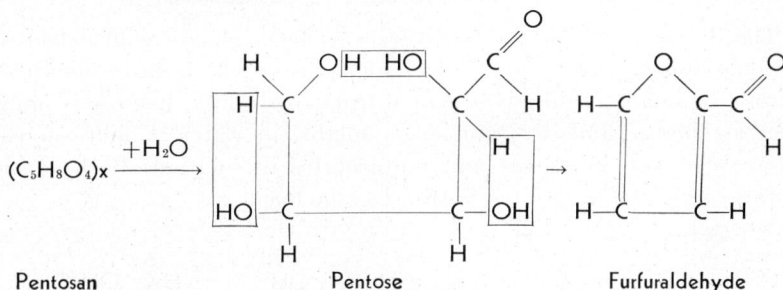

Pentosan   Pentose   Furfuraldehyde

Disaccharides

Of the several disaccharides, the four of outstanding importance are sucrose, lactose, maltose, and cellobiose. Disaccharides are identified by their source and by the hydrolysis products which they yield.

Disaccharides of importance ($C_{12}H_{22}O_{11}$):

    A. Sucrose: Cane or beet sugar, yields glucose and fructose upon hydrolysis.

    B. Lactose: Milk sugar, yields glucose and galactose upon hydrolysis.

    C. Maltose: Obtained from sprouting grain and from starch, yields glucose upon hydrolysis.

    D. Cellobiose: Obtained from cellulose (wood), yields glucose upon hydrolysis.

Sucrose is found in many plants. Sugar beets, sugar cane, and sugar maple are the chief sources of sucrose. Refined sucrose, table sugar, is one of the cheapest and purest chemical compounds. Sucrose is not a reducing sugar. In solution it is dextrorotatory (specific rotation +66.5°) and it does not mutarotate. By acid hydrolysis sucrose solution forms a mixture of glucose and fructose which is slightly sweeter than the original sucrose solution. The mixture is also levorotatory.[4] Because of this opposite change or inversion of rotation, the mixture is called invert sugar. The formula for sucrose may be written in the lactol or pyranose-furanose forms.

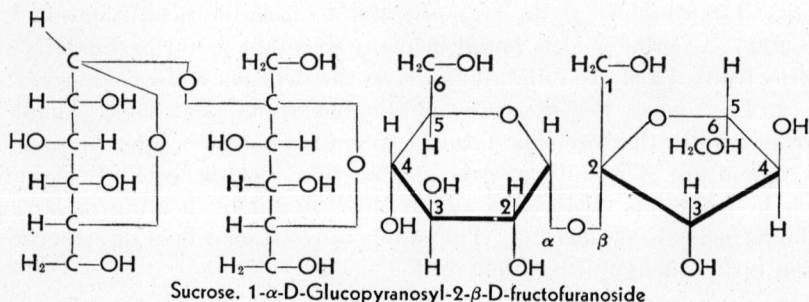

Sucrose. 1-α-D-Glucopyranosyl-2-β-D-fructofuranoside

The disaccharide **lactose** hydrolyzes to form glucose and galactose. The only substantial natural source of lactose is the milk of mammals. Lactose for infant feeding is prepared from cow's milk. Lactose is not as sweet as sucrose and it is much less soluble in water. Unlike sucrose, lactose is a reducing sugar and mutarotates in solution. Both α- and β-forms of glucose are united with β-D-galactoside.

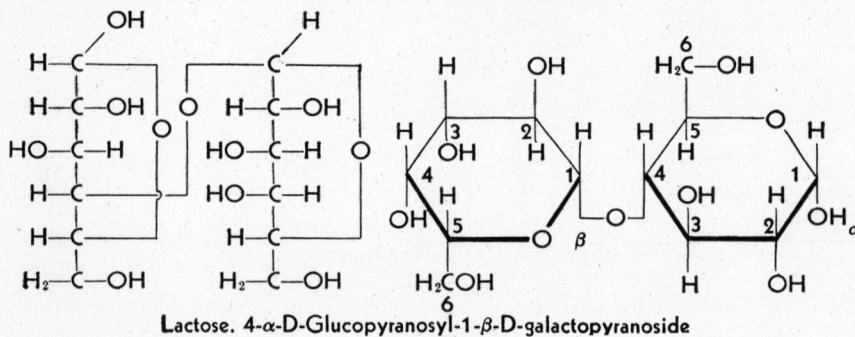

Lactose. 4-α-D-Glucopyranosyl-1-β-D-galactopyranoside

Maltose is produced by the acid or enzymatic hydrolysis of starch. When maltose is hydrolyzed, two glucose units are produced. Maltose

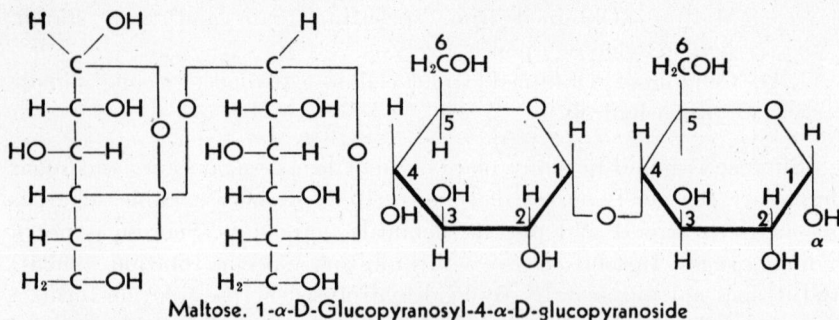

Maltose. 1-α-D-Glucopyranosyl-4-α-D-glucopyranoside

[4] The specific rotation of glucose is $+52.5°$ and of fructose is $-92°$, leaving the solution a difference of $-20.5°$ in specific rotation.

is a glucose-α-glucoside. It mutarotates in solution. Maltose is also a reducing sugar. When maltose is hydrolyzed, since 2 molecules of glucose are formed, it readily ferments with yeast forming ethanol and carbon dioxide.

Cellobiose is produced by the hydrolysis of cellulose. Like maltose it yields two molecules of glucose when hydrolyzed. It is unlike maltose in its ether linkage as it is a glucose-β-glucoside. This small difference in

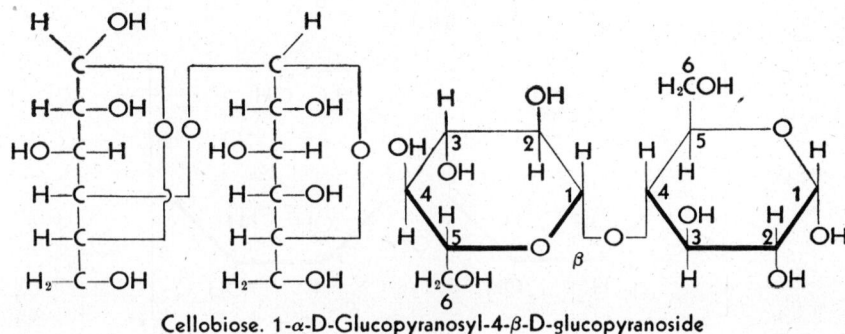

Cellobiose. 1-α-D-Glucopyranosyl-4-β-D-glucopyranoside

structure is one reason why carnivorous animals do not digest cellulose. Maltose can be fermented by the intestinal enzyme maltase. The enzyme cellobiase is present in many animals which can digest cellulose, but not in man.

Trisaccharides

Raffinose ($C_{18}H_{32}O_{16}$) is the best known trisaccharide. Upon hydrolysis it yields one molecule each of glucose, fructose, and galactose. It is found in molasses from beet sugar, and in cottonseed. Partial enzymatic hydrolysis yields fructose and melibiose, a disaccharide, which on further hydrolysis yields glucose and galactose.

Polysaccharides

Polysaccharides are molecules containing many monosaccharide units bound together by ether linkages. Since molecular weights may vary from around 2,000 to 500,000, they are macromolecules. In the acid hydrolysis of polysaccharides, molecules of varying sizes are formed, but the ultimate hydrolysis of the two most common polysaccharides, starch and cellulose, yields glucose. The chemical difference between starches and celluloses is in the type of glucoside units present in the molecule. Starches are formed by α-glucoside units whereas cellulose is made up of the same elements arranged as β-glucosides. The exact number of glucoside units in a molecule may vary from 20 to more than 3000. On partial hydrolysis starches yield maltose and cellulose yields cellobiose. The respective formulas of these compounds, repeated indefinitely presents the accepted configuration of starch and cellulose.

Cellulose

Starch

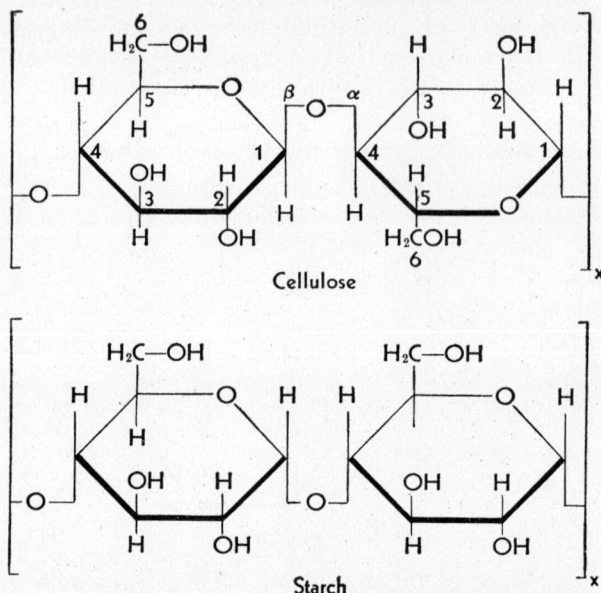

A. Starch. Starches, as ordinarily obtained from vegetable sources, contain in addition to monosaccharide units very small percentages of higher fatty acids. They are believed to be composed of from 26 to 28 monosaccharide units and to have a molecular weight of about 5000. Depending upon the complexity of the molecule, the solubility of starches ranges from sparingly soluble to insoluble. On boiling suspensions of starch, sufficient hydrolysis and physical hydration occurs to permit the formation of colloidal dispersions which form gels upon cooling. When starch is hydrolyzed in acid solution, the ultimate hydrolysis product is glucose, but varying degrees of hydrolysis produce intermediate products known as *dextrins* and maltose. Dextrins are water soluble starches of lower molecular weight. They are called dextrins because they are dextrorotatory in their optical activity. They are formed by the partial hydrolysis of starch and by heating starch, as in the crust of bread.

A simple test for starch consists of treating it with a few drops of a solution of iodine in potassium iodide. A deep blue color appears if starch is present. This test is also made use of in quantitative analysis wherein iodine compounds are used as titrating agents; in this case starch serves as an indicator, turning blue as soon as free iodine forms.

Glycogen is an "animal starch." In man it is stored in the liver and muscles. On hydrolysis it yields glucose. The molecular weight of glycogen is about 2000.

Inulin, found in the Jerusalem artichoke, is a starch formed of fructosidal units. Upon hydrolysis fructose is formed.

B. Cellulose. Cellulose is insoluble in water and organic solvents. The molecules contain as many as 300 β-glucoside units. It is soluble in Schweitzer's reagent (cupric ammonium complex ion and ammonium hydroxide) and in zinc chloride-hydrochloric acid solution. Relative solubility of cellulose in varying concentrations of sodium hydroxide solution leads to its classification as α-, β-, and γ-cellulose. The α-cellulose is important as the raw material for making cellulose plastics. Hydrolysis of cellulose is much slower than that of starch. One of the purest natural sources of cellulose is cotton.

Cellulose differs from hemicellulose in that the latter may have pentose units condensed with those of glucose in the molecule. Hemicellulose is more readily hydrolyzed than cellulose.

The cellulose industries embrace the utilization of lumber, cotton, and many agricultural waste products together with their conversion into paper, rayon, plastics, explosives, and allied materials.

Cellulose does not dissolve in organic solvents but it forms esters with inorganic acids, acetic anhydride, strong alkali, and carbon bisulfide. Guncotton is a nitrated cellulose. Pyroxylin, another nitrated cellulose, is used in making lacquers, celluloid, collodion, and artificial leather. Celanese is made from cellulose acetate. The first safety glass for automobiles was made by placing cellulose acetate between two panes of glass and bonding by applying high pressure. Viscose rayon and cellophane are made by the reaction of cellulose with carbon bisulfide and sodium hydroxide, followed by the regeneration of celluloselike material in a sulfuric acid bath.

The importance of carbohydrates to life can probably be expressed best by saying that they are the storehouses of energy from the sun. They have fixed within their molecules about one third of all the carbon dioxide on our earth. They are easily available, easy to utilize. Without them man would not be able to exist.

Study and Review Questions

1. Why was the name "carbohydrate" selected for the class of compounds treated in this chapter?
2. Give an accurate definition of "carbohydrate."
3. What is meant by the term "simple sugar"?
4. What suffix is used to represent the general class of carbohydrates?
5. Write the formula of an aldotetrose. Of a ketopentose. How many asymmetric carbon atoms are in each?
6. Write the graphic formula of D-galactose. Of L-galactose.
7. Using Kiliani's synthesis, show a series of equations for transforming D-lyxose into D-galactose.
8. What is meant by D(+)-galactose? By D(−)-fructose?

9. Show by a series of equations how to transform D-glucose into D-xylose.

10. What is meant by the term "reducing sugar"?

11. Show by graphic formulas the relationship between sorbose and vitamin C (p. 271).

12. Write the graphic formulas of three tautomers of glucose.

13. Write the formula β-D-glucopyranose. Write the δ-lactol formula of β-D-glucose.

14. Explain the phenomenon called "mutarotation."

15. What is a "glucoside"? Write a graphic formula of β-methyl glucoside.

16. Show in steps the reaction of glucose with phenylhydrazine. Name the final product.

17. How is the reaction of simple sugars with phenylhydrazine used in determining specific sugars?

18. By a series of equations show how an aldohexose may be converted to a ketohexose.

19. What specific compound is meant by the name levulose?

20. From Fig. 27 on p. 152 select two aldopentoses. Show with graphic formulas how furfural is the dehydration product of each.

21. Name four disaccharides. What does each yield upon being hydrolized?

22. By what type linkage are monosaccharides joined in the formation of disaccharides?

23. Explain the structural differences between maltose and cellobiose.

24. Why does the human body digest starches but not cellulose?

25. What difference exists between dextrose and dextrin?

26. What is glycogen?

27. Name five commercial products of cellulose.

Chapter 12

NOMENCLATURE

CHAPTER PROLOGUE

Acute "polynomenclaturitis" is a disease peculiar to organic chemists. It is an insidious affliction seldom noticed by the patient but annoying and at times distressing to the patient's associates. In the organic-chemistry-neophyte the disease often begets bewilderment at first and then discouragement.

For more than a century "polynomenclaturitis" has been congenital—handed down from teacher to student. More than 60 years ago chemists in different countries recognized the existence and nature of the disease. In 1892 at Geneva, Switzerland, the International Union of Chemistry prescribed a remedy—a scientific system of nomenclature for organic chemicals. Like most good remedies, however, the Geneva or I.U.C. system has not received the widespread usage it deserves, despite the repeated urging of the International Union of Chemistry (1923 and 1930) and of the American Chemical Society (1933).

In the first 11 chapters of this book we have tried to use consistently only the simplest recommended system of naming compounds. Since students will be exposed in later life and in collateral reading to other systems of nomenclature, this chapter will introduce them to the confusion they have thus far been spared. It is our hope that with a secure, scientific foundation in the naming of compounds they will be less terrified or discouraged by the inconsistency of their elders.

During the first 11 chapters of this book a single consistent method of nomenclature has been followed. The names of compounds have in all instances been those officially approved in 1930 by the International Union of Chemistry at Liège, Belgium. Rule 1 on nomenclature adopted at that meeting is an interesting departure from the first rule of the Geneva meeting in which a single official name was to be assigned to each compound: "As few changes as possible will be made in terminology universally adopted."[1] This rule is the justification for many acceptable names of compounds found in this chapter.

To facilitate cross-reference with systems of nomenclature enjoying variable degrees of universal adoption, each preceding class of compounds will be discussed separately.[2]

Alkanes

In the naming of unbranched alkanes the names of compounds are the same as outlined in Chapter 2. In the naming of branched chain isomers the prefix *iso-* is used to designate two methyl groups attached at the same end of the chain. Thus, isopentane, $(CH_3)_2CHCH_2CH_3$, is often used for 2-methylbutane, and the radical $(CH_3)_2CHCH_2CH_2$—

---

[1] J.A.C.S., 55, 3907 (1933). All references to rule numbers will be found in this article.

[2] Students will be able to interpret more clearly and to interrelate the explanations herein given by writing the graphic formula of every example cited, naming it by I.U.C. and then studying the example to see the reason for the substitute name. Each section should be mastered before proceeding to the next, for, as an example, the naming of alkanes may, by the use of the derived radical, be applied to alkenes, halogen derivatives, and so on.

is the isopentyl radical. Misuse of this basic system has led to a few examples of commercial interest wherein well-known materials such as 2,2,4-trimethylpentane (called isooctane) are named; as such cases are met in industry they will be readily recognized.

Another method of naming branch-chain hydrocarbons is based upon methane as a parent compound. In this system 2-methylbutane would be called dimethylethylmethane. The inherent weakness of so naming higher branch-chain homologues is gradually diminishing the use of this system of nomenclature.

Alkenes

In naming unsaturated hydrocarbons containing one double bond, such as $CH_3CH:CHCH_3$, ethylene may be considered to be the parent compound. Alkyl substitution for the hydrogen atoms of ethylene leads to the naming of the hydrocarbon as ethylene derivatives, such as symmetrical (*sym-*) dimethyl ethylene for 2-butene, the example cited above. *Unsym*-dimethyl ethylene is then the name for 2-methylpropene $(CH_3)_2$-$C:CH_2$ and ethyl ethylene for 1-butene $CH_3CH_2CH:CH_2$.

In naming the derivatives of such hydrocarbons as the chloro-derivatives for example, an ambiguity is introduced for the derivative is named as an addition product. Thus 1,2-dichloroethane, CH_2ClCH_2Cl, is known as ethylene chloride, ethylene being the unsaturated compound to which the chlorine was added.

The I.U.C. explicitly sanctions (Rule 56) the use of $CH_2\diagup^{\diagdown}$ as the methylene radical. By the same rule $CH_3CH\diagup^{\diagdown}$ becomes ethylidine, $CH_3CH_2CH\diagup^{\diagdown}$ becomes propylidine, and so on. From this sanction, for example, dichloromethane, CH_2Cl_2, becomes methylene chloride and 1,1-dichlorobutane, $CH_3CH_2CH_2CHCl_2$, becomes butylidine chloride.

Alkynes

In the I.U.C. system the name "acetylene" is highly acceptable, ranking with propane and butane. In naming acetylenic derivatives, the parent compound system uses $-C\equiv C-$ as the basis for naming these compounds, as dimethyl acetylene for 2-butyne, $CH_3C:CCH_3$. Halogen additions, as described under alkenes, are named for the parent compound. Thus 1,1,2,2-tetrachloroethane is called acetylene tetrachloride.

Halogen Compounds

Halogen derivatives of alkanes are often named like binary inorganic compounds, such as ethyl chloride (CH_3CH_2Cl), and isopropyl chloride ($(CH_3)_2CHCl$). In some few instances the type designation of a carbon atom is used in naming halogen derivatives, such as secondary butyl bromide ($CH_3CHBrCH_2CH_3$) or tertiary butyl bromide ($(CH_3)_3$-CBr. The naming of complex halogen derivatives of this system is most unsatisfactory. The naming of chloro-derivatives of unsaturated hydrocarbons is given under alkenes and alkynes.

Alcohols

Other systems of naming alcohols are based (a) upon the type of carbon atom to which an hydroxy group is attached or (b) upon the groups substituted for hydrogen in the $H—CH_2OH$ called "carbinol."[3] Five-carbon atom alcohols are called *amyl* alcohols in the former system. The names of other alcohols are derived from the names used for acids in that system after myristic acid (p. 170). Thus the tetradecanols, alcohols with 14-carbon atoms, are myristyl alcohols. The names of the eight pentanols in each system are shown below with structural formulas.

$CH_3CH_2CH_2CH_2CH_2OH$	n-amyl alcohol	butyl carbinol
$(CH_3)_2CHCH_2CH_2OH$	isoamyl alcohol	isobutylcarbinol
$CH_3CH_2CH_2CH(OH)CH_3$	sec. amyl alcohol	methylpropylcarbinol
$CH_3CH_2CH(OH)CH_2CH_3$	sec. amyl alcohol	diethylcarbinol
$(CH_3)_2CHCH(OH)CH_3$	sec. amyl alcohol	isopropylmethylcarbinol
$CH_3CH_2C(OH)(CH_3)_2$	sec. amyl alcohol	dimethylethylcarbinol
$(CH_3)_3CCH_2OH$	neopentyl alcohol	*tert*-butylcarbinol
$CH_3CH_2CH(CH_3)CH_2OH$	active amyl alcohol	*sec*-butylcarbinol

The carbinol system of naming alcohols is widely used in industry and the other system still finds many adherents especially in the field of drugs.

Ethers

The I.U.C. system of naming ethers has undoubtedly deterred many organic chemists from adopting it. The name ethoxyethane seems too involved for what has been long known by the simple name "ether." The naming of more complex ethers, however, becomes almost impossible by the system that places so much emphasis on the location of the oxygen. Simple ethers are named in accord with the groups on each side of the oxygen atom such as $(CH_3)_2CHOCH_3$, methylisopropyl ether. The more common simple ethers of this system are entitled to I.U.C. recognition under Rule 1. Internal ethers such as epoxyethane, H_2C———CH_2, with bridging O, are called oxides, in this case "ethylene oxide."

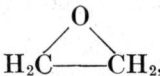

[3] At this point note the general rule for naming compounds, using a root name: Such compounds are considered to be formed by substituting for hydrogen on the root compound. Accordingly the name shows this derivation, as: CH_3CH_2OH is methyl *carbinol;* CH_3I is iodo*methane;* $(CH_3)_4C$ is tetramethyl *methane;* $CH_3COC_2H_5$ is methyl ethyl *ketone.*

Aldehydes

In nomenclature systems other than the I.U.C., names of aldehydes are derived from the acid formed by more complete oxidation (below). In this way the aldehyde whose oxidized product is acetic acid would become acetaldehyde. Propionaldehyde and butyraldehyde are further examples of this method of naming. The names acetaldehyde and formaldehyde are directly acceptable in the I.U.C. system since the use is so universal.

Ketones

Ketones are often named by indicating the hydrocarbon radicals attached to the carbonyl group, $=C=O$. Accordingly butanone, the ketone whose structural formula is $CH_3COCH_2CH_3$, is known as methylethyl ketone.[4] Acetone is the commonly accepted name for propanone or dimethyl ketone.

Acids

The greatest single departure in nomenclature by the I.U.C. system is in the naming of acids and the result has been a rewarding simplification for neophyte and veteran chemist alike. Two systems of naming acids have survived the years since the introduction of the Geneva system.

In one system the name "carboxylic acid" refers to the —COOH group. All acids are regarded as hydrocarbon *additions* to this group. Thus propanoic acid, CH_3CH_2COOH, is called ethane carboxylic acid. Multiples of the carboxyl group are "dicarboxylic," "tricarboxylic" acid, etc. Thus, butanedioic acid, more familiarly called "succinic acid," $HOOCCH_2CH_2COOH$, is known as "ethane dicarboxylic acid."

Another system of naming acids is less systematic. It might be called the "isolationist system" since the chemist must memorize both the formula and the unique name given to each compound. For convenience only the number of carbon atoms for normal compounds is listed below. The carbon in the —COOH is counted in the total shown at the left of each compound.

C 1	Formic acid	C 12	Lauric acid
C 2	Acetic acid	C 14	Myristic acid
C 3	Propionic acid	C 16	Palmitic acid
C 4	Butyric acid	C 17	Margaric acid
C 5	Valeric acid	C 18	Stearic acid
C 6	Caproic acid	C 20	Arachidic acid
C 7	Enanthic acid	C 22	Behenic acid
C 8	Caprylic acid	C 26	Cerotic acid
C 9	Pelargonic acid	C 31	Melissic acid
C 10	Capric acid	C 32	Lacceric acid

[4] It is interesting to note at this point that many industries evolve their own language (or perhaps "slanguage"). It is a common practice in industries using butanone as a solvent

Among the unsaturated or alkenoic acids, the following of commercial importance have the names:

Acrylic acid	$CH_2{:}CHCOOH$
Crotonic acid	$CH_3CH{:}CHCOOH$
Oleic acid	$CH_3(CH_2)_7CH{:}CH(CH_2)_7COOH$
Linoleic acid	$CH_3(CH_2)_4CH{:}CHCH_2CH{:}CH(CH_2)_7COOH$
Linolenic acid	$CH_3CH_2CH{:}CHCH_2CH{:}CHCH_2CH{:}CH(CH_2)_7COOH$

For the names of other unsaturated acids in this system, students should consult any larger textbook. Each acid has a specific name, usually unassociated with the names of similar acids.

The alkanedioic (dicarboxylic) acids in this system also have specific names. Using the same method of representing them as in the foregoing list, these acids are:

C 2	Oxalic acid		C 7	Pimelic acid
C 3	Malonic acid		C 8	Suberic acid
C 4	Succinic acid		C 9	Azelaic acid
C 5	Glutaric acid		C 10	Sebacic acid
C 6	Adipic acid			

Many names of these acids may be considered as a part of the I.U.C. system because of the latitude allowed by Rule 1 in retaining names enjoying universal usage.

Acyl Halides

The difference between naming acyl halides by the I.U.C. system and by other systems is very slight. A contributing factor is that these compounds are man-made substances of comparatively recent discovery and use. The I.U.C. system uses the suffix -*oyl* on the root name of the acid and other systems use -*yl*. Thus ethanoyl chloride is acetyl chloride and butanoyl chloride is butyryl chloride.

Acid Anhydrides

In simple anhydrides the I.U.C. and other systems use the same basic method of naming, using of course their own acid names. Thus propanoic anhydride is propionic anhydride. In mixed anhydrides, other systems of nomenclature use the names of both acids and the words "anhydride with" between them. Thus ethanoic propanoic anhydride is "acetic acid, anhydride with propionic acid."

Amides

The names of amides are derived by dropping the "ic" of the acid and adding "amide." Propanamide is thus called "propionamide."

---

to simply call it M.E.K. The public has accepted many of these short-cut, letter names for complex substances as illustrated by TNT, DDT, and 2,4-D.

Salts and Esters

I.U.C. Rule 31 states: "The existing conventions will be retained for salts and esters." In Chapter 8 many acceptable I.U.C. names were intentionally avoided such as "tristearin" for "glyceryl trioctadecanoate." The suffix "in" is used with the root of the word describing the acid to designate the glycerol ester. Triolein is the ester of oleic acid with glycerol. The name "glyceride" is also used to designate esters of glycerol.

Other Names

One of the outstanding contributions of I.U.C. nomenclature has been the systematic use of numbers in designating the location of groups in various compounds. Older systems of nomenclature utilized letters of the Greek alphabet usually calling the carbon atom attached to the carbon of the principal functional group, the alpha (α) carbon. A contrast of the two systems applied to the naming of an acid follows.

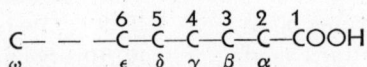

$$\underset{\omega}{C}- \ - \ \underset{\epsilon}{\overset{6}{C}}-\underset{\delta}{\overset{5}{C}}-\underset{\gamma}{\overset{4}{C}}-\underset{\beta}{\overset{3}{C}}-\underset{\alpha}{\overset{2}{C}}-\overset{1}{COOH}$$

Thus, 3-hydroxyhexanoic acid, $CH_3CH_2CH_2CH(OH)CH_2COOH$ is known as β-hydroxycaproic acid. The end carbon atom farthest from the functional group, regardless of number, is frequently designated as the omega (ω) carbon after the last letter in the Greek alphabet.

Greek lettering systems as described are commonly applied to acids, aldehydes, ketones, alcohols, and thio (sulfur) compounds. In ethers, thio ethers, and in aromatic compounds, the lettering starts with the carbon atom attached directly to the key atom. One of the most interesting examples is the chemical warfare agent "mustard gas," $ClCH_2CH_2SCH_2CH_2Cl$ called β,β'-dichlorodiethyl sulfide. This system applied to the nomenclature of aromatic (cyclic) compounds follows the same pattern.

The problem of nomenclature during the remainder of this book is largely solved by using the simplest of approved names for various compounds. Extremely complex organic substances are not usually described in elementary texts. Where cyclic compounds are shown, the numbers for each point of reaction will be noted. Nomenclature is not a *simple* branch of Organic Chemistry. In choosing the methods of naming compounds in this book, the student has received primary consideration, but he must in turn remember that other approved names exist, names that have served well in the development of the science.

Chapter 13

CYCLIC COMPOUNDS OF CARBON
WITH HYDROGEN

CHAPTER PROLOGUE

Chemical compounds now well known and produced in large tonnages have often been the enigmas of an earlier generation. A century ago benzene was such a compound. Some of the most active and agile minds of science have wrestled with its problems. Faraday discovered benzene. Liebig gave it the name "benzol" by which it is still known in most of Europe. Hofmann isolated it as a product of coal tar and Perkin commercialized one of its derivatives in the form of the dye "mauve." August Kekulé, who had simplified the idea of the carbon atom's structure, gave the world the present structure of the benzene ring 40 years after the compound was discovered. Nevertheless it took another 60 years after Kekulé before Bragg and Lonsdale proved by X-ray spectroscopy the ring structure and size of the molecule.

Carbon forms ring compounds of varying size and structure. Often the carbon atoms are joined by single bonds or by irregularly spaced double and single bonds. Most important, however, is the special structure of the benzene molecule.

In an earlier chapter we saw that open-chain compounds were not straight-chain compounds. Now we shall see that the formation of ring compounds is largely due to the bond direction of the carbon atom. We shall also see how these rings are formed and how they will react—how their utility has improved the world in which we live.

In addition to open-chain compounds of carbon and hydrogen discussed in earlier chapters (Chapters 2 and 3), carbon and hydrogen also form cyclic (G. *cyclos*, circle) or ring compounds. There are two general types of carboxylic ring compounds, *alicyclic* and *aromatic*. Aromatic ring compounds are those whose ring structure has a *special* kind of bonding, usually represented by alternate double bonding in a six carbon ring, the benzene ring. Alicyclic compounds are those ring structure compounds of carbon and hydrogen which do not have the special bonding structure of the benzene ring.

Alicyclic Compounds

Names and Kinds of Alicyclic Compounds. Alicyclic compounds of carbon and hydrogen are also called "naphthenes." They are classified in accord with the bonds between the carbon atoms. If single bonds unite the carbon atoms in a single ring, the compound is called a *cycloalkane*. When one double bond is in the ring, the generic name of the compound is *cycloalkene*. A cycloalkadiene has two double bonds in the carbon ring. Regular alkane names with the prefix *cyclo-* are used to designate specific compounds. Thus, cyclopropane, cyclobutane, and cyclopentane refer to ring compounds of three, four, and five carbon atoms respectively, bonded together by single bonds. These compounds have the type formula C_nH_{2n} instead of the usual alkane formula C_nH_{2n+2} because there are no terminal carbon atoms requiring three hydrogens each, as in the case of open-chain compounds. Alicyclic compounds of hydrogen and carbon having as many as 30 carbon atoms have been found in nature.

Structure and Bonding. Carbon atoms are united to each other and to hydrogen atoms by covalent bonding. In the formation of cyclo-alkanes the number of carbons in the ring is important. The normal bond angle (p. 4) of carbon is 109°28′. When three carbon atoms form cyclo-propane, the bond angle is reduced to 60°. Bond angles in cyclobutane and in cyclopentane are 90° and 108° respectively. For cyclohexane (to

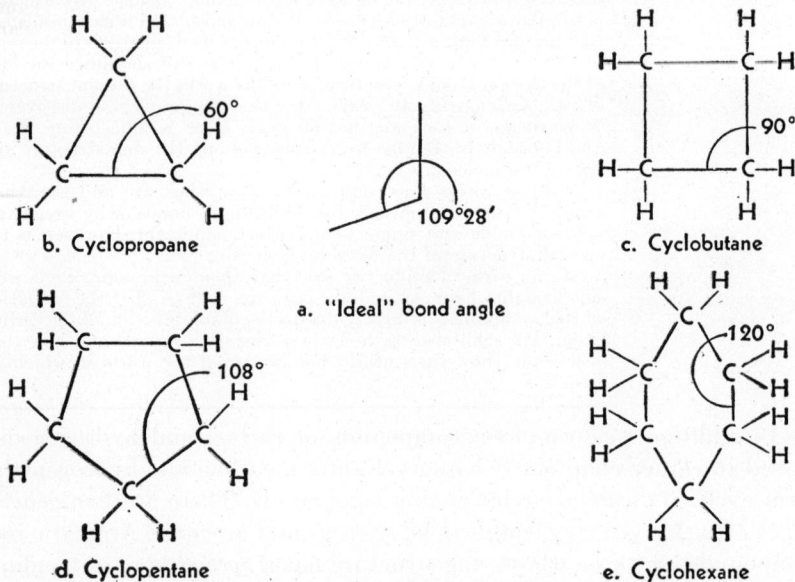

Fig. 28. Bond angles in carbocyclic structures: (a) ideal angle; (b), (c), and (d) subject to strain; (e) angle too great so the molecule in space becomes "puckered."

be discussed later) shown as a plane figure the angle is 120°. Baeyer explained the increased activity of cycloalkanes over their alkane counter-parts by the strain upon the bonds in attaining ring structure. This strain necessitates that the molecule containing abnormal bond angles also requires abnormally high energy content to maintain stability of structure; cleavage of the ring releases this energy and therefore the molecule is more reactive in proportion to the strain. Cyclopropane reacts readily with bromine, breaking the ring and adding bromine to form 1,3-dibromopropane. Baeyer, in accord with this theory and on the erroneous assumption that all carbon atoms were in the same plane, deduced that large cycloalkanes would not be found in nature because of strain upon bonds. Sachse and Mohr later explained the existence of large cycloalkanes in apparently strain-free condition by showing that carbon atoms need not lie in a plane. It is now known that cyclohexane can exist in two molecular shapes, the bed-type (C-form) and the chair-type (Z-form), based upon normal tetrahedral angles of the carbon bonds.

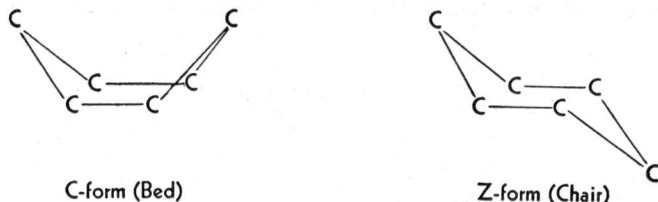

C-form (Bed)   Z-form (Chair)

Fig. 29. Bed and chair shapes of cyclic compounds.

Preparation. Cycloalkanes may be prepared by the reaction of zinc dust upon dibromoalkanes where the bromine atoms are located on different carbon atoms. This reaction of zinc on 1,2-dibromoethane has even led some authors to place ethene in the general classification of alicyclic compounds as a ring compound of two carbon atoms.

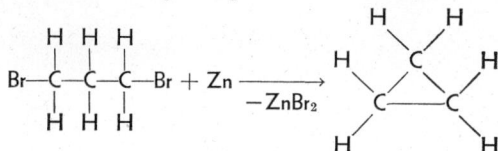

Reactions. Like ordinary alkanes, cycloalkanes are inactive. Cyclopropane reacts with bromine to break the ring, but higher homologues react with bromine to form substitution products as do the alkanes.

Cycloalkenes and cycloalkadienes are unsaturated compounds capable of addition reactions similar to those of ordinary alkenes and alkadienes. Terpenes and carotenes are notable examples of unsaturated alicyclic compounds. The hydroxy derivative of β-carotene is known as Vitamin A.

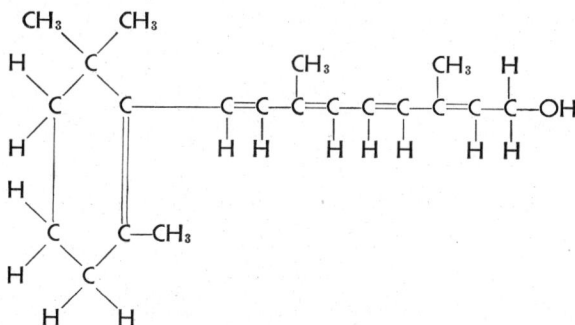

Vitamin A $(C_{20}H_{30}O)$

Aromatic Compounds

A. Benzene (C_6H_6). Benzene is the simplest of aromatic compounds. Michael Faraday first discovered benzene in 1825 as a component of illuminating gas. The name "phenyl" used to designate the univalent radical C_6H_5— is derived from the Greek "pheno," meaning "I bear light," and honors the source of its discovery. Benzene is a liquid whose vapors are somewhat toxic. It is produced in large tonnage as a by-product of the coke industry.

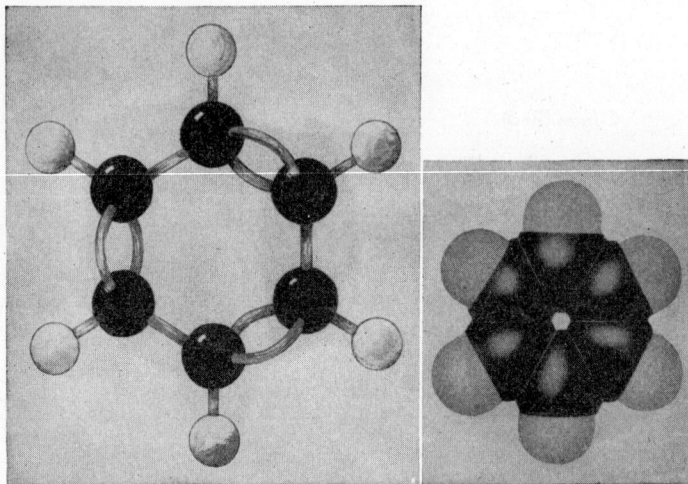

Fig. 30. The benzene molecule.

STRUCTURE AND BONDING. For many years the structure of benzene was the subject of widespread discussion by some of the foremost chemists of Europe. In 1865 Friederich August Kekulé suggested the structure of benzene now most generally used to designate the compound—the alternate double bond structure—adding, a year later his idea that the double bonds were in a state of dynamic oscillation between adjoining

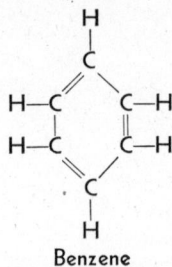

Benzene

pairs of carbon atoms. Kekulé's oscillating double bond was thus a precursor of the modern explanation of benzene structure based upon resonance. From X-ray methods of measuring molecular size, the length of bonds between carbon-carbon atoms in benzene is known to be 1.40 Å. This distance is slightly less than the average length of carbon-carbon single bonds (1.54 Å) and slightly more than carbon-carbon double bonds (1.33 Å). The true structure of benzene is shown by the electronic formula. The double bonds shown in formulas are not true double bonds since additions to the double bonds are rather difficult to produce. Substitution reactions quite readily occur in which groups replace one or more of the hydrogen atoms of benzene forming derivatives. All carbon and hydrogen atoms lie in the same plane.

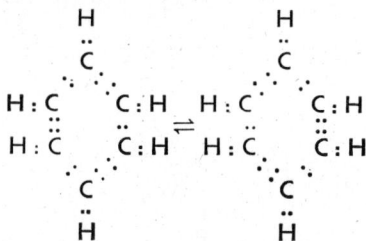

NAMING BENZENE DERIVATIVES. In naming derivatives of benzene the preferred I.U.C. system is to number the carbon atoms in clockwise fashion using the lowest numbers possible. The substituted group name is then used as a prefix. Thus:

Methylbenzene   Chlorobenzene   Nitrobenzene   1,3-Dimethyl
                                                          benzene

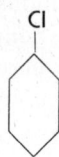

Also acceptable because of general usage is that in which positions 2 and 6 are called ortho- (in relation to #1 carbon), positions 3 and 5 are called meta-, and position 4 is called para-. They are represented by small letters followed by a hyphen o-, m-, and p-. The ortho- position is thus adjacent to the substituent on the ring, the para- position is opposite, and the meta- position has one carbon atom lying between it and the substituent.

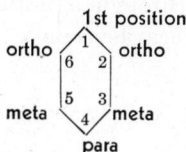

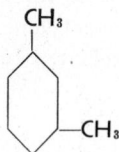

m-Dimethylbenzene   p-Dichlorobenzene

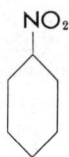

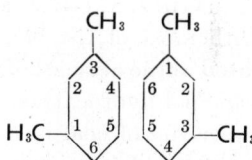

Two other usages of importance require brief explanation: (1) the group C_6H_5— may be designated as the phenyl group. Accordingly

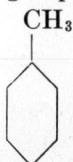

becomes phenylmethane. (2) Many fundamental aromatic compounds are so very important that I.U.C. permits the use of the names of these compounds as root names. Most important of these and the approved names of corresponding substituted products are:

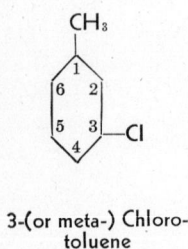

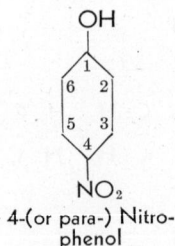

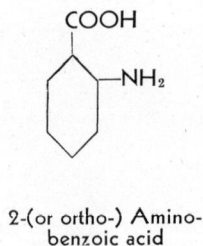

3-(or meta-) Chloro-   4-(or para-) Nitro-   2-(or ortho-) Amino-
        toluene   phenol   benzoic acid

The names of other individual important aromatic hydrocarbons are given in Table 9 on p. 182. The rules by which the formation of these substituted products may be predicted are given on p. 185.

PREPARATION: A. COMMERCIAL. 1. The coke industry has supplied the world with most of the benzene of commerce. Some of the gases from coal, heated in the absence of air, are collected as a crude light oil. Benzene is the principal hydrocarbon recovered from the crude oil. The volume of benzene per ton of coal is small, 2.0 to 2.5 gallons. The total volume of benzene produced from coal is large because of the enormous tonnage of coal converted to coke for the steel industry.

2. During World War II, the demand for benzene and especially for toluene (used in making T.N.T.) required the aid of the petroleum industry in the production of these two essential compounds. Products of the cracking industry, C1–C6 alkyl molecules, heated in the presence of Cr_2O_3, are united to form aromatic hydrocarbons including benzene.

B. ACADEMIC. Most benzene for commercial use is prepared by the two methods just outlined. Some of the methods listed below as academic have been used to produce benzene when demand has been strong.

1. From Acetylene. When acetylene is passed through a hot iron pipe or over finely divided nickel, it polymerizes to form benzene.

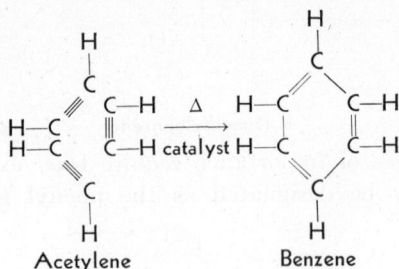

        Acetylene   Benzene

2. From Sodium Benzoate. Decarboxylation of sodium benzoate fused with soda lime ($NaOH + Na_2CO_3$) produces benzene.

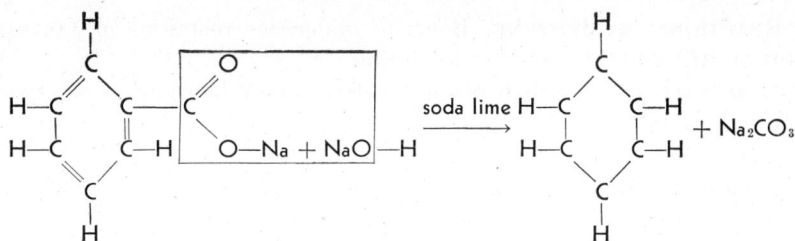

3. **From Phenol.** Phenol and zinc dust when heated produce benzene and zinc oxide.

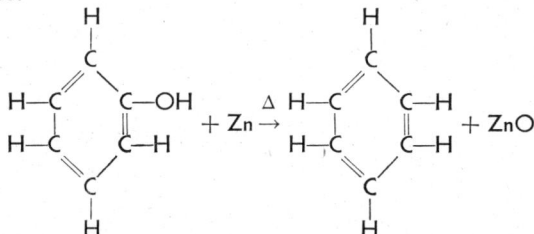

4. **From Cyclohexane.** Cyclohexane when heated in the presence of sulfur or selenium catalyst is dehydrogenated to form benzene.

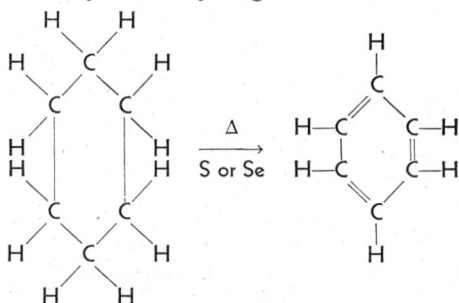

5. **By Diels-Alder[1] Reaction and Dehydrogenation.** Ethene adds to 1,3-butadiene, forming cyclohexene. The reaction occurs without any catalyst and at ordinary temperatures. Cyclohexene is then dehydrogenated by passing while hot over sulfur, selenium, or palladium.

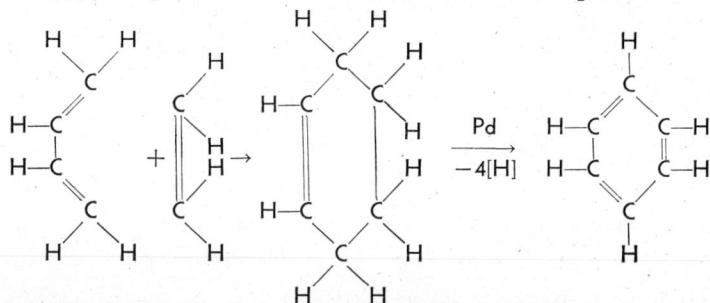

[1] In December 1950, the Nobel Prize in Chemistry was awarded to Otto Diels and Kurt Alder for their discovery of this reaction in 1928.

REACTIONS OF BENZENE. Benzene undergoes reactions of (a) substitution, (b) addition, and (c) oxidation.

A. SUBSTITUTION. Substitution reactions with benzene occur much more readily than with alkanes.

1. Halogenation. Halogenation by substitution occurs when a suitable carrier catalyst is present such as iron.

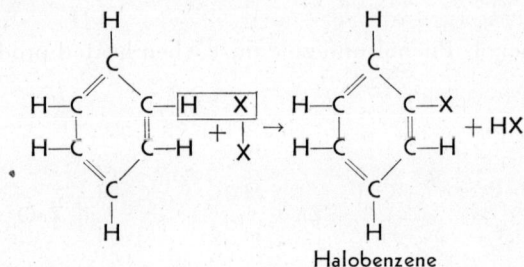

Halobenzene

2. Sulfonation. Concentrated sulfuric acid reacts with benzene to form benzenesulfonic acid. Sulfuric acid serves as its own dehydrating agent.

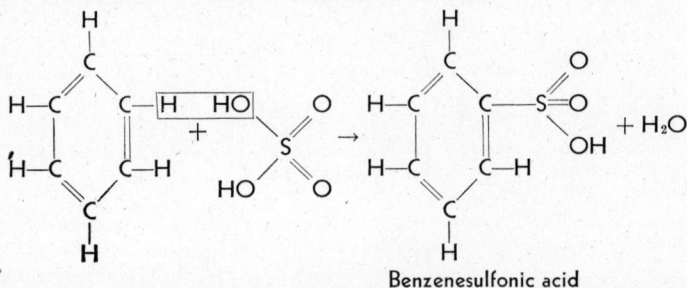

Benzenesulfonic acid

3. Nitration. Nitrobenzene is formed by the reaction of benzene and a mixture of concentrated sulfuric and nitric acids. Sulfuric acid serves as the dehydrating agent.

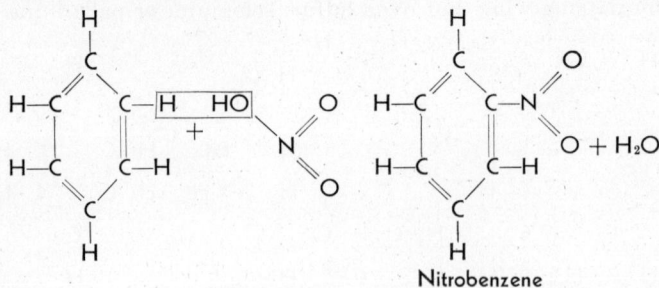

Nitrobenzene

4. Alkylation. Benzene reacts with chloromethane or chloroethane in the presence of anhydrous aluminum chloride, forming an alkyl

benzene (toluene or ethyl benzene). This type of reaction is called the Friedel-Crafts Reaction.

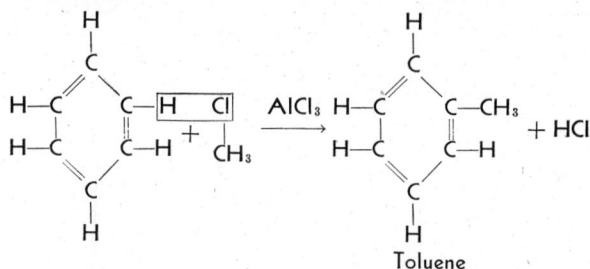

Toluene

B. ADDITION. 1. Benzene in the vapor phase adds hydrogen in the presence of nickel catalyst to form cyclohexane. This is called the Sabatier-Senderens reaction.

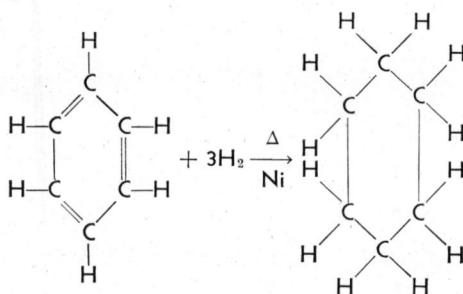

2. In the presence of sunlight chlorine or bromine add directly to benzene to form hexahalocyclohexane.

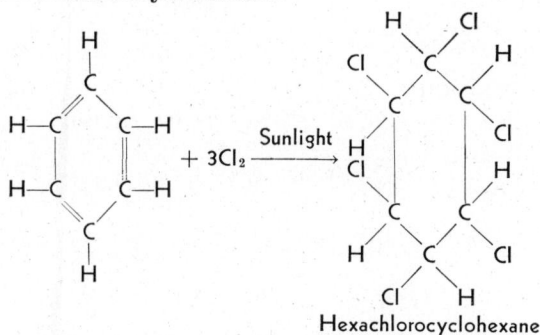

Hexachlorocyclohexane

C. OXIDATION. In air, benzene burns as do other hydrocarbons producing H_2O and CO_2.

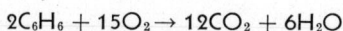

$$2C_6H_6 + 15O_2 \rightarrow 12CO_2 + 6H_2O$$

B. Benzene Derivatives. Many hydrocarbon derivatives of benzene can be obtained in small quantities from the fractional distillation of coal

Table 9

AROMATIC HYDROCARBONS WITH ACCEPTED NAMES

Name	Formula	M.P. C.	B.P. C.	Sp. Gr.	Properties and Uses
Benzene		5.5	80	0.879	
Toluene	—CH₃	−95	110.8	0.866	
Ethylbenzene	—C₂H₅	−93.9	136.15	0.867	Physical Properties: Colorless liquid or white solid. Odors characteristic.
o-Xylene	CH₃ —CH₃	−29	144	0.875	Solubility: Water Insoluble / Alcohol Soluble / Ether Soluble
m-Xylene	—CH₃ —CH₃	−53.6	138.8	0.864	Chemical Properties: Halogen: Substitutes readily for hydrogen.
p-Xylene	CH₃ —CH₃	13.2	138.5	0.861	Nitration: One to three nitro groups. Sulfonation: One or more
Biphenyl		70	254	1.180	Alkylation: Groups usually ortho directing
Naphthalene		80.2	217.9	1.145	Uses: Solvents for fats, resins, oils and
Anthracene		217	354	1.25	various organic compounds. Synthesis.
Phenanthrene		100	340.2	1.025	

tar. Most of them, however, are now synthesized commercially from simple coal tar products by reactions listed here and in more comprehensive texts on organic chemistry. Most of the simple hydrocarbon derivatives of benzene are liquids at ordinary temperatures with boiling points higher than that of water. They are less dense than water and all are inflammable. Many of these compounds have pleasant odors or aromas, a fact that earned for them the classification name, "aromatic" hydrocarbons. The pleasant odors, however, shield a toxicity that is often dangerous.

PREPARATION: 1. THE FRIEDEL-CRAFTS METHOD of preparing methyl and ethyl derivatives of benzene was discussed on p. 181. For Friedel-Crafts reactions anhydrous aluminum chloride is usually shown as the catalyst. It is the strongest catalyst in a series of anhydrous compounds capable of carrying out the same type of reaction. Ferric chloride, stannic chloride, boron fluoride, and zinc chloride, all in the anhydrous state, serve the same purpose with decreasing vigor of reaction. An increased concentration of strong catalyst can cause the substitution of an alkyl group in additional positions in the ring.

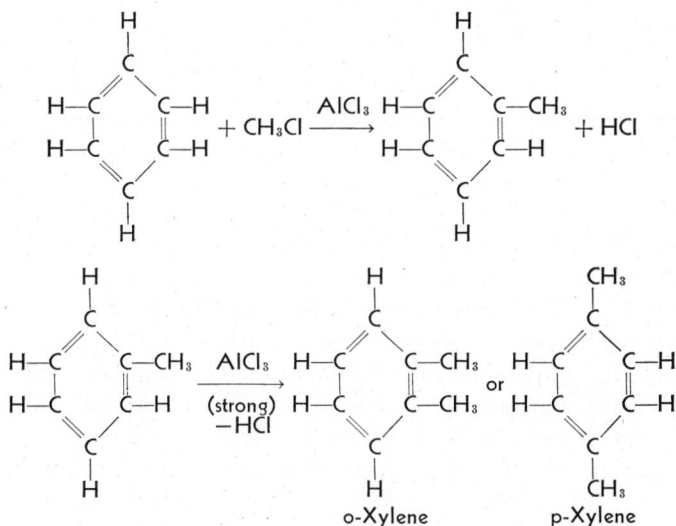

o-Xylene   p-Xylene

2. THE WURTZ-FITTIG METHOD of alkylation is based upon the Wurtz synthesis of hydrocarbons (p. 22). In the synthesis of alkanes only one haloalkane is used with sodium because of the difficulty in separating the alkanes formed. When a haloalkane and a halobenzene are used, the separation of the final products is not so difficult. The other possible combination of the reactants will result, of course, in the formation of some biphenyl ⟨◯⟩—⟨◯⟩ and some alkane.

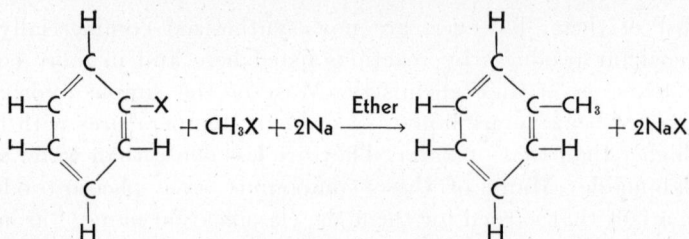

3. DEHYDROGENATION OF ALKANE. During World War II, the n-hep-tane fraction of the petroleum industry (octane rating = 0) was passed over a catalyst (Cr_2O_3 and Al_2O_3) at temperatures of 500–600°C. Toluene formed by this reaction was used for making the explosive, trinitro-toluene (T.N.T.).

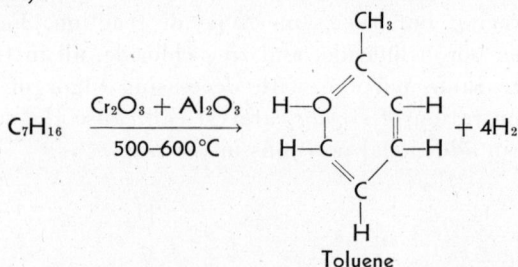

Toluene

REACTIONS OF BENZENE DERIVATIVES. Since hydrocarbon derivatives of benzene are composed of two distinct groups, the aryl and the alkyl group, reaction may take place on either or both groups.

1. REACTION OF THE ARYL GROUP. The aryl group reacts independ-ently by substitution at other points on the benzene ring. In benzene itself each carbon and hydrogen atom is equal to the others in the com-pound. Once a single substitution is made, however, the carbon atom where substitution occurs, becomes the 1-carbon atom of the molecule. In writing the graphic formula of such a compound, the top carbon atom is usually represented as the carbon atom with the substituted group. If it is placed elsewhere the carbon atom to which it is attached is designated as the 1-carbon atom. For uniformity of graphic repre-

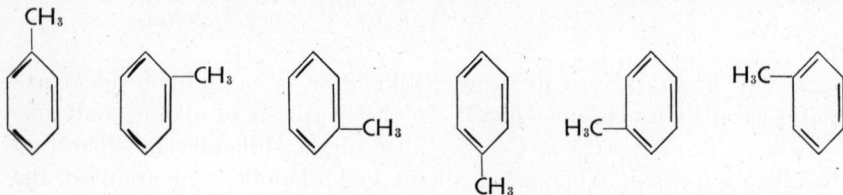

sentation, we shall use the top carbon atom as number one, unless some reaction is better illustrated by one of the other formulas. Other carbon atoms in the ring are counted clockwise from the starting point. For

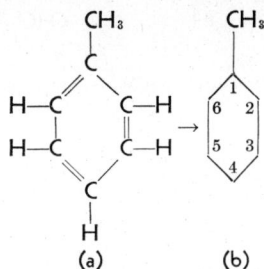

(a)   (b)

facility in representing the benzene ring, the single line hexagon (b) is used except in cases of those polycyclic compounds where one ring is not of the benzene type.

Substitution in the Benzene Ring. Simple substitution on the benzene ring was shown in the reactions of benzene on p. 180. Further substitution occurs in different places, varying with the nature of the substituent already on the ring. Certain rules of substitution have been formulated, based upon the observed place of substitution under varying conditions. Only common groups will be considered in this text. For a more complete list of substituents, students should consult one of the larger texts on organic chemistry.

Rule I. When one of the following groups or elements is on the ring, $-NH_2$, $-OH$, $R-$, $Cl-$, $Br-$, $I-$, the next substitution of a group or element will occur in the 2-, 4-, or 6- position, (para- and ortho-) in varying amounts.

Rule II. When one of the following substituents is already on the ring, $-COOH$, $-CHO$, $-NO_2$, $-SO_3H$, $-CN$, the next substitution will be directed to the 3-position (meta-).

Rule III. When two different types of directing substituents are on the ring, those listed in Rule I generally exert a more powerful directing influence than those listed in Rule II. The nature of the groups also influence the ease with which further substitution occurs. Once a substitution is made on the benzene ring, further substitution becomes easier in the presence of o- p- directing groups except halogens, and usually more difficult with m- directing groups.

Many new students of organic chemistry make the serious mistake of thinking that substitution takes place *only* at the positions designated by these rules. As a matter of fact substitution usually occurs in several positions. The rules simply designate those positions preponderantly occupied in the resulting equilibrium mixture.

2. REACTION OF THE ALKYL GROUP. When an alkyl group is already on the ring as in toluene, oxidation affects the alkyl group, producing successively benzyl alcohol, benzaldehyde, and benzoic acid. Each product can be made separately by varying the conditions of the reaction.

$$
\underset{\text{Toluene}}{\overset{\text{CH}_3}{\bigcirc}} \xrightarrow{[O]} \underset{\text{Benzyl alcohol}}{\overset{\text{CH}_2\text{OH}}{\bigcirc}} \xrightarrow{[O]} \underset{\text{Benzaldehyde}}{\overset{\text{CHO}}{\bigcirc}} \xrightarrow{[O]} \underset{\text{Benzoic acid}}{\overset{\text{COOH}}{\bigcirc}}
$$

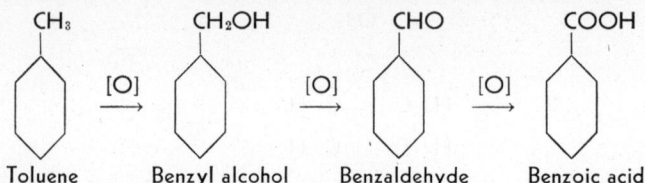

Alkanes could not be partially oxidized directly. The aryl group thus serves as a modifying influence on an alkyl group in oxidation. When a longer alkyl group is present, oxidation also occurs resulting in the formation of benzoic acid and an aliphatic acid.

Halogenation of the aryl or alkyl group is also dependent upon modifying influences. When toluene is chlorinated at low temperatures in the presence of a catalyst (Fe), substitution occurs in the ring. When no catalyst is used and at elevated temperatures (with or without sunlight), substitution occurs in the side-chain.

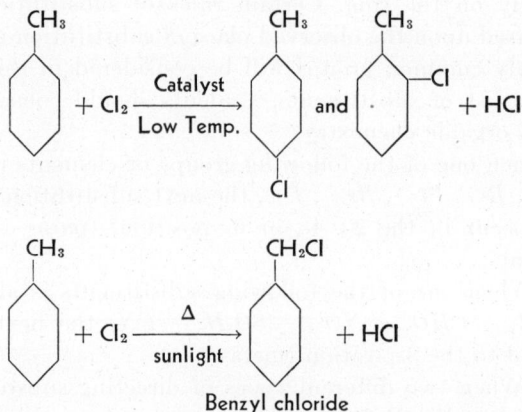

$$
\overset{\text{CH}_3}{\bigcirc} + \text{Cl}_2 \xrightarrow[\text{Low Temp.}]{\text{Catalyst}} \overset{\text{CH}_3}{\underset{\text{Cl}}{\bigcirc}} \quad \text{and} \quad \overset{\text{CH}_3}{\bigcirc}{-}\text{Cl} + \text{HCl}
$$

$$
\overset{\text{CH}_3}{\bigcirc} + \text{Cl}_2 \xrightarrow[\text{sunlight}]{\Delta} \underset{\text{Benzyl chloride}}{\overset{\text{CH}_2\text{Cl}}{\bigcirc}} + \text{HCl}
$$

The alkyl part of the molecule is not affected by the action of nitric or sulfuric acid.

C. Polycyclic Aromatic Hydrocarbons. Several polycyclic aromatic hydrocarbons are produced by the distillation of coal tar. Three of these are naphthalene, anthracene, and phenanthrene. Naphthalene and anthracene have been used extensively in the making of dyes. Phenanthrene is important because it serves as the nucleus of several complex compounds found in the human body.

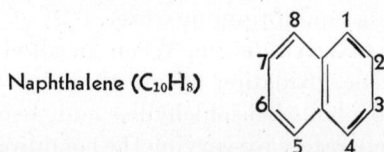

$$
\text{Naphthalene (C}_{10}\text{H}_8\text{)} \qquad \overset{8\quad 1}{\underset{5\quad 4}{\underset{6\qquad 3}{\bigcirc\!\bigcirc}}}\!\!{}^{2}_{7}
$$

Anthracene (C₁₄H₁₀)

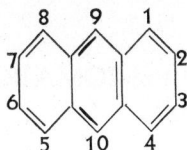

Phenanthrene (C₁₄H₁₀)

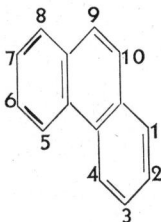

Study and Review Questions

1. What is the principal difference between alicyclic and aromatic compounds?
2. Write the graphic formula of a cycloalkane. Name your compound.
3. Write the graphic formula of a cycloalkene. Name your compound.
4. Explain the Baeyer "strain theory" of alicyclic compounds.
5. In what way did Sachse and Mohr explain the formation of cyclic compounds?
6. Write an equation with graphic formulas showing how to prepare cyclopropane.
7. Write an equation showing how 1 mole of cyclohexane reacts with bromine under favorable conditions.
8. How does the nuclear distance in C-C linkage of benzene compare with the same linkage of ethane and of ethene?
9. How is benzene prepared commercially?
10. Write the equations for two laboratory methods of preparing benzene.
11. Discuss comparatively: substitution of halogen in benzene, toluene, and cyclohexane.
12. What is a "carrier catalyst"?
13. What catalyst is used in the Sabatier-Senderens reduction of benzene to cyclohexane?
14. Compare the Friedel-Crafts method of preparing alkyl-benzene with the Wurtz-Fittig method.
15. Write three rules for substitution in the benzene ring and illustrate the rules by writing graphic formulas of compounds formed.
16. Summarize with equations oxidation reactions of the alkyl part of benzene derivatives.
17. What modifying influence does the benzene ring have on the activity of hydrogen atoms of a side-chain?
18. Showing numbers of substitution positions write graphic formulas of benzene, naphthalene, anthracene, and phenanthrene.

Chapter 14

AROMATIC COMPOUNDS WITH HALOGENS AND SULFUR

CHAPTER PROLOGUE

One of the problems of applied science is getting the correct substance for a job in the right place at the proper time. Many such practical problems of applied science arise in times of war. During the early stages of World War II there arose the problem of reducing illness spread by various insects, especially in tropical climates. Malaria and tyhpoid fever had decimated the armies of other generations. Why not find an insecticide capable of destroying the chief carriers of these dreadful diseases? To find an effective insecticide researchers turned back the historical pages of chemical synthesis to the year 1874 before they found their answer in a compound of carbon, hydrogen, and chlorine. Zeidler's compound, dichlorodiphenyltrichloroethane, better known as DDT, had waited nearly 70 years to become one of our heroes of war.

In this brief story of DDT there also lies an answer to that oft-asked American question regarding pure scientific research: "So what?" Very often our pragmatic evaluation of effort suffers from the myopia of dollar-overconsciousness. True, scientific progress for its very life frequently depends upon its ability to pay its own way, but for its health and vitality it depends more upon a love of new knowledge and a desire to penetrate the fascinating recesses of a world unknown. Among the compounds cited in this chapter there may be another substance, useless now, for which future generations will be extremely grateful.

Aromatic hydrocarbons form many compounds with the halogens. The chief sulfur compounds are the sulfonic acids and their derivatives. In a brief course such as this only those few compounds which illustrate principles of reaction or whose use has become relatively commonplace can be considered.

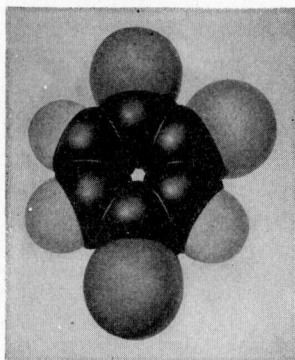

Fig. 31. 1,2,4-Trichlorobenzene molecule.

I. Aromatic Compounds with Halogens

Aromatic compounds react with all the halogens, but reactions with chlorine and bromine are more common than with fluorine and iodine. Fluorine and iodine require different preparation procedures because fluorine is so extremely reactive and iodine so very slow in reacting. The order of activity among the halogens decreases with the increase in atomic weight of the halogen.

Preparation. Halogen compounds can be made (1) by direct addition and (2) by substitution of a hydrogen by a halogen atom.

1. BY ADDITION. Halogen will add directly to an aromatic compound at low temperature in the presence of sunlight, provided no catalyst favorable to substitution is present in the reaction medium. This reaction

188

was discussed briefly under reactions of alicyclic hydrocarbons (p. 181). The compound formed is an alicyclic polyhalide. The reaction is slow even in the case of the more active halogens. With chlorine the reaction is:

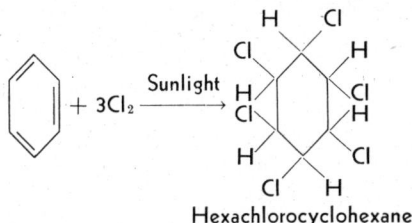

Hexachlorocyclohexane

2. By Substitution. When the aromatic compound is an alkyl derivative of benzene, substitution may occur either (a) in the ring or (b) in the alkyl side-chain.

A. IN THE RING. Substitution of a hydrogen atom by a chlorine atom in the ring always requires the presence of a catalyst of the carrier type.[1] Iron is frequently used but other catalysts such as anhydrous aluminum chloride or bromide will also effect the same kind of reaction. Neither sunlight nor more than ordinary temperature are required for this reaction. As a rule substitution of the chlorine in benzene itself is rather slow. The location of substitutions on the ring is in accord with the general rules of substitution (p. 185). Upon reading the rules for substitution, one may be led to believe that substitution occurs in only one place on the ring. The rules actually designate the predominant position to be found substituted in the equilibrium mixture that forms. Thus substitutions may occur at different places in different molecules of the same compound, making the products of the reaction a mixture of ortho- and para- with a very small per cent of the meta- chlorides, rather than just ortho- or just para- chloride alone.

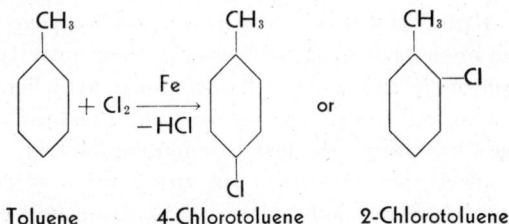

Toluene   4-Chlorotoluene   2-Chlorotoluene

B. IN ALKYL GROUP. Substitution in the alkyl side-chain of a benzene derivative such as toluene requires the absence of moisture and of any catalyst used in ring substitutions. Higher temperatures and the presence of sunlight or ultraviolet light are required for the reaction. In an excess of chlorine, two or more of the alkyl hydrogens may be replaced.

[1] Carrier catalysts liberate Cl^+ by forming complex ions with Cl^-.

$$CH_3 + Cl_2 \xrightarrow[\text{Sunlight}]{\Delta} CH_2Cl \quad \text{Benzyl chloride}$$

$$CH_3 + 2Cl_2 \xrightarrow[\text{Sunlight}]{\Delta} CHCl_2 \quad \text{Benzal or Benzylidene Chloride}$$

$$CH_3 + 3Cl_2 \xrightarrow[\text{Sunlight}]{\Delta} CCl_3 \quad \text{Benzotrichloride}$$

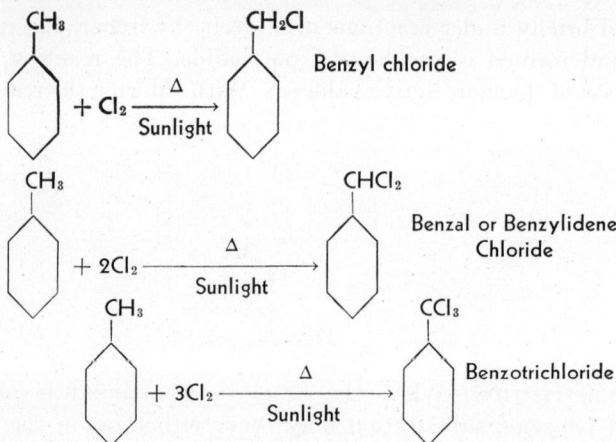

When alkyl substituents of benzene are halogenated, the halogen replaces hydrogen atoms on the carbon atom attached to the benzene ring, under the usual conditions of temperature, sunlight, and absence of catalysts. In naming this type of compound by one system of nomen-

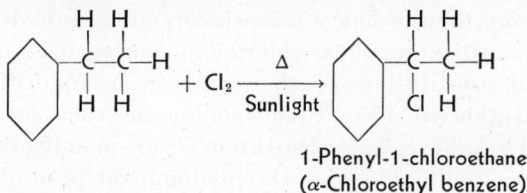

$$\begin{array}{c} \underset{H\ H}{\overset{H\ H}{C-C-H}} + Cl_2 \xrightarrow[\text{Sunlight}]{\Delta} \underset{Cl\ H}{\overset{H\ H}{C-C-H}} \end{array}$$

1-Phenyl-1-chloroethane
(α-Chloroethyl benzene)

clature, the carbon atom nearest the ring is called the α-carbon and the others are called in sequence β, γ, δ, etc. The I.U.C. name for the compound formed in the preceding equation is 1-phenyl-1-chloroethane.

Reactions. Reactions of halogenated aromatic compounds depend upon the location of the halogen.

1. IN THE RING. Halogens substituted for hydrogen atoms of the ring or benzene nucleus are very different in their activity from halogens in the alkyl part of the molecule. A halogen once attached to the benzene nucleus often requires high temperature and pressure for substitution and it is difficult to remove the last halogen completely.

A. IN THE FITTIG MODIFICATION OF THE WURTZ SYNTHESIS, a halogenated benzene nucleus reacts with a haloalkane and sodium to form sodium halide and alkylbenzene. Biphenyl is made by the same type of reaction. In both cases the reaction is slow.

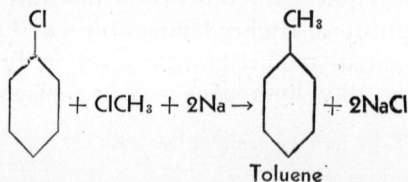

$$Cl + ClCH_3 + 2Na \rightarrow CH_3 + 2NaCl$$

Toluene

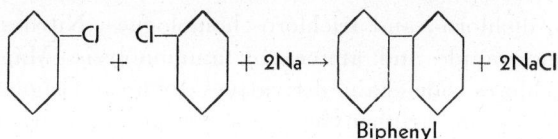

Biphenyl

B. HYDROLYSIS. Chlorobenzene reacts with sodium hydroxide or aqueous sodium carbonate only at high temperatures and pressures, forming phenol.

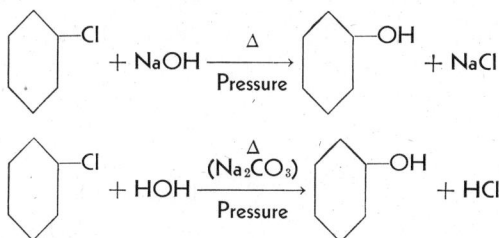

C. AMMONOLYSIS. Ammonia reacts with chlorobenzene at high temperature and pressure to form aniline and ammonium chloride.

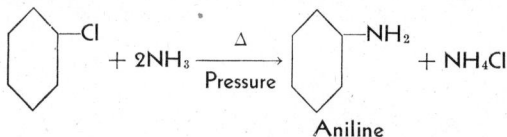

Aniline

D. NITRATION. Chlorobenzene reacts with nitric acid in the presence of concentrated sulfuric acid to produce the 2- and 4-nitrochlorobenzene. The latter (p-) is produced in larger quantity by this reaction. With an excess of nitric acid further substitution occurs.

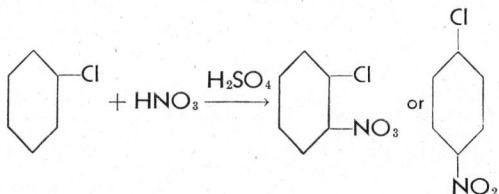

E. GRIGNARD REACTION. Grignard reagents similar to alkyl Grignard compounds are formed with bromo- and iodobenzene.

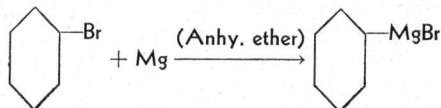

2. IN SIDE-CHAIN. Halogens attached to side-chain alkyl groups are very much more active than in aliphatic halides and than halogens on the ring. Their manner of reacting is fundamentally like that of haloalkanes. By hydrolysis, side-chain alcohols, aldehydes, and acids are formed

from chloro-, dichloro-, and trichloro- homologues. Nitriles are formed by potassium cyanide and amines by ammonolysis. Magnesium also reacts with chloro- side-chain derivatives to form Grignard reagents. Diphenylmethane and triphenylmethane are produced by Wurtz-Fittig reactions. Some triphenylmethane derivatives are brilliant dyes.

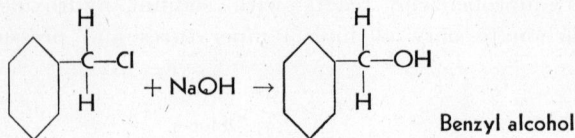

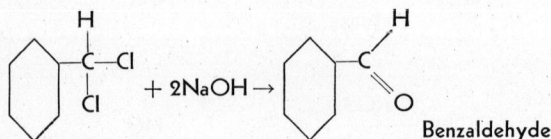

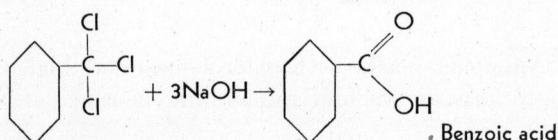

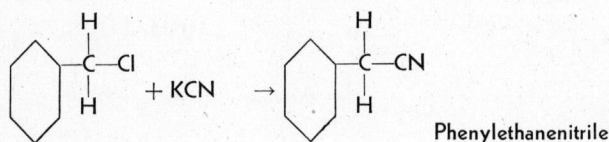

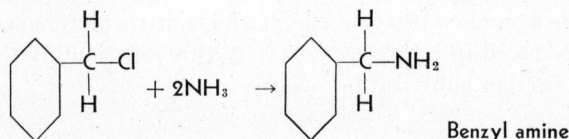

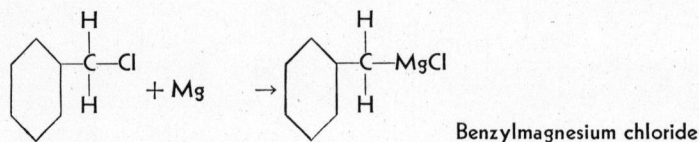

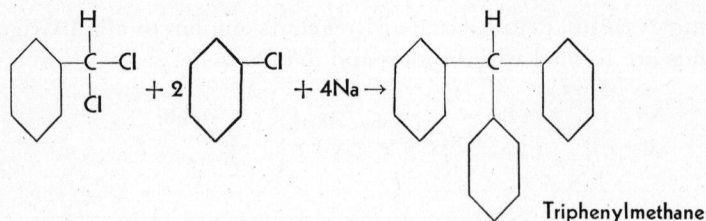

Benzyl alcohol

Benzaldehyde

Benzoic acid

Phenylethanenitrile

Benzyl amine

Benzylmagnesium chloride

Triphenylmethane

Important Chloro-Aromatic Compounds. Two compounds developed during the past 10 years have gained widespread acceptance in national life and welfare. Since World War II, DDT (p, p'-dichlorodiphenyl-trichloroethane) has become a household weapon against various insects. In very dilute solutions 2,4-D (2,4-dichlorophenoxyacetic acid) is a specific eradicator of broadleaf plants such as dandelions and poison ivy.

2,2,2-Trichloroethanal   Chlorobenzene   1,1,1-Trichloro-2,2'-bis-
(Chloral)   (4-chorophenyl) ethane
                                                   (DDT)

2,4-Dichlorophenoxyacetic acid
(2,4-D)

II. Sulfonic Acids

The sulfonic acids or aromatic hydrocarbons are colorless, crystalline substances, insoluble in common organic solvents such as ether but soluble in both water and alcohol. They are strong acids possessing many characteristics of sulfuric acid.

Structure and Bonding. The sulfonic acid group is —SO_3H. The sulfur atom is always bonded directly to a carbon atom. Aryl sulfonic acids are very stable ionic compounds. They are also capable of reacting in much the same manner as the carboxyl group, resulting in the formation

of sulfonic amides,   , and sulfonyl chlorides   .[2]

The chlorosulfonic acids are extremely reactive agents for synthesis.

---

[2] Many authors prefer the use of an arrow to designate bonding of coordinate bonds as

in the sulfonic acids —S—OH, rather than the designation —S—OH which may be mis-

interpreted as a double covalent bond.

Preparation. Sulfonic acids are prepared by the reaction of concentrated sulfuric acid at slightly elevated temperatures with an alkyl or an aryl group. Strong sulfuric acid serves as its own dehydrating agent for the water formed by the reaction.

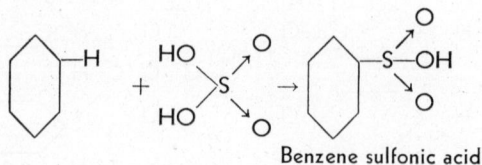

Benzene sulfonic acid

When an excess of sulfuric acid is present, disulfonic acid and trisulfonic acids are formed. The sulfonic acid group is meta- directing.

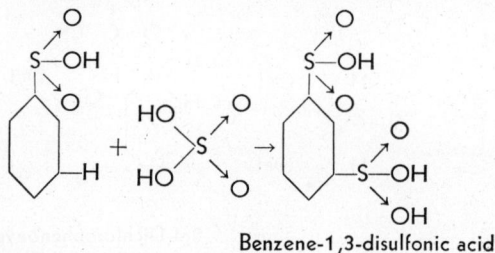

Benzene-1,3-disulfonic acid

Sulfonation of aryl compounds occurs more readily than in the case of alkyl compounds.

Reactions. Aryl sulfonic acids are very active. For commercial purposes they are converted to the sodium salt. The sodium salt may be prepared by the reaction of the acid with a large excess of sodium chloride.

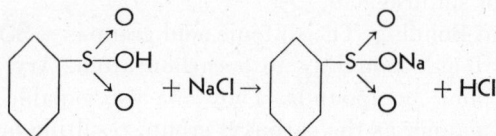

The common ion effect of an excess of sodium ion causes the sodium sulfonate to precipitate.

Certain direct reactions of benzene (or aryl) sulfonic acid are usually accomplished by fusion with suitable reactants. Fusion with sodium hydroxide is used as a commercial procedure for making phenol. Fusion with sodium cyanide produces a nitrile which can be readily hydrolyzed to benzoic acid.

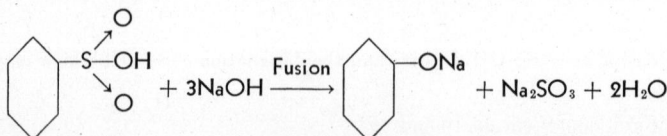

Phenol

Benzoic acid

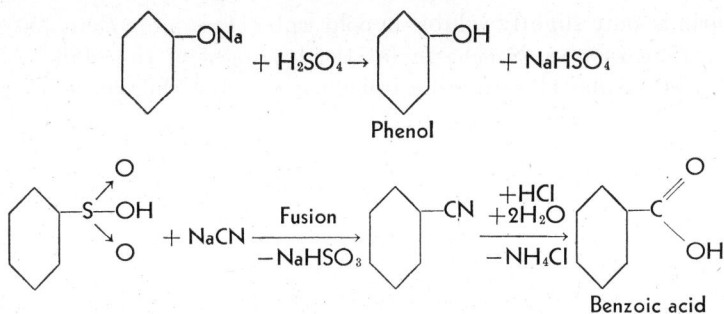

Benzenesulfonic acid and its sodium salt react with phosphorus pentachloride to form benzenesulfonyl chloride. This reaction is analogous to the reaction of carboxylic acids to give acyl halide.

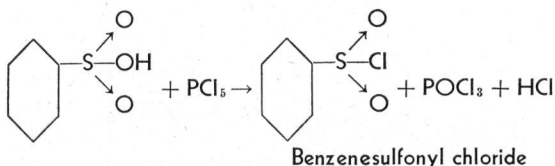

Benzenesulfonyl chloride

Benzenesulfonyl chloride, analogous to acyl chloride, is used in preparing esters and amides. With methanol for example the methyl ester of benzenesulfonic acid is formed.

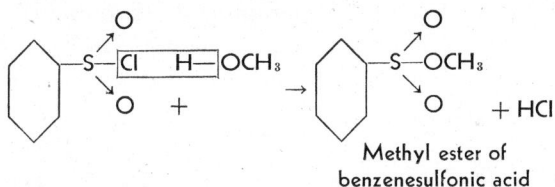

Methyl ester of
benzenesulfonic acid

Benzenesulfonyl chloride reacts with ammonia to form water insoluble benzenesulfonamide.

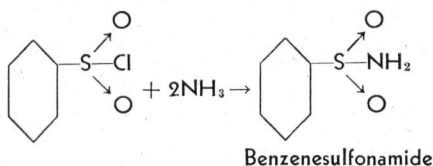

Benzenesulfonamide

Interesting preparations of three well-known substances—saccharin, chloramine T, and halazone illustrate the use of benzene sulfonamide in synthesis. In the sulfonation of toluene, both ortho- and para-toluenesulfonic acids are formed. When ortho-toluenesulfonic acid is converted to the amide, and the methyl group is oxidized to a carboxyl group by potassium permanganate solution, o-benzosulfonimide, better known as saccharin, is formed in weakly basic solution with loss of water.

Saccharin is only slightly soluble in cold water. It is more than 500 times sweeter than sucrose. Saccharin (at the hydrogen of the imide) is acid in its reaction due to activating influence of double-bonded oxygens on nearby carbon and sulfur atoms. The sodium salt of saccharin is readily soluble in water.

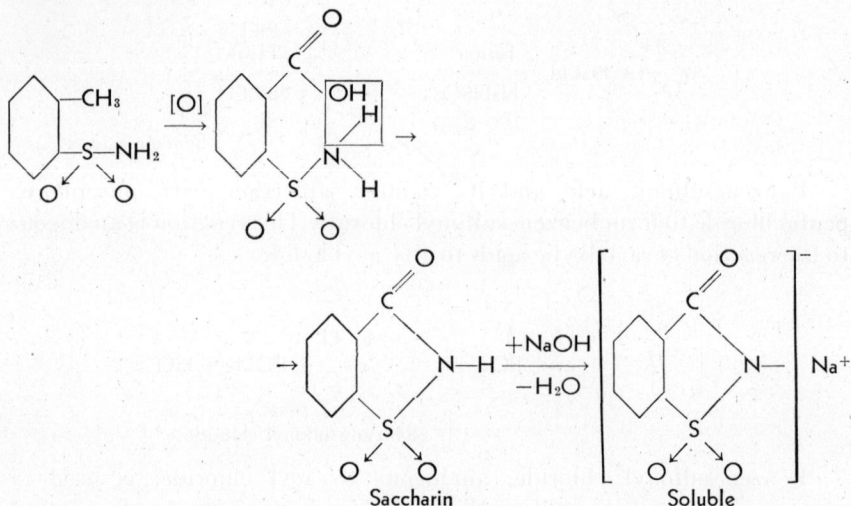

Saccharin   Soluble

The p-toluenesulfonic acids have been used in the preparation of antiseptics and disinfectants. "Chloramine T" is prepared by the reaction of sodium hypochlorite on p-toluenesulfonamide. Hypochlorous acid is liberated slowly when Chloramine T is dissolved in water.

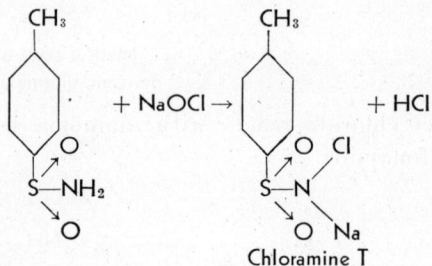

Chloramine T

Further chlorination by sodium hypochlorite yields dichloramine T, a strong antiseptic.

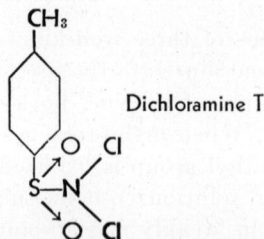

Dichloramine T

When p-toluenesulfonamide is oxidized to p-benzosulfonamide the amide may be converted to dichloroamide with sodium hypochlorite. The resulting compound is called halazone. During World War II, halazone was used in the purification of drinking water.

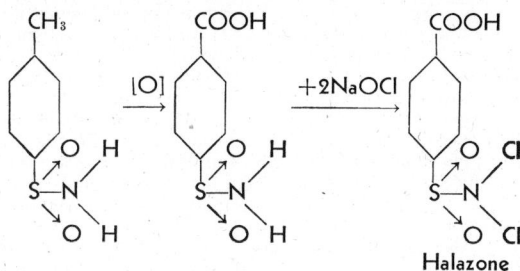

Study and Review Questions

1. Show by a series of equations with special emphasis on conditions of reaction, how halogens react with benzene by addition and by substitution on the ring or on a substituent alkyl group.
2. Name every organic compound used in equations in question 1.
3. Show by equations how phenol and also aniline may be prepared from benzene.
4. Compare the activity of chloromethane with the activity of benzyl chloride ($C_6H_5CH_2Cl$).
5. Compare the activity of trichloromethane (chloroform) with the activity of benzotrichloride. How does the benzene ring influence the activity of trichloro- derivatives?
6. Write the graphic formula of DDT and of 2,4-D and give their correct chemical names.
7. Compare benzene sulfonic acid with sulfuric acid from the point of view of structure and activity.
8. How is benzene sulfonic acid usually stored?
9. How does sulfonic acid respond to hydrolysis?
10. In what type of reaction is sulfonic acid used in preparing phenol and benzoic acid? Write equations for these reactions.
11. Show by equations how to prepare the methyl ester of benzene sulfonic acid from sulfonic acid.
12. Benzene sulfonamide is similar to sulfanilamide. Write the formula of each for comparative purposes. (See Chapter 20 for sulfanilamide.)
13. Compare the structure of Saccharin with that of Chloramine T.
14. How do Chloramine T, Dichloramine T, and Halazone achieve their water purifying results?

CHAPTER PROLOGUE

Next to carbon, nitrogen is the most versatile of the elements. Like carbon, nitrogen is lazy and inactive at ordinary temperatures. When nitrogen is heated, however, it becomes very active. In this active condition it unites with just about anything, even with other nitrogen, though in a limited sort of way. When nitrogen unites with carbon and other elements, the nature of these other elements is very important. Nitrogen and carbon are like well-behaved lone children from different families. They play together reasonably well, but when others join them, what happens depends upon their new associates and the circumstances surrounding them. For example, if they unite with hydrogen as their only companion, they are the gentle lambs of the dye industry, serving as a foundation for soft, subdued shades so soothing to the eyes. With oxygen as an associate, however, they easily become irresponsible rascals, littering their path with destruction.

Both types of compounds have added immeasurable wealth to our way of life. The dye industry has inexpensively exalted the common man to the status of a prince of Tyre in fine purple. Alfred Nobel's dynamite has accomplished in a moment tasks unmatched by the slaves of tyrants. And withal, the surface of progress has hardly been scratched. A century has not passed since young Perkin discovered mauve or Nobel first controlled a nitro compound. In the century immediately before us lies the real future, unlimited, almost uncharted, beckoning for you.

The aromatic compounds of nitrogen treated in this chapter are those in which nitrogen is attached to an aromatic ring. The ring compound may be either of the simple benzene ring type or of the polycyclic type. When nitrogen becomes one of the elements within a ring such as in pyridine, the compound is called heterocyclic (Chapter 17). In this chapter the three principal kinds of derivatives to be considered are: (I) nitro, (II) amino, and (III) diazo derivatives.

I. Nitro Aromatic Derivatives

Structure and Bonding. Nitro aromatic compounds always have a direct C—N bonding of the nitro group ($-N{\overset{\nearrow O}{\underset{\searrow O}{}}}$) with the benzene ring.

They are prepared by the dehydration reaction of nitric acid through substitution for a hydrogen atom attached directly to a carbon of the ring (p. 180). Nitric acid is a strong oxidizing agent.[1] The nitro group is also a strong acceptor of electrons. This electro-negative character of the nitro group attached to a benzene ring tends to attract electrons from the adjacent carbon atoms causing them to become electro-positive. The effect extends around the ring in turn causing the carbon atoms to

---

[1] In inorganic chemistry, an oxidizing agent is described as a substance that causes an increase in the positive valence of the material oxidized by taking electrons from it. Thus an oxidizing agent is an acceptor of electrons.

be respectively more positive and more negative than in the unsubstituted ring.

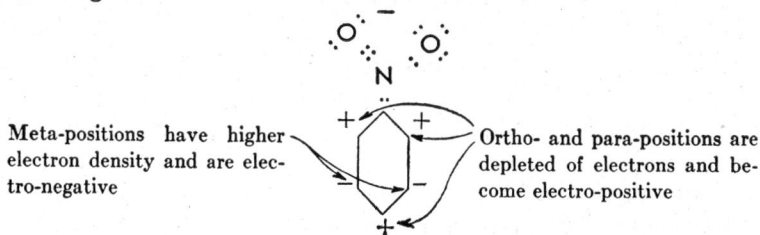

Meta-positions have higher electron density and are electro-negative

Ortho- and para-positions are depleted of electrons and become electro-positive

There is also the opportunity for resonance within the nitro group and this can extend itself to include the resonance of the benzene ring itself. This results in the possibility of at least eight resonance forms, three of which are shown here.

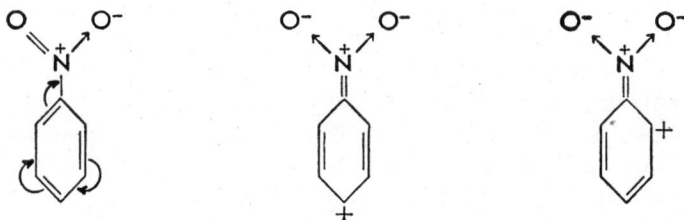

Since reactivity is dependent upon the opportunity for electron bonding to occur, it follows that the most reactive positions on a nitrobenzene molecule are the two meta- positions. It is for this reason that ring structures carrying a nitro group tend predominantly to permit substitution in the meta- position. This conforms to the rules for substitution presented on p. 185.[2]

As in the chloroacetic acids (p. 128) a second effect of the presence of the electro-negative group is to induce a greater tendency toward ionization of hydrogen atoms. Just as trichloroacetic acid was found to be a very strong (highly ionized) acid, so the ring-structured acids with electro-negative nitro groups are much stronger than the unnitrated molecules. This effect is illustrated in the case of the ionization constants for nitrophenols.

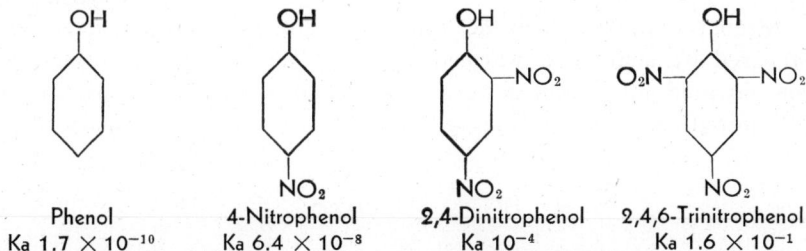

Phenol	4-Nitrophenol	2,4-Dinitrophenol	2,4,6-Trinitrophenol
Ka 1.7×10^{-10}	Ka 6.4×10^{-8}	Ka 10^{-4}	Ka 1.6×10^{-1}

[2] Similar interpretation of the electro-positive and electro-negative character of substituted groups may be used to explain further the rules for substitution reactions.

In compounds like aniline [benzene ring with —NH$_2$] the effect is such that the basicity

of the amino group is decreased. This polarity also serves as a limiting factor for the number of nitro groups possible on a benzene ring, preventing the combination of more than three nitro groups to any one ring.

Physical Properties. With the exception of nitrobenzene and o- and m- nitrotoluenes, nitro aromatic compounds are solids at room temperature. They have high boiling points and are insoluble or at best only slightly soluble in water. They can be separated from their reaction mixtures by steam distillation. Their color range is yellow through orange to red, depending upon the number of nitro groups and their acidity or basicity. Some nitro aromatic compounds are used as indicators because of their change in color at various pH ranges.

As previously noted nitro groups have a powerful activating influence on elements or groups adjacent to them on the benzene nucleus. Chlorine, for example, is difficult to remove from the ring in a chloro aromatic. When a nitro group is ortho- or para- to it, the chlorine will respond to hydrolysis when heated with sodium hydroxide—a reaction requiring heat and pressure when no nitro group is present. This activating effect of nitro groups is also cumulative. Trinitrochlorobenzene can be converted to trinitrophenol by boiling with water or in cold sodium carbonate solution.

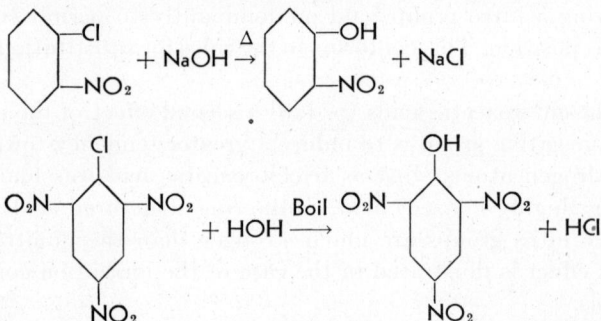

Preparation. Nitrobenzene, the simplest nitroaromatic compound, is prepared by shaking a warmed (30°–50°C) mixture of benzene with concentrated nitric and sulfuric acids. The concentrated sulfuric acid is required as a dehydrating agent. Nitro derivatives of other aromatic

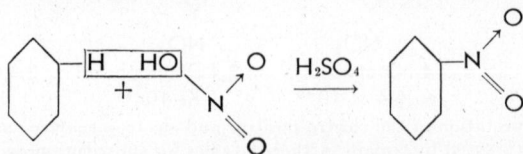

homologues may be prepared similarly, the concentration of nitric acid varying with substituents already on the ring. Nitration of benzene derivatives is usually promoted by the presence of ortho- and para-directing substituents except the halogens. Nitration is definitely hindered or retarded by the presence of halogens and of meta-directing substituents.

Among the better known nitroaromatic compounds are the explosives trinitrotoluene (TNT) and trinitrophenol (picric acid).

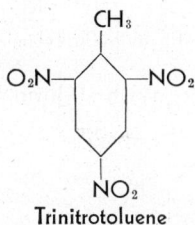

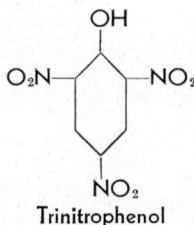

Trinitrotoluene   Trinitrophenol

Reactions. An important reaction of nitroaromatic compounds is their reduction. The reduction of the nitro group may be complete or partial, resulting in the formation of an amino group or some intermediate compound.

1. REDUCTION TO AMINO GROUP. The most important of these reduction reactions is the reduction of a nitro group to an amino group in the presence of hydrochloric acid reacting with tin or stannous chloride. This type of reaction is used commercially for producing amino benzene (aniline) from nitrobenzene, except that for industrial production iron is used in the place of tin or stannous chloride. The reduction reactions

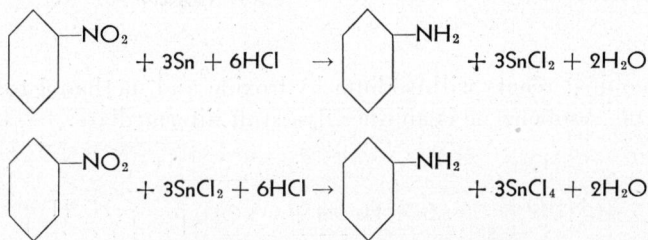

for the production of aniline should be remembered. The other reduction reactions presented here are for reference.

2. PARTIAL REDUCTION WITH ZINC. Depending upon the reagents used with zinc and upon the concentration of nitrobenzene, partially reduced compounds of nitrobenzene are formed.

a. Zinc dust with water will reduce nitrobenzene to nitrosobenzene.

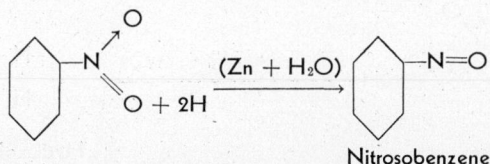

Nitrosobenzene

b. The hydrolysis of ammonium chloride produces sufficient hydrogen ion to react with zinc dust and reduce nitrobenzene to β-phenyl-hydroxylamine. With hydrochloric acid β-phenylhydroxylamine rearranges to

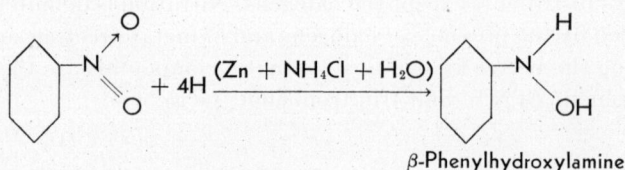

β-Phenylhydroxylamine

form p-aminophenol, a photographic developer. With sodium hydroxide

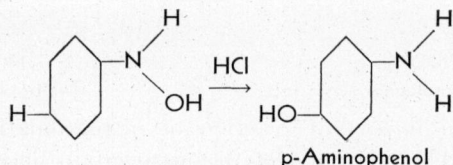

p-Aminophenol

β-phenylhydroxylamine condenses with nitrosobenzene to form azoxybenzene. Azoxybenzene, a dye intermediate, can also be made directly by reducing nitrobenzene with potassium hydroxide and ethanol.

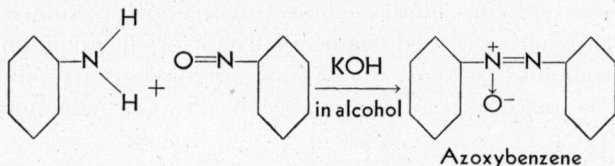

Azoxybenzene

c. Zinc dust reacts with sodium hydroxide and methanol to produce azobenzene. Azobenzene is another dye stuff intermediate.

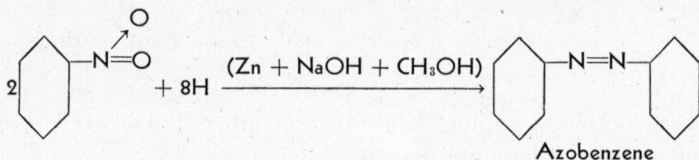

Azobenzene

d. Zinc dust reacts with sodium hydroxide and water to form hydrazobenzene which rearranges to benzidine on heating. Benzidine is an important dye intermediate.

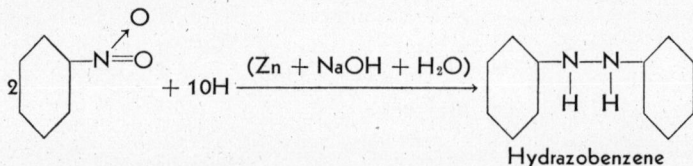

Hydrazobenzene

Benzidine

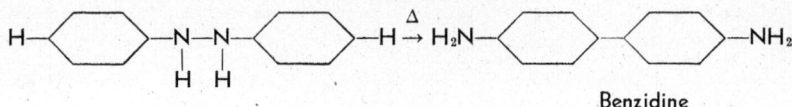

II. Aromatic Amines

Kinds of Aromatic Amines. Alkyl amines are of three types: primary, secondary, and tertiary (p. 143). Aromatic amines also exist in these three types. Alkyl radicals may also be present in place of aryl radicals. The essential characteristic of an aromatic amine is that it have at least

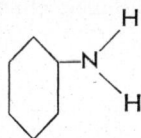

Aniline   Diphenylamine   Triphenylamine   Dimethylaniline

one aryl group directly attached to the nitrogen of an amine. Benzyl amine, , is not an aromatic amine. It is more correctly a benzene derivative of methyl amine. Moreover the benzene nucleus may have more than one amino group attached to it, forming o-, m-, and p-phenylenediamines.

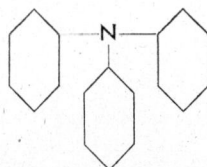

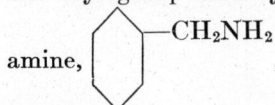

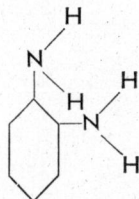

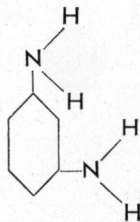

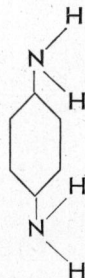

o-Phenylenediamine   m-Phenylenediamine   p-Phenylenediamine

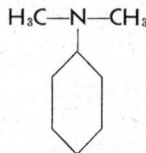

The simplest of all aromatic amines is aniline. It is also the most important. Aniline is the parent compound for the colors produced by the aniline dye industry. All mononuclear (benzene structure only) aromatic amines may be considered as derivatives of aniline.

Structure and Bonding. Because of the unshared electron pair on the nitrogen in aniline, the molecule readily forms addition products. This electron also makes possible several resonance forms, three of which are shown below. As in nitrobenzene (p. 199) reactions are dependent upon these structures:

Resonance forms
(ortho- and para- positions activated)

More stable
ionic form

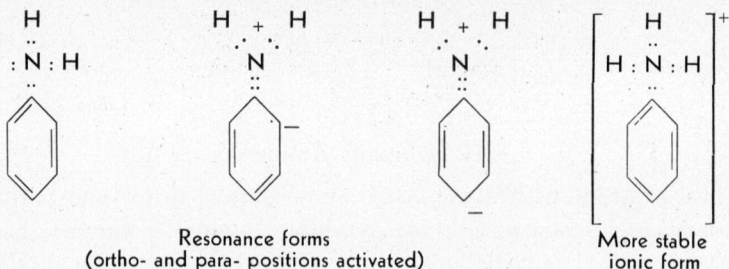

Preparation. Aniline is made commercially in large tonnages, chiefly by two methods.

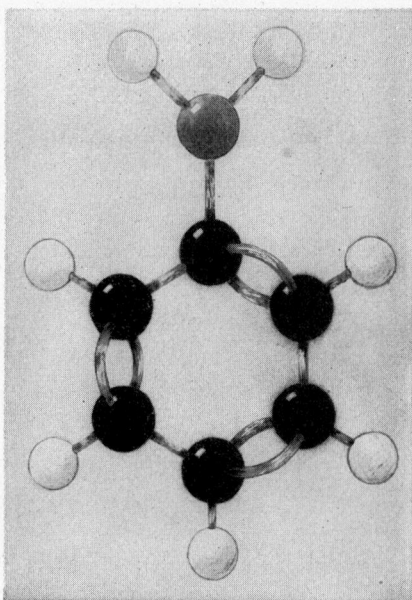

Fig. 32. The aniline molecule.

1. REDUCTION OF NITROBENZENE. Aniline is produced when nitrobenzene is reduced by the action of hydrochloric acid on scrap iron.

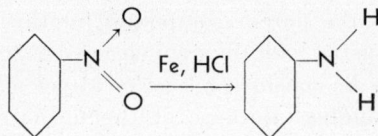

2. AMMONIZATION OF CHLOROBENZENE. Aniline is also produced commercially by the reaction of ammonia under pressure with heated chlorobenzene and cuprous oxide, using cuprous chloride as a catalyst.

$$2 \; \bigbone{Cl} + 2NH_3 + Cu_2O \xrightarrow[\text{Pressure}]{\substack{Cu_2Cl_2 \\ 200\,°C}} 2 \; \bigbone{N\substack{H \\ H}} + Cu_2Cl_2 + H_2O$$

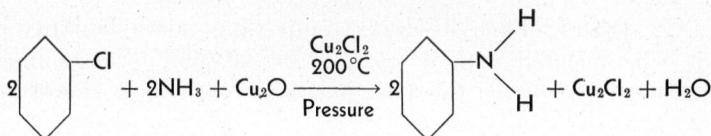

3. LABORATORY METHODS OF REDUCING NITROBENZENE by tin or stannous chloride with hydrochloric acid have been discussed on p. 201.

Reactions. Aniline reacts in two ways: the amino group or the ring part may react separately.

1. BY ADDITION OF AN ACID. Aniline is basic but its degree of basicity is considerably less than that of a primary alkyl amine due to the electronegative influence of the benzene ring. Reactions with inorganic acids and haloalkanes are additive, after the manner of ammonia in forming ammonium salts. Compounds of this type are salts of aniline and like

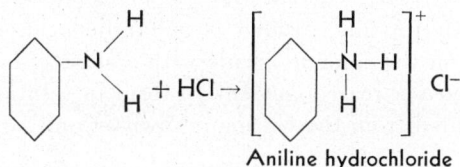

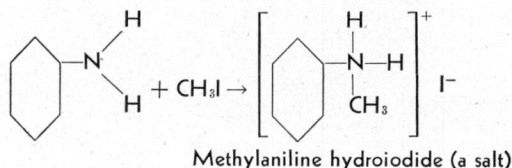

$$\bigbone{N\substack{H \\ H}} + HCl \rightarrow \left[\bigbone{N\substack{H \\ | \\ H}-H} \right]^+ Cl^-$$

Aniline hydrochloride

$$\bigbone{N\substack{H \\ H}} + CH_3I \rightarrow \left[\bigbone{N\substack{H \\ | \\ CH_3}-H} \right]^+ I^-$$

Methylaniline hydroiodide (a salt)

ammonium salts, are ionic in character. When these salts of aniline are treated with sodium hydroxide, aniline and methylaniline are formed respectively. Many organic acids and aromatic amines are identified

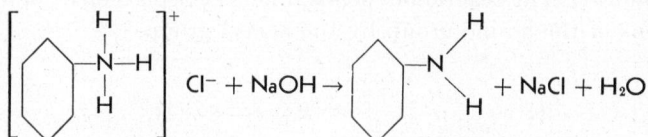

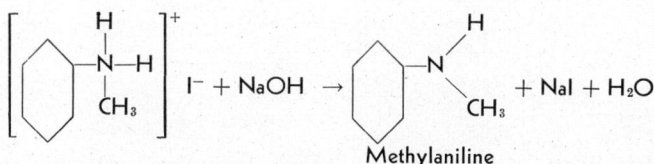

$$\left[\bigbone{N\substack{H \\ | \\ H}-H} \right]^+ Cl^- + NaOH \rightarrow \bigbone{N\substack{H \\ H}} + NaCl + H_2O$$

$$\left[\bigbone{N\substack{H \\ | \\ CH_3}-H} \right]^+ I^- + NaOH \rightarrow \bigbone{N\substack{H \\ CH_3}} + NaI + H_2O$$

Methylaniline

by converting them to their anilides, since they are easily purified, crystalline solids.

Several additive salts of aniline on heating undergo rearrangement with hydrogen from the ring. This rearrangement is due to the resonance structure of aniline and to the para-directing influence of the —NH_2 group (p. 185). β-Phenylhydroxylamine and methylaniline hydro-chloride when heated with a small excess of acid form aminophenol and toluidine, chiefly in the para-position. With aniline and sulfuric

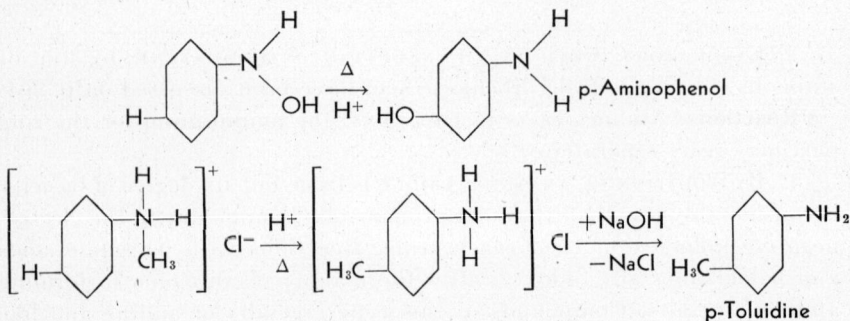

p-Aminophenol

p-Toluidine

acid the addition product formed is sulfanilic acid, in which an acid group is present in the same molecule with a slightly basic amino group. This salt undergoes a rearrangement to form in solution a dipolar ionic characteristic called from the German "Zwitterion."

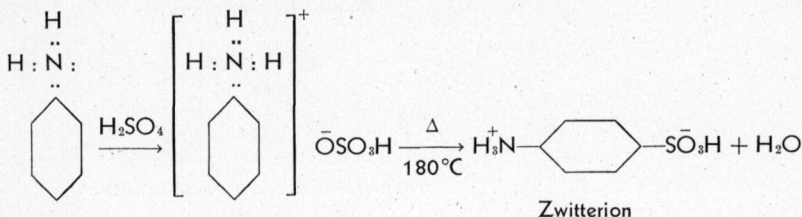

Zwitterion

2. REACTION WITH ACYL HALIDE. Acetanilide, used in headache remedies, is an important derivative of aniline that is prepared by using acetyl chloride. The reaction depends upon the replacement of one of the hydrogens of the amino group by the acetyl group.

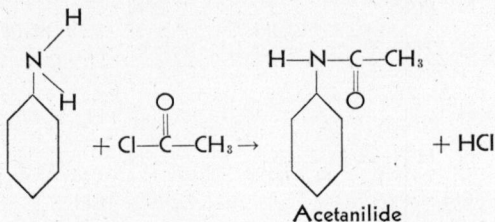

Acetanilide

3. REACTIONS WITH THE RING. Since the amino portion of the aniline molecule is so reactive, it is often necessary in preparing ring substitution products to first mask the amino part of the molecule by converting aniline to a compound like acetanilide. Substitution may then be accomplished without interference from the amino group, and the acetanilide part can be reconverted to an amino group by hydrolysis. Nitration and partial bromination of aniline are accomplished in this manner. This "masking" technique is necessary since if concentrated nitric acid

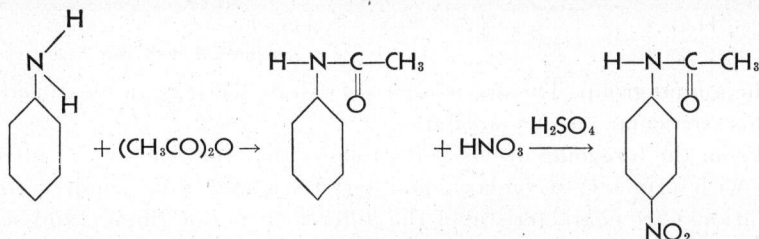

is added directly to aniline, many oxidation products of the —NH$_2$ group are obtained.

4. WITH NITROUS ACID.[3] Aniline reacts with nitrous acid in two ways, depending upon the temperature of the reactants. At room temperature aniline reacts like primary alkyl amines, forming nitrogen gas, water, and phenol. At the temperature of melting ice, aniline hydrochloride reacts with nitrous acid to form benzenediazonium chloride. Diazo compounds will be discussed later in this chapter (p. 208).

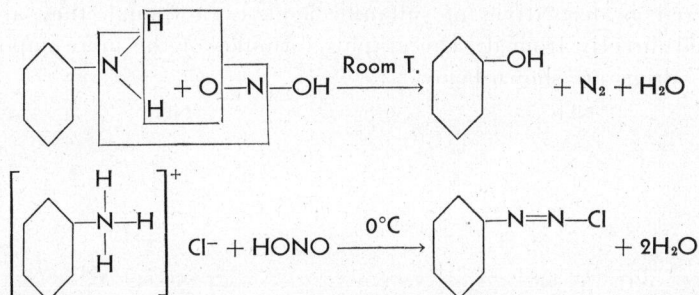

Secondary aromatic amines react with nitrous acid in the same way as secondary alkyl amines, forming nitroso compounds.

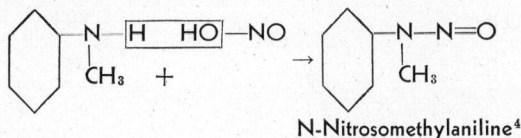

N-Nitrosomethylaniline[4]

Tertiary alkyl amines react with nitrous acid to give the nitrite salt. Tertiary aryl amines react even at low temperatures with nitrous acid by substituting a nitroso group in the ring always para- (when available)

[3] At this point reference to the reactions of alkyl amines with nitrous acid is recommended.

[4] "N—" in such a name is read as "Nitrogen" or as "N—." It signifies that the nitroso group is attached directly to the nitrogen rather than to some carbon on the ring. The latter is sometimes called C-nitroso. These designations are used only when necessary to avoid ambiguity.

$$H_3C \diagdown N - \bigcirc - \boxed{H + HO}NO \rightarrow H_3C \diagdown N - \bigcirc - NO$$

p-Nitrosodimethylaniline

to the amino group. The green-colored crystals serve as an identification of this particular tertiary aryl amine.

From the foregoing discussion it follows that the reactions of nitrous acid with primary, secondary, and tertiary amine salts (under proper conditions) are characteristic of the different types of amines, and serve as a valuable means of identifying them. (a) Primary aryl amines yield diazonium compounds $C_6H_5N_2X$. (b) Secondary aryl-alkyl amines yield N-nitroso compounds C_6H_5N—NO. (c) Tertiary aryl-dialkyl amines yield p-nitroso substitution compounds if the p-position is available.

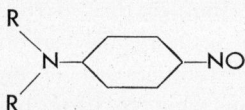

$$R \diagdown N - \bigcirc - NO$$

Sulfa Drugs. The sulfa drugs of modern medicine (p. 282) may all be considered as derivatives of sulfanilic acid, even though they are not prepared directly from it. The graphic formulas of the more important "sulfa" drugs are shown below.

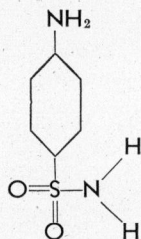

Sulfanilamide

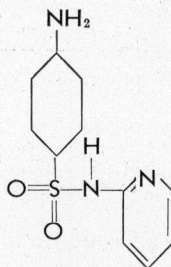

Sulfapyridine

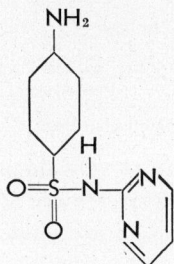

Sulfadiazine

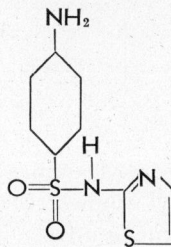

Sulfathiazole

III. Diazo Derivatives

Diazo compounds have two nitrogen atoms bonded to each other, one nitrogen being also bonded to an aryl group and the other nitrogen bonded to some element or group other than an aryl group. When a second aryl group is bonded to the second nitrogen atom, the product is called an azo compound.

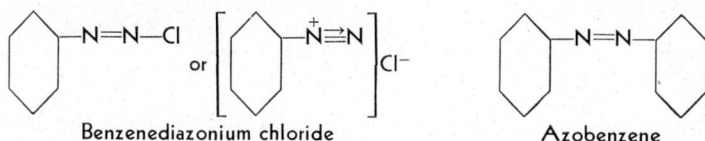

Benzenediazonium chloride   Azobenzene

Diazonium salts are soluble in water and are ionic in character, due to the similarity of one nitrogen atom as in ammonium salts. Diazonium salts are nearly always used in aqueous solution, making unnecessary their preparation in the easily explosive crystalline form.

Preparation. When aniline reacts with nitrous acid, two types of compounds are formed according to the temperature of the reaction (p. 207). At the temperature of melting ice nitrogen is not liberated. Two atoms of nitrogen (French *azote*, nitrogen) are bonded together forming a diazonium salt. This is called the "diazotization reaction."

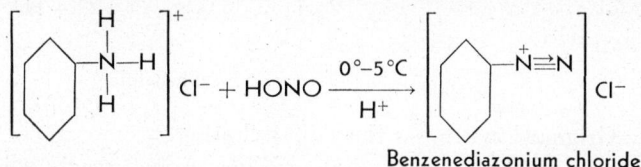

Benzenediazonium chloride

Reactions: 1. PARTIAL REDUCTION. Partial reduction of benzene-diazonium chloride results in the formation of phenylhydrazine, Emil Fischer's useful reagent in identifying carbohydrates. Sodium sulfite is used as a reducing agent. Phenylhydrazine hydrochloride is converted

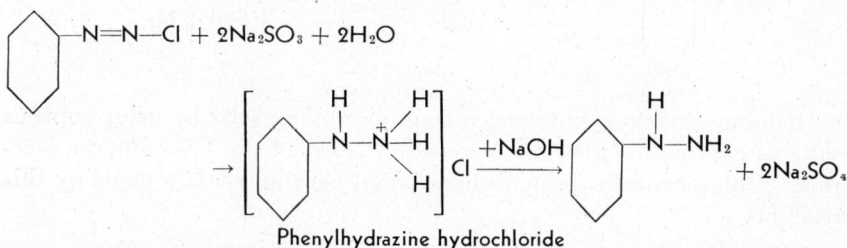

Phenylhydrazine hydrochloride

to phenylhydrazine by sodium hydroxide. The hydrochloride is a more stable form for storage.

2. DENITRIFICATION. Removal of nitrogen (denitrification) of aryl nitrogen compounds occurs with complete reduction to benzene or by a replacement of the —N≡N— group by another group or element.

A. COMPLETE REDUCTION. Complete reduction of a diazonium compound to benzene is accomplished by the use of a strong reducing agent such as sodium stannite in alkaline solution or sometimes by formaldehyde in alkaline solution. Nitrogen gas is always liberated.

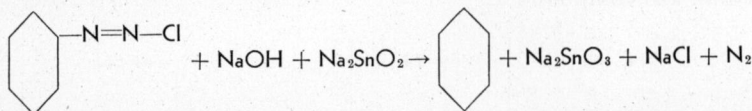

$$\text{C}_6\text{H}_5\text{—N}\!\!=\!\!\text{N—Cl} + \text{NaOH} + \text{Na}_2\text{SnO}_2 \rightarrow \text{C}_6\text{H}_6 + \text{Na}_2\text{SnO}_3 + \text{NaCl} + \text{N}_2$$

Ethanol may also serve as a denitrifying agent. Some of the diazonium salt is reduced to benzene and an equivalent amount of alcohol is oxidized to acetaldehyde. Most of the reaction however results in the liberation of nitrogen and the formation of phenetole (ethyl phenyl ether). The two reactions occur simultaneously. Phenetole is a safer medium to use

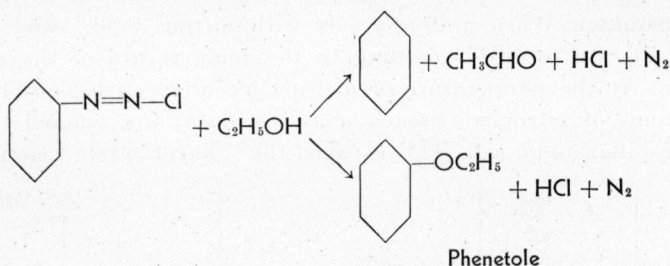

$$\text{C}_6\text{H}_5\text{—N}\!\!=\!\!\text{N—Cl} + \text{C}_2\text{H}_5\text{OH}$$

$$\nearrow \quad \text{C}_6\text{H}_6 + \text{CH}_3\text{CHO} + \text{HCl} + \text{N}_2$$

$$\searrow \quad \text{C}_6\text{H}_5\text{—OC}_2\text{H}_5 + \text{HCl} + \text{N}_2$$

Phenetole

for many Grignard syntheses than diethyl ether.

B. BY REPLACEMENT OF —N=N—. When aniline reacts with nitrous acid at room temperature, nitrogen is evolved and phenol is formed. Similarly, when a diazonium salt solution (formed from aniline and nitrous acid) is warmed, nitrogen is liberated and phenol is formed.

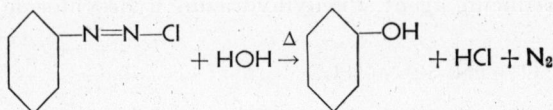

$$\text{C}_6\text{H}_5\text{—N}\!\!=\!\!\text{N—Cl} + \text{HOH} \xrightarrow{\Delta} \text{C}_6\text{H}_5\text{—OH} + \text{HCl} + \text{N}_2$$

Sandmeyer removed nitrogen from diazonium salts by using cuprous salts as catalysts. This type of reaction is called the "Sandmeyer reaction." Chlorobenzene, bromobenzene, and benzonitrile are made by this reaction.

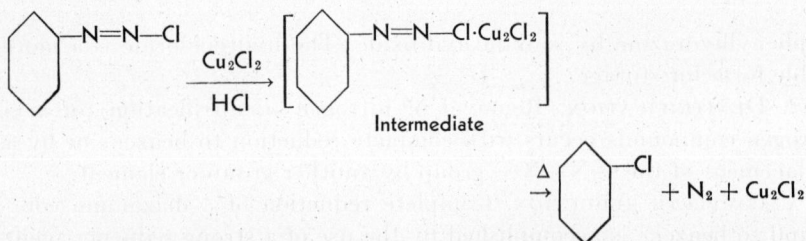

$$\text{C}_6\text{H}_5\text{—N}\!\!=\!\!\text{N—Cl} \xrightarrow[\text{HCl}]{\text{Cu}_2\text{Cl}_2} \left[\text{C}_6\text{H}_5\text{—N}\!\!=\!\!\text{N—Cl}\cdot\text{Cu}_2\text{Cl}_2 \right]$$

Intermediate

$$\xrightarrow{\Delta} \text{C}_6\text{H}_5\text{—Cl} + \text{N}_2 + \text{Cu}_2\text{Cl}_2$$

$$\Delta \atop \rightarrow \quad \text{Br} + N_2 + Cu_2Br_2$$

$$\Delta \atop \rightarrow \quad \text{CN} + N_2 + Cu[CN]Cl$$

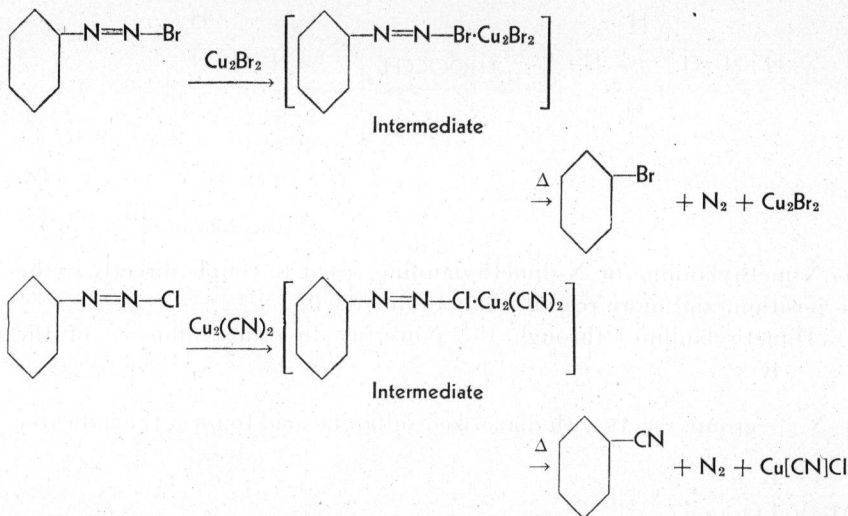

Diazonium salts react directly with potassium iodide without cuprous ion catalyst, to form iodobenzene.

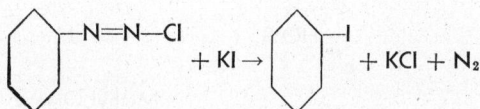

3. COUPLING REACTIONS. Coupling reactions of diazonium compounds are of great importance to the dye industry (p. 253). Many useful dyes are of the azo type.

A. WITH PHENOLS. Phenolic compounds readily react in alkaline solution with diazonium chloride in the para- position when available, otherwise in the ortho- position.

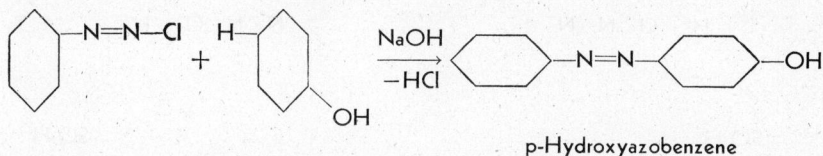

p-Hydroxyazobenzene

B. WITH AROMATIC AMINES. Diazonium salts also couple with aromatic amines (primary, secondary, and tertiary) in the presence of sodium acetate, forming diazoamino compounds. Sodium acetate is used to maintain the degree of basicity most favorable for reaction, pH 5–pH 8. The resulting compound when warmed with aniline hydrochloride or H$^+$ rearranges to p-aminoazobenzene. Secondary and tertiary amines, such

p-Aminoazobenzene

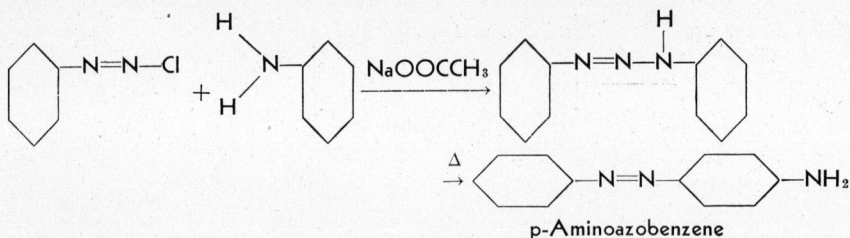

as N-methylaniline or N-dimethylaniline, seem to couple directly in the p- position, and more readily than aniline itself.

Dimethylaniline, through the powerful directing influence of the

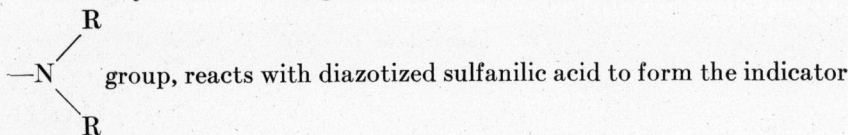

$$-N\begin{smallmatrix} R \\ \\ R \end{smallmatrix}$$

group, reacts with diazotized sulfanilic acid to form the indicator Methyl Orange.

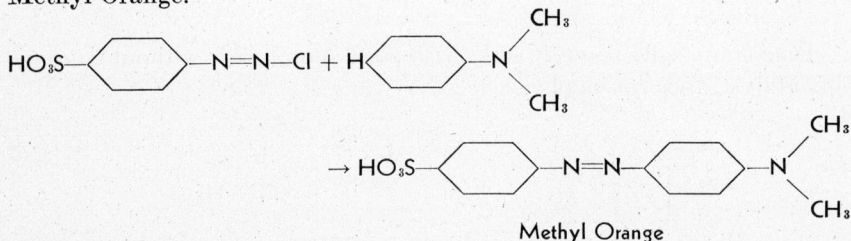

Methyl Orange

The preceding examples of diazotization and azo derivatives have all involved only single diazotization and single-ring aromatic compounds. The same reactions apply to more complex compounds. When benzidine hydrochloride (p. 203) reacts with nitrous acid both ends of the molecule are diazotized. Coupling with naphthionic acid occurs at both ends of the molecule, resulting in the dye and indicator Congo Red.

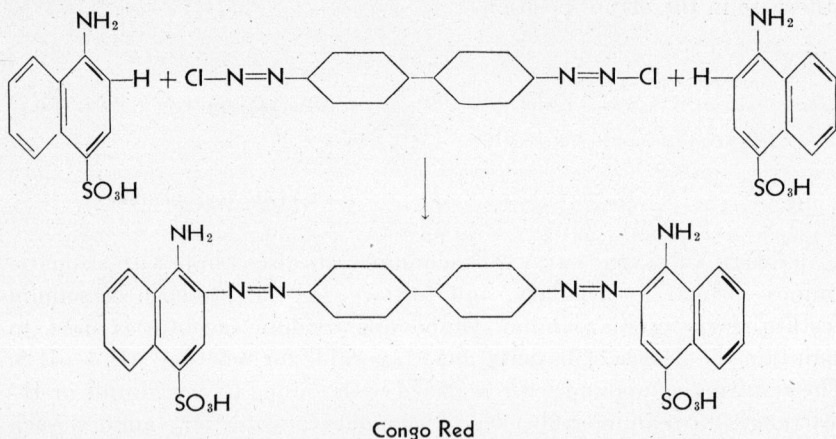

Congo Red

Study and Review Questions

1. How does a nitro group affect a hydroxy substituent on the same benzene ring? Give examples of cumulative effect.
2. How does a nitro group affect a chlorine substituent on the same benzene ring? Give examples of cumulative effect.
3. Write the equation for the complete reduction of nitrobenzene. Name the product.
4. Zinc and various reagents are used in obtaining different end products from the reduction of nitrobenzene. Make a short table showing reagents and end products.
5. Show what is meant by the benzidine rearrangement.
6. What is required in the structure of amines to classify them as aromatic amines?
7. Write the equations for two commercial methods of making aniline.
8. Show by an equation how aniline forms a salt with sulfuric acid. With hydrochloric acid. Name the salts.
9. Show by a series of equations how to prepare methylaniline from chlorobenzene.
10. Show by a series of equations how to prepare acetanilide from nitrobenzene.
11. Show by a series of equations how p-toluidine may be prepared from benzene.
12. Show by a series of equations how p-sulfanilic acid may be prepared from benzene.
13. Show by a series of equations how m-sulfanilic acid may be prepared from benzene.
14. Define "zwitterion" and explain by example.
15. Explain the "masking" of a compound—why and how it is accomplished. Give example.
16. Compare the reaction of nitrous acid with primary, secondary, and tertiary amines, aliphatic and aromatic.
17. Name four "sulfa" drugs.
18. What is the structural difference between "diazonium" and "azo" compounds? Give example.
19. By a series of equations show how phenylhydrazine may be made from nitrobenzene.
20. Show by equations two ways of denitrifying an azo compound.
21. Discuss the Sandmeyer reaction.
22. Coupling reactions are especially useful in some dye syntheses. Starting with benzene and reagents of your own choice, devise a method for making methyl orange.

Chapter 16

AROMATIC COMPOUNDS WITH OXYGEN

CHAPTER PROLOGUE

Oxygen is the most abundant element on earth and it is just about the "nosiest." It is the ubiquitous infiltrator getting into molecules often by insinuating itself between atoms like carbon and hydrogen which are so peaceful side by side. Once oxygen gets into a side-chain of an aromatic compound, other oxygen atoms enter more easily and soon take over by having an oxygen atom entice two hydrogens away to form water. When oxygen edges in between a ring carbon and a hydrogen, it forms one of our most useful compounds, phenol. Phenol is an acid, not an alcohol. Oxygen on the ring excites the hydrogens to such an extent that its own hydrogen may ionize part of the time and three other hydrogens on the ring get ready to go off and form some other molecule if favorable circumstances arise.

Aromatic compounds with oxygen are, however, among the very beneficial materials used by man. Phenol with formaldehyde has given us the Bakelite plastics. Bacteria die in the presence of phenol and kindred compounds. Sodium benzoate preserves our foods. Other compounds serve as photographic developers. Mankind in general and some organic chemistry students have enjoyed the refreshing relief of aspirin. The dye fluorescein, saving messenger of aviators forced down at sea, finds a new use in the labeling of clouds by the professional rainmaker. These are some of the aromatic compounds with oxygen to be studied in the following pages.

Aromatic compounds containing oxygen may be considered from two points of view: (I) those in which the oxygen is attached directly to the benzene ring and (II) those in which the oxygen is attached to the side-chain of some benzene derivative. In the first case the oxygen modifies the properties of the ring whereas in the second case the ring may be considered as a modifier of the side-chain itself. In this chapter our first consideration will be the modifying influence of oxygen on the ring. Later the modifying influence of the benzene ring on a side-chain will be considered.

I. Oxygen on the Benzene Ring

The simplest and most important aromatic compound of oxygen is phenol. When freshly distilled it is a colorless, crystalline substance of low melting point, 41°C. When exposed to air and light, phenol darkens, due to oxidation to various complex compounds, often to a reddish purple color. It is only slightly soluble in water but is readily soluble in many organic solvents such as alcohol and ether. Unlike the alcohols phenol is definitely, though weakly, acidic ($Ka = 10^{-10}$). Its common name is "carbolic acid." Phenol is the oldest known organic disinfectant. Its bactericidal action on *B. typhosus* in comparison with that of other compounds determines the phenol coefficient of bactericides (p. 286). Phenol must be handled carefully because it is caustic and is absorbed by the skin on contact. It is highly toxic, as little as 15 g. causing death.

Preparation. Large tonnages of phenol are prepared from coal tar residues as a by-product of the coke industry. Since World War I, three important commercial methods of preparing phenol have also been used, all starting with benzene.

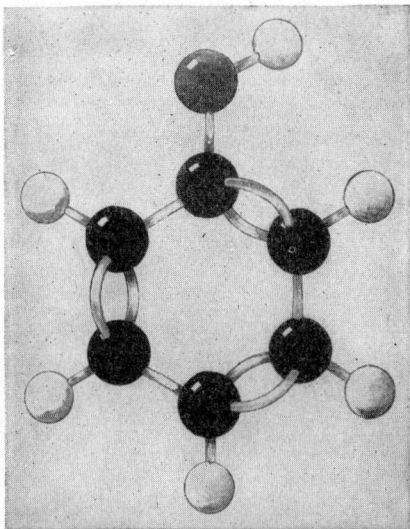

Fig. 33. The phenol molecule.

1. Benzenesulfonic acid, converted to the sodium salt, is fused with sodium hydroxide. The resulting sodium phenate is converted to phenol by hydrochloric or sulfuric acid. This is the typical reaction of a salt of any

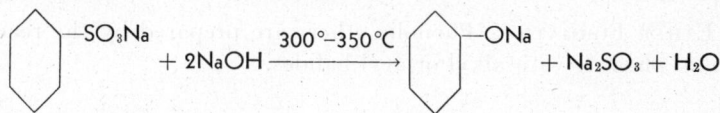

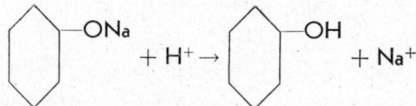

$$\text{C}_6\text{H}_5-\text{SO}_3\text{Na} + 2\text{NaOH} \xrightarrow{300°-350°C} \text{C}_6\text{H}_5-\text{ONa} + \text{Na}_2\text{SO}_3 + \text{H}_2\text{O}$$

$$\text{C}_6\text{H}_5-\text{ONa} + \text{H}^+ \rightarrow \text{C}_6\text{H}_5-\text{OH} + \text{Na}^+$$

weak acid when treated by a strong acid.

2. The Dow method of producing phenol consists of the hydrolysis of chlorobenzene. A dilute solution of sodium hydroxide is used at high temperature (300°C) and high pressure (3000 p.s.i.).

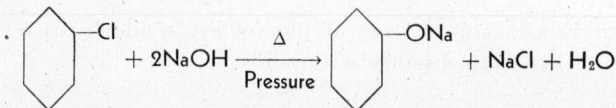

$$\text{C}_6\text{H}_5-\text{Cl} + 2\text{NaOH} \xrightarrow[\text{Pressure}]{\Delta} \text{C}_6\text{H}_5-\text{ONa} + \text{NaCl} + \text{H}_2\text{O}$$

3. In Germany the Raschig method of converting benzene into phenol is based upon the catalytic formation of chlorobenzene from benzene, hydrochloric acid and oxygen from air, followed by the catalytic hydrolysis of chlorobenzene to phenol. Hydrochloric acid is regenerated by the latter reaction and used again in subsequent formation of chlorobenzene.

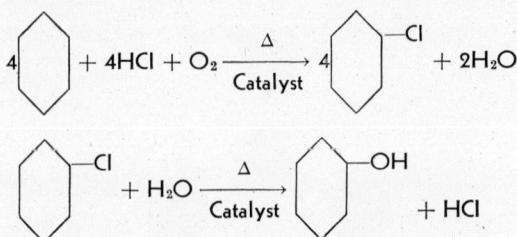

$$4\,\bigcirc + 4HCl + O_2 \xrightarrow[\text{Catalyst}]{\Delta} 4\,\bigcirc\!\!-Cl + 2H_2O$$

$$\bigcirc\!\!-Cl + H_2O \xrightarrow[\text{Catalyst}]{\Delta} \bigcirc\!\!-OH + HCl$$

Reactions. Reactions of phenols are of two types. The —OH group may react in whole or in part and reaction may occur with another part of the benzene nucleus.

1. SALT FORMATION. As an acid, phenol forms salts with sodium and potassium hydroxides. It does not form sodium phenate with sodium

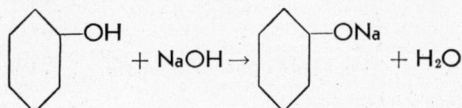

$$\bigcirc\!\!-OH + NaOH \rightarrow \bigcirc\!\!-ONa + H_2O$$

carbonate because phenol is a weaker acid than carbonic acid. The addition of CO_2 to sodium phenate solution will cause the precipitation of phenol.

2. ETHER FORMATION. Phenolic ethers are prepared by the reaction of sodium phenate with alkyl or aryl halides.

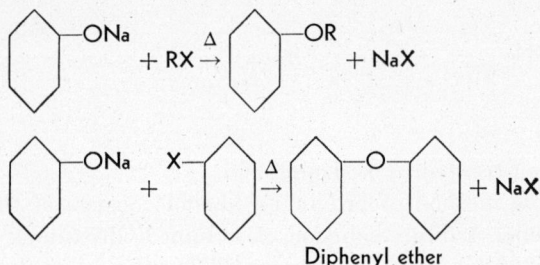

$$\bigcirc\!\!-ONa + RX \xrightarrow{\Delta} \bigcirc\!\!-OR + NaX$$

$$\bigcirc\!\!-ONa + X\!\!-\!\bigcirc \xrightarrow{\Delta} \bigcirc\!\!-O\!\!-\bigcirc + NaX$$

Diphenyl ether

3. ESTER FORMATION. Esters of phenol are made by reaction with acyl halides rather than directly with acids.

$$\text{C}_6\text{H}_5\text{—ONa} + R\text{—C}\overset{O}{\underset{Cl}{<}} \rightarrow \text{C}_6\text{H}_5\text{—O—C}\overset{O}{\underset{}{<}}\text{—R} + NaCl$$

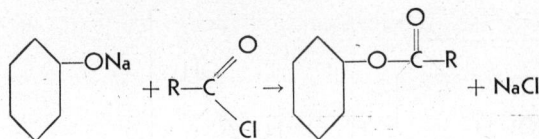

4. REPLACEMENT OF —OH. The replacement of —OH groups for halogen and other groups, such as occurs in alkyl compounds with PCl_3 or PCl_5, results in such small yields that their mention here is justified only in showing the difference between phenol and the alcohols. Replacement of —OH by H may be regarded as a reduction.

5. REDUCTION. Depending upon conditions phenol may be reduced to two very different products, i.e. to benzene or to cyclohexanol. Zinc dust reacts with phenol to form benzene and zinc oxide. Hydrogen in the presence of nickel adds to the double bonds of the ring, reducing phenol to cyclohexanol. The latter reaction is used in the production of adipic acid for nylon intermediates.

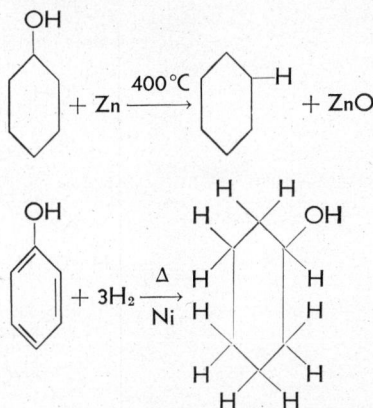

$$\text{(phenol)} + Zn \xrightarrow{400\,°C} \text{(benzene)} + ZnO$$

$$\text{(phenol)} + 3H_2 \xrightarrow[Ni]{\Delta} \text{(cyclohexanol)}$$

6. ADDITION AND CONDENSATION REACTIONS. Phenol forms an addition product with aldehydes. With formaldehyde the reaction is:

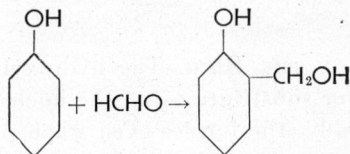

$$\text{(phenol)} + HCHO \rightarrow \text{(o-hydroxybenzyl alcohol)}\text{—CH}_2\text{OH}$$

By condensation phenol formaldehyde resin (Bakelite) is produced. The exact mechanism of the reaction is not definitely known but the final structure, consisting of many crosslinkages, is thought to be the following:

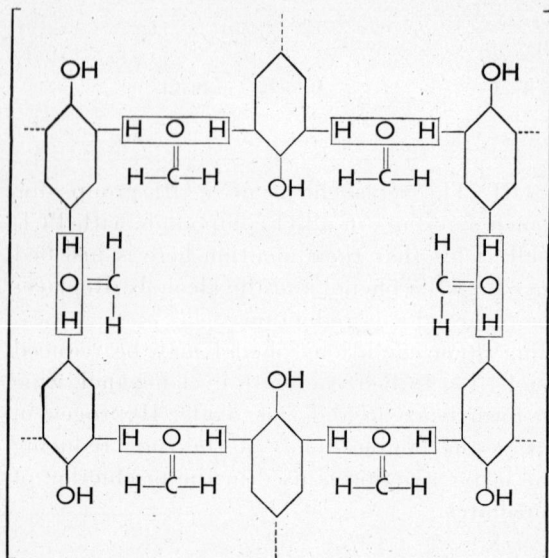

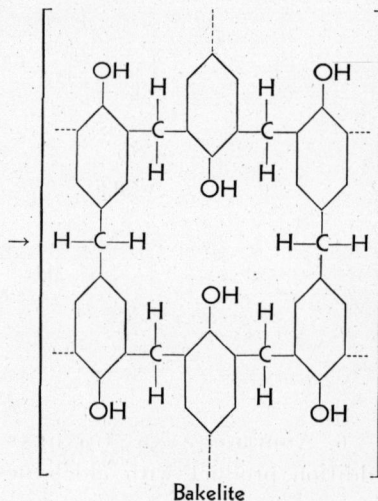

Bakelite

7. Substitutions on the Ring. The hydroxyl group of phenol is ortho-para- directing for substitutions on the nucleus. Bromination and chlorination occur readily, the former even when present in very small aqueous concentrations (1:100,000 parts). This reaction is used effectively in manufacturing bromine from sea water. Phenol is nitrated by dilute nitric acid, in contrast to the strong concentration required for nitrating benzene. Sulfonation occurs with cold sulfuric acid.

8. Test for Phenols. Of the tests available for detecting the presence of phenol the ferric chloride test and Millon's test are most often used.

Ferric chloride in dilute solution will turn violet, blue, green, or red with dilute solutes of phenol. Color is due to complex ions such as $Fe(OC_6H_5)_6^=$. Nonphenolic substances such as oximes and acetates will also give a red color. Millon's test for the hydroxy phenyl group is very sensitive. Millon's solution is made by dissolving mercurous and mercuric nitrates in nitric acid. A brick-red color indicates the presence of the phenol group.

Phenol Derivatives: 1. WITH —OH. Phenolic compounds with two hydroxy groups are called benzenediols, with numbers to designate the location of the hydroxy groups. All three dihydric phenols are acidic

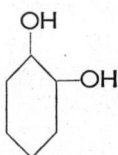

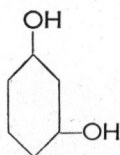

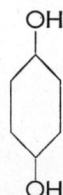

Catechol
Benzene-1,2-diol

Resorcinol
Benzene-1,3-diol

Hydroquinone
Benzene-1,4-diol

and have antiseptic properties. Trihydric phenols are called pyrogallol, phloroglucinol, and hydroxyhydroquinone. Their graphic formulas are

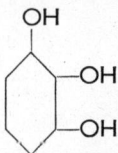

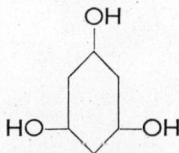

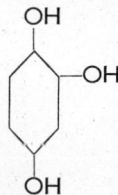

Pyrogallol
Benzene-1,2,3-triol

Phloroglucinol
Benzene-1,3,5-triol

Hydroxyhydroquinone
Benzene-1,2,4-triol

shown above. Naphthols are important in dye and perfume synthesis.

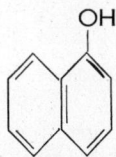

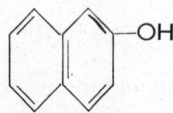

1-Naphthol
(α-Naphthol)

2-Naphthol
(β-Naphthol)

2. WITH —CH₃. Phenolic compounds having a methyl substituent are called cresols. They exist in ortho-, meta-, and para- isomers. Cresols are less toxic to man and animals and yet are more powerful as germicides

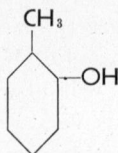

o-Cresol

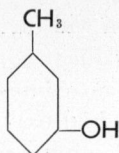

m-Cresol

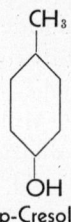

p-Cresol

(p. 214) than phenol. Lysol contains a mixture of all three cresols. Xylenols have two methyl groups and one hydroxy group on a benzene nucleus. Thymol, used in tooth paste, is 1-hydroxy-2-isopropyl-5-methylbenzene.

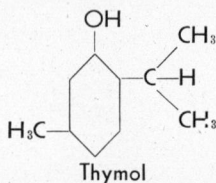

Thymol

3. WITH OTHER GROUPS AND ELEMENTS. Aminophenols and nitrophenols have been mentioned in Chapter 15. Nitro groups activate the

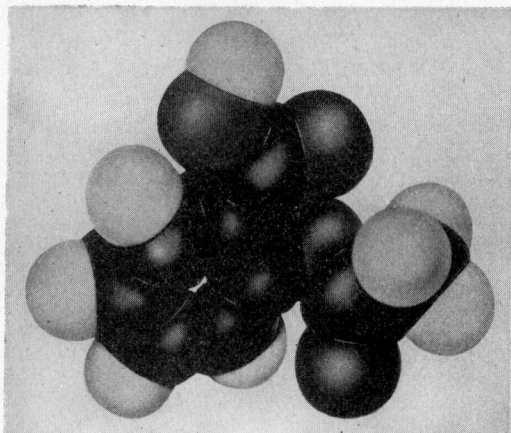

Fig. 34. The aspirin molecule showing steric proportion.

—OH group making it more acidic. When a carboxyl group is ortho- to an hydroxy group, the substance is called salicylic acid. Phenylsalicylate, called salol, is an intestinal antiseptic, insoluble in the acid gastric juices of the stomach but soluble in the basic mixtures in the intestines. Acetyl salicylic acid, known as aspirin, is widely used for relief from headache and fever. Methyl salicylate is known as oil of wintergreen.

Salicylic acid   Salol   Aspirin   Oil of wintergreen

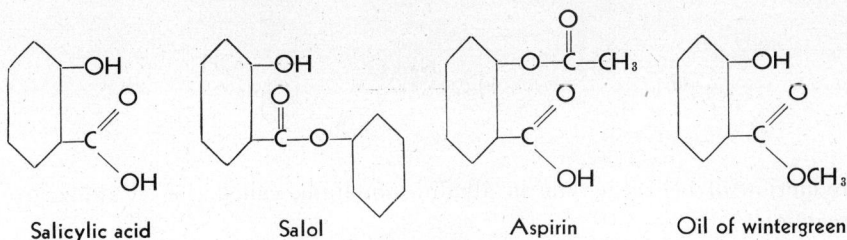

II. Oxygen in Aromatic Side-chains

A. Aromatic Alcohols. Aromatic alcohols are similar in structure to aliphatic alcohols, in that one hydrogen of the chain is replaced by a benzene nucleus. The simplest aromatic alcohol is benzyl alcohol, also called phenyl carbinol. It is methanol with a benzene ring substituted for a hydrogen atom. If the remaining hydrogen atoms are replaced by

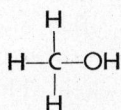

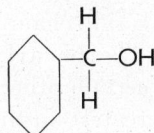

Methanol   Benzyl alcohol[1]

phenyl groups, diphenylcarbinol and triphenylcarbinol are formed.

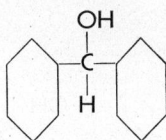

Diphenyl carbinol   Triphenyl carbinol

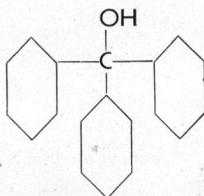

A phenyl group may take the place of hydrogen from either the C_1 atom or the C_2 atom in ethanol. These compounds are accordingly named 1-phenylethanol (α-phenylethanol) or 2-phenylethanol (β-phenylethanol).

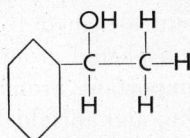

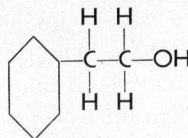

1-Phenylethanol   2-Phenylethanol

PREPARATION. Benzyl alcohol may be prepared by the hydrolysis of benzyl chloride. It is also one of the products of simultaneous oxidation-

---

[1] In naming compounds of this typ benzyl refers to $C_6H_5CH_2$—, benzal or benzylidene to $C_6H_5CH=$, and benzo- to $C_6H_5C\equiv$.

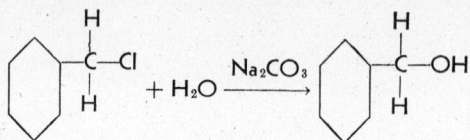

reduction of benzaldehyde in alkaline medium, called the "Cannizzaro reaction."

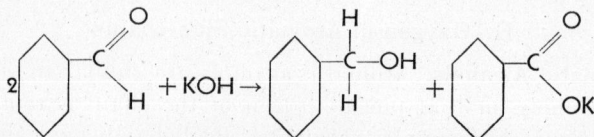

REACTIONS. The phenyl radical acts like a relatively electronegative substituent, and either "activates" or "deactivates" other groups attached to a side-chain in the same way as any other electronegative substituent. The phenyl radical also has capacity to resonate, which may modify certain other substituents on the side-chain. This activating influence is greatest on those elements or groups nearest the ring. Benzyl alcohol reacts like aliphatic alcohols in the formation of esters and ethers. It is easily oxidized to benzaldehyde and benzoic acid.

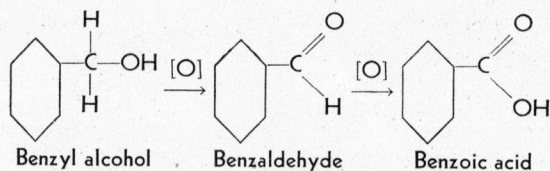

Benzyl alcohol ． Benzaldehyde ． Benzoic acid

B. Aromatic Aldehydes. Among aromatic compounds used in synthesis, the aldehydes rank next to phenols and amines. Aliphatic aldehydes are among the most active and versatile alkyl compounds in forming new products (Chapter 6). Aromatic aldehydes show similar activity and versatility.

Aromatic aldehydes have an aldehyde group attached to the ring or to a side-chain. The ring part may be either a phenyl group or some phenyl derivative. Among the commercially important aromatic aldehydes are benzaldehyde, salicylaldehyde, vanillin, and anisaldehyde.

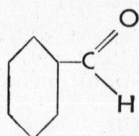

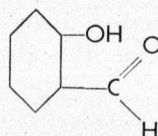

Benzaldehyde ． Salicylaldehyde

OH

—OCH₃ → $-OCH_3$

O

C—H

Vanillin

OH

O

C—H

Anisaldehyde

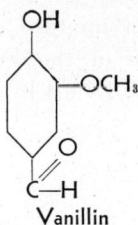

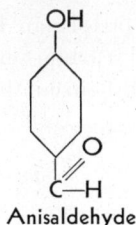

Preparation. Benzaldehyde, the simplest aromatic aldehyde, is an oily liquid called "oil of bitter almonds," owing to its odor. It can be prepared in several ways. A few of the more common methods are the following:

1. Oxidation of Toluene. Toluene may be called phenyl methane or methyl benzene (p. 177). We have called attention to the fact that the methyl group *activates* the o-, p- H atoms in the ring in methyl benzene (p. 185).

As phenyl methane (phenyl being an electronegative substituent on methane) the activation of the methane residue toward *oxidation* is probably due to the resonance effect, there being *conjugation* of the double

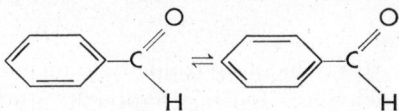

bonds of the —C=O and the ⬡ . This tendency for oxidation exemplifies the activating influence of the phenyl group on carbon atoms attached to it. It will be recalled that methane and other alkanes cannot be oxidized directly.

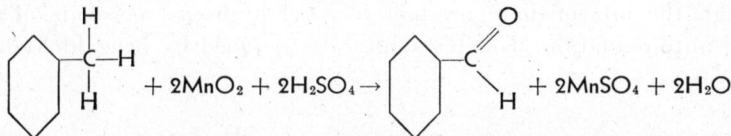

$$+ 2MnO_2 + 2H_2SO_4 \rightarrow \qquad + 2MnSO_4 + 2H_2O$$

The most important industrial method of producing benzaldehyde consists of passing a mixture of toluene vapor and O₂ over heated V₂O₅.

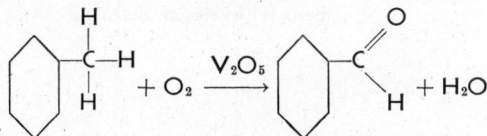

$$+ O_2 \xrightarrow{V_2O_5} \qquad + H_2O$$

2. By hydrolysis benzal chloride can be converted into benzaldehyde.

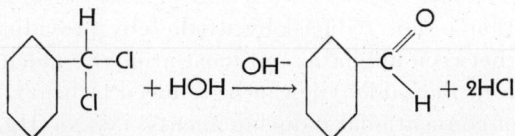

$$+ HOH \xrightarrow{OH^-} \qquad + 2HCl$$

3. The Gatterman-Koch method, utilizing a mixture of hydrogen chloride and carbon monoxide, is an extension of the Friedel-Crafts reaction with benzene in the presence of $AlCl_3$ and Cu_2Cl_2. Possibly the

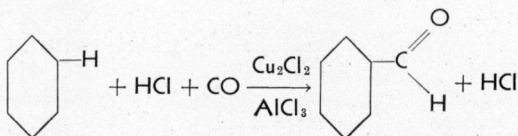

unstable methanoyl chloride, $H—C$, is an intermediate in the reaction, since alkanoyl halides with $AlCl_3$ react with aromatics to give alkyl-aryl ketones.

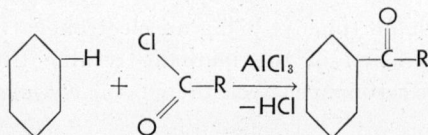

REACTIONS. The reactions of benzaldehyde include most of the reactions of aliphatic aldehydes cited in Chapter 6. Students should review those reactions, using benzaldehyde as a reactant with Grignard reagent and as an addition reagent with hydrogen cyanide (HCN), sodium bisulfite ($NaHSO_3$), hydroxylamine (H_2NOH), and phenylhydrazine ($H_2NNHC_6H_5$).

1. OXIDATION. Benzaldehyde oxidizes in air to benzoic acid, passing through the intermediate product benzoyl hydrogen peroxide. This is called auto-oxidation. Tollen's solution is reduced by benzaldehyde but

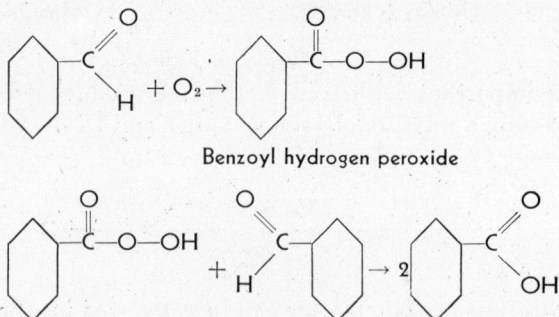

Benzoyl hydrogen peroxide

Fehling's solution is not reduced by it, thereby providing a method of distinguishing between aliphatic and most aromatic aldehydes.

2. REDUCTION. Benzaldehyde can be reduced to benzyl alcohol with a great variety of conventional reducing agents, i.e. Na/Hg, etc.

The aluminum isopropoxide of Meerwein, Pondorf, and Verley (p.

91) is a *very specialized* reducing agent, valuable because it can reduce aldehydes to alcohols without effecting ordinarily reactive double and triple bonds in the same molecule. Thus

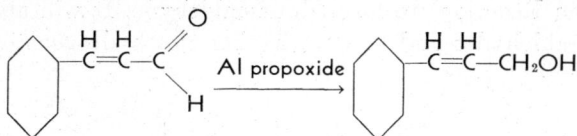

3. SIMULTANEOUS OXIDATION-REDUCTION. Cannizzaro discovered the reaction by which benzaldehyde is converted to benzyl alcohol and benzoic acid. This reaction is effective only when —CHO is directly attached to the ring.

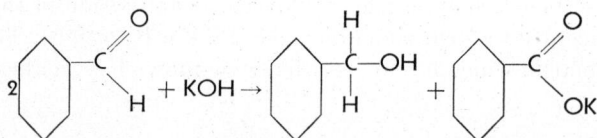

4. WITH HALOGENS. The phenyl group has so great an activating influence on the carbonyl group hydrogen that halogens substitute for the hydrogen of the group directly producing benzoyl chloride. The

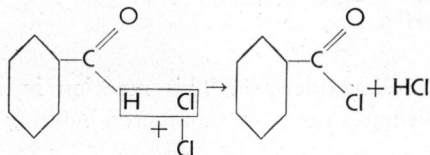

contrast of this reaction with that observed in halogenation of aliphatic aldehydes (p. 87) is striking. In the latter class substitution occurs for hydrogen atoms on the alpha carbon atom.

5. WITH AMMONIA. Aliphatic aldehydes, except formaldehyde, form aldehyde-ammonia addition products insoluble in ether (p. 96). Aromatic aldehydes, with —CHO attached directly to the ring, react more like formaldehyde with ammonia in the forming of urotropin (p. 94). Benzaldehyde condenses with ammonia to form hydrobenzamide, $(C_6H_5CH)_3N_2$.

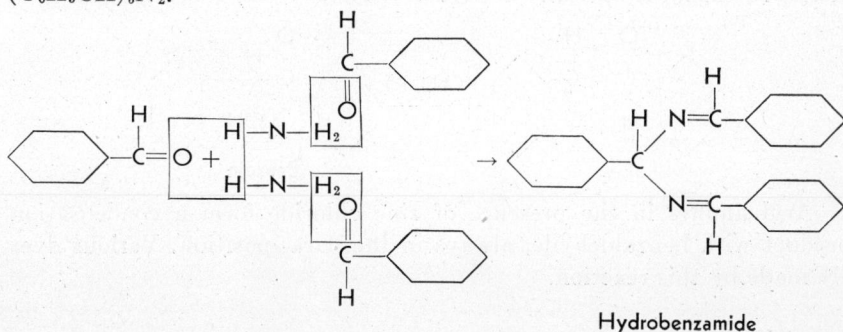

Hydrobenzamide

6. ADDITIONAL CONDENSATION REACTIONS. Benzaldehyde undergoes several condensation reactions, but it cannot condense with itself in the manner of the aldol condensation (p. 89). It can condense with acetaldehyde (Claisen Reaction) to form cinnamaldehyde. It will also condense with higher aldehydes and ketones having two hydrogens on an alpha carbon.

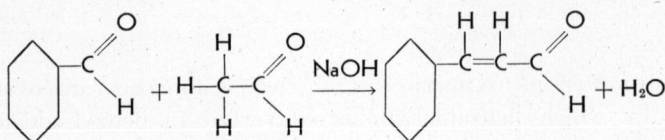

Similarly, benzaldehyde reacts with acetic anhydride in the presence of sodium acetate to form cinnamic acid (Perkin Reaction). The production of Coumarin, which has an odor like new-mown hay, makes use of the

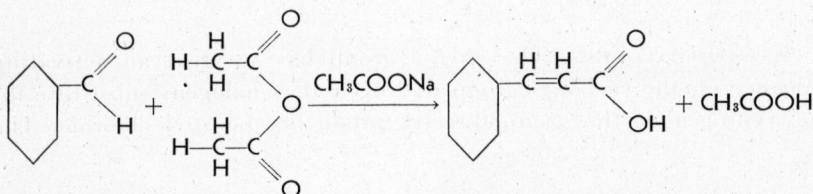

Perkin reaction of salicylaldehyde. This reaction is like many others that result in the formation of a six-membered heterocyclic ring.

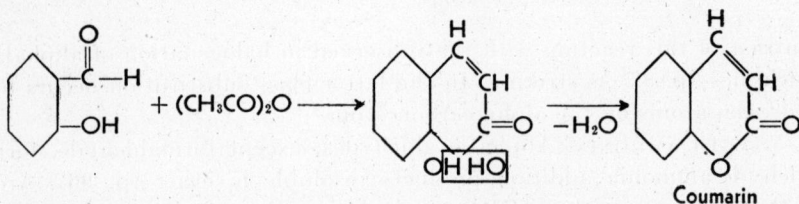

In the presence of alcoholic potassium cyanide catalyst benzaldehyde (other aldehydes also) condenses with itself to form benzoin, a keto-alcohol. Benzoin is optically active.

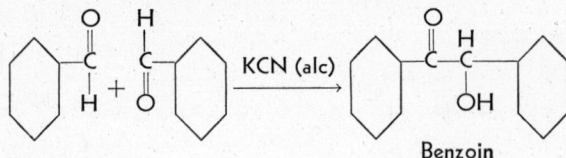

Aryl amines in the presence of zinc chloride form a condensation product with benzaldehyde, always in the para- position. Various dyes are made by this reaction.

p,p'-Diaminotriphenylmethane

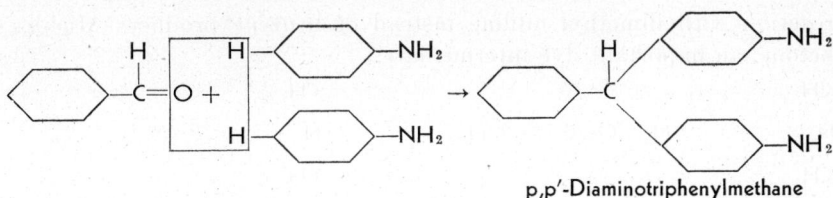

C. Aromatic Ketones. Aromatic ketones are ketones in which some aryl group is present. The aryl group takes the place of one or both alkyl groups in the formula R—CO—R'. Acetophenone is the simplest mono-aryl ketone and benzophenone is the simplest diaryl ketone.

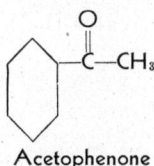

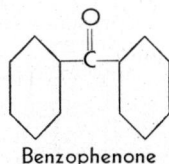

Acetophenone

Benzophenone

1. Acetophenone is prepared by the Friedel-Crafts reaction of benzene

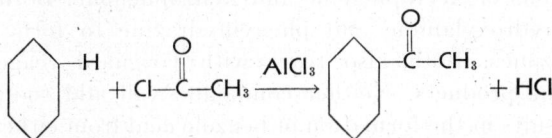

with ethanoyl chloride. It also may be prepared by the oxidation of 1-phenylethanol. Homologues of acetophenone are prepared by the same methods.

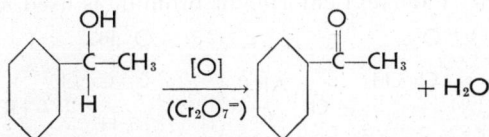

2. Benzophenone may be prepared by the Friedel-Crafts reaction of benzene and benzoyl chloride. It may also be prepared by the Friedel-

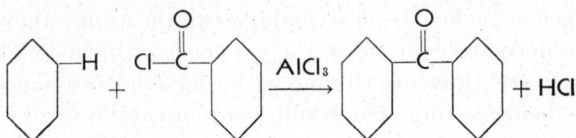

Crafts reaction of benzene with phosgene ($COCl_2$). The latter type

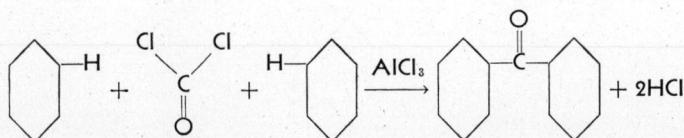

reaction with dimethyl aniline instead of benzene produces Michler's ketone, an important dye intermediate.

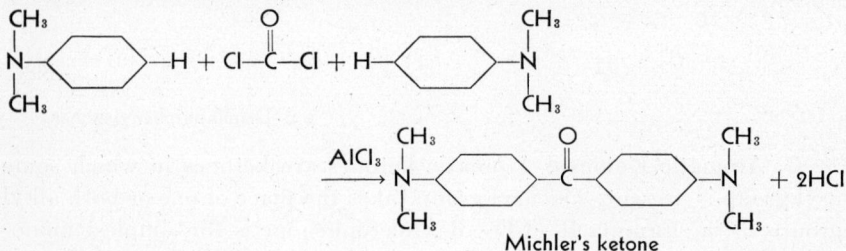

Michler's ketone

Benzophenone may also be prepared by the oxidation of diphenylmethane. The presence of two phenyl groups activates the hydrogen for easy oxidation.

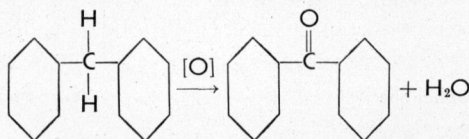

3. Reactions of Acetophenone and Benzophenone. Both compounds react with hydroxylamine and phenylhydrazine to form oximes and phenylhydrazones. Both also react with Grignard reagent to form regular ketone products. Neither compound will add sodium bisulfite. Oxidation results in the formation of benzoic acid from either compound, one ring breaking in the case of benzophenone. Both can be reduced to secondary alcohols or completely to the hydrocarbon state.

Acetophenone, when chlorinated in the presence of AlCl₃, forms phenacyl chloride. Phenacyl chloride or bromide is used as a tear gas.

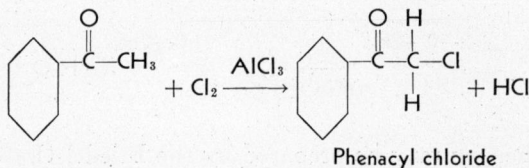

Phenacyl chloride

D. Quinones. Quinones are like ketones because they have a double-bonded oxygen attached to a secondary carbon atom; they are unlike true benzene derivatives because the oxygen is attached to a carbon of the ring structure thus interfering with the true resonance character of the simple benzene ring. The result is a tendency toward tautomerism and to the formation of colored compounds.

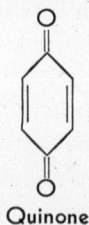

Quinone

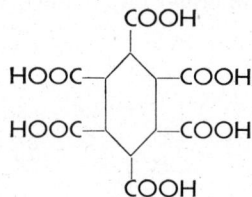

E. Aromatic Acids. Aromatic acids have one or more carboxyl groups attached to an aromatic nucleus. Mellitic acid has six carboxyl groups attached to a benzene nucleus. Benzoic acid is the simplest aromatic acid. Phthalic acid, especially in the anhydride form, is extensively used

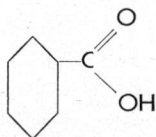

Benzoic acid   Mellitic acid

in the preparation of plastics, dyes, and indicators. Salicylic acid (p. 220) is used in preparing aspirin and other useful pharmaceuticals.

1. BENZOIC ACID: PREPARATION. The activating influence of a benzene nucleus upon hydrogen of a side-chain makes possible the direct oxidation of toluene to benzoic acid. Other aryl acids, as phthalic acid, may be prepared in the same way from selected hydrocarbons.

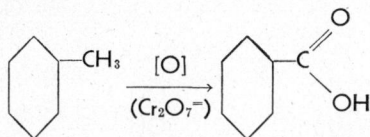

Benzoic acid may also be prepared by the hydrolysis of benzo-trichloride or the hydrolysis of benzonitrile. The Cannizzaro reaction

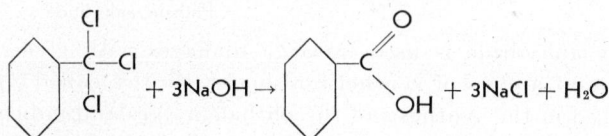

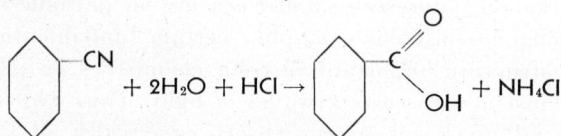

(p. 93) of an aldehyde also produces benzoic acid. Phenyl Grignard reagent adds carbon dioxide (dry ice) to form benzoic acid on hydrolysis.

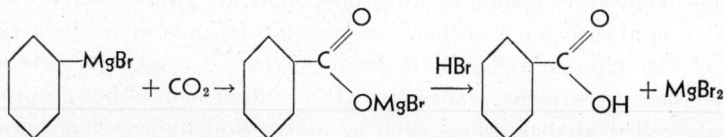

REACTIONS. Benzoic acid forms salts with sodium hydroxide, esters with alcohols, benzoyl chloride with PCl$_5$, benzamide with ammonia, benzanilide with aniline, and in general follows the reactions of aliphatic acids. Benzoic acid is a weaker acid than acetic acid. The sodium salt is used as a food preservative.

2. PHTHALIC ACIDS. There are three phthalic acids but the term "phthalic acid" is used for the ortho- acid because it is the only one to attain commercial importance.

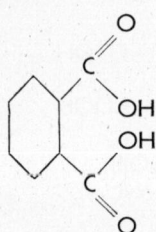

o-Phthalic acid   m-Phthalic acid   p-Phthalic acid

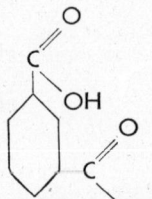

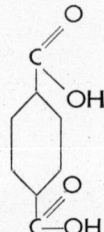

The phthalic acids may be prepared by the oxidation of the corresponding xylene. Commercial o-phthalic acid is made by the air oxidation of naphthalene at elevated temperatures in the presence of vanadium oxide and is dehydrated to produce phthalic anhydride.

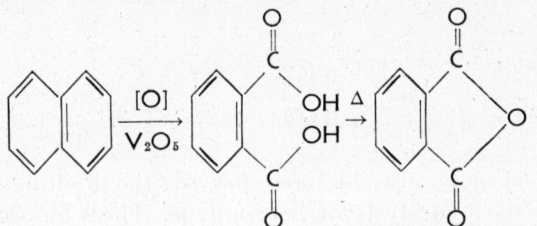

Phthalic anhydride

Phthalic anhydride is used in large tonnages with glycerol in the manufacture of a class of synthetic resins or plastics called "glyptals." It is also used in the synthesis of the phthalein dyes and indicators.

Phenolphthalein, fluorescein, and eosin are all phthalic acid derivatives. Their changes in color exemplify certain fundamental principles in molecular structure important in color chemistry. To produce color a substance must absorb certain waves of light. Thus white substances reflect light and black substances absorb practically all light. It has been found that many chemical groups possessing double bonds have selective absorption of various vibration bands of light. Groups having a double bond and capable of absorbing light are called chromophores. Variation in absorption and therefore of color depends upon the arrangement of the groups within a molecule. When conjugated systems of double bonds are present, colors are further intensified. Other groups and elements called auxochromes, such as amino and hydroxyl groups and the halogens, assist chromophores in intensifying color.

In the formula of phenolphthalein given below the red color of the basic solution is attributable to the conjugated double bonds of the quinone part of the molecule. Fluorescein and eosin form similar chromophoric arrangements.

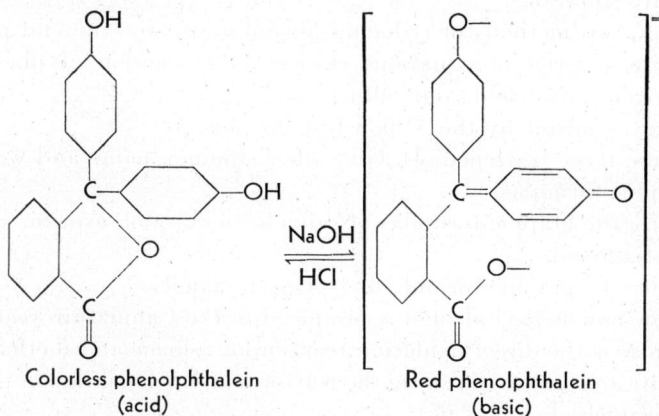

Colorless phenolphthalein
(acid)

Red phenolphthalein
(basic)

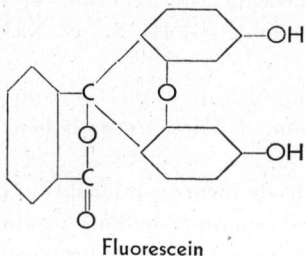

Fluorescein

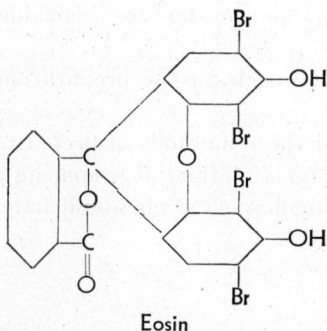

Eosin

Study and Review Questions

1. List the physical properties of phenol.
2. Contrast the Dow and Raschig methods of producing phenol.
3. Why is it impossible to produce sodium phenoxide by the reaction of phenol with sodium carbonate?
4. Show by an equation how to prepare anisol (methoxybenzene).

5. Show by an equation how to prepare phenyl ethanoate (acetate) from phenol and other chemicals of your choice.
6. Contrast the reaction of ethanol and phenol with PCl_3 or PCl_5. Write equations.
7. Show two methods of reducing phenol and name the end products.
8. Write a series of equations showing the reaction of phenol with halogen, nitric acid, and sulfuric acid.
9. What is meant by the Millon test for phenol?
10. Name three benzenediols. Give their common name and write their graphic formulas.
11. Write the graphic formulas of salicylic acid, salol, aspirin, and oil of wintergreen.
12. Write the graphic formula of 1-phenylethanol.
13. Show how benzyl alcohol is produced in the Cannizzaro reaction.
14. Contrast the direct oxidation reaction of toluene and methane.
15. Write an equation showing the Gatterman-Koch method of preparing benzaldehyde.
16. Write equations showing the reaction of benzaldehyde with (a) methyl Grignard reagent, (b) HCN, (c) $NaHSO_3$, (d) $HONH_2$, and (e) $H_2NNHC_6H_5$.
17. Give a test for distinguishing aliphatic from aromatic aldehydes.
18. Contrast the reaction of halogens with benzaldehyde and aliphatic aldehydes.
19. Show how benzaldehyde forms condensation products with ammonia.
20. Cinnamaldehyde and cinnamic acid are produced from benzaldehyde. Contrast the methods used for their preparation.
21. Show the condensation product of benzaldehyde with dimethyl-aniline.
22. Contrast one method each for the preparation of acetophenone and acetanilide.
23. Write equations for three methods of preparing benzoic acid.
24. Write an equation for a method of preparing phthalic anhydride.
25. Account for the color changes of phenolphthalein with a change in pH.

CHAPTER PROLOGUE

Heterocyclics are the last compounds to be studied in most organic chemistry textbooks but their importance is far from the least. More than one third of all known organic compounds are heterocyclic. Small packages of heterocyclic glucose rings carry the bargain names of starch and cellulose. Animal and plant organisms are profusely supplied with heterocyclics. Chlorophyll in plants and hemin in blood are complex structures with nitrogen in several rings. A roll call of heterocyclic compounds would reveal such distinguished company as most members of the vitamin and hormone families, antibiotics like penicillin, drugs like atabrine, caffeine, nicotine, and cocaine.

The presentation of heterocyclic compounds in this text will not emphasize many of the compounds cited above because most of them merit further consideration under the general heading of biological compounds. Presented in this chapter, however, and we believe for the first time, will be a new approach to the field of heterocyclics. Just as we can compare alkanes and cyclohydrocarbons, so also can we compare simple aliphatic compounds, with heterocyclics. The mere fact that an oxygen or a nitrogen has tied itself into a knot should hardly deprive it of recognition from the family of its earlier possible origin. We hope this method of presentation is helpful to the student and useful to the teacher.

In Chapter I (p. 11) the divisions of organic chemistry were listed. The final division was that dealing with cyclic compounds wherein one or more atoms of elements other than carbon are present in ring structures. This division represents the most complex of organic compounds. However, keeping in mind the relationship of every class of compound as we have studied it—the existence of compounds containing atoms other than carbon in a ring structure does not add many complications. In much the same manner that aromatic compounds differ from aliphatic—so the existence of oxygen, sulfur, and nitrogen (the three elements most frequently forming heterocyclic compounds) in the ring causes slight differences in the physical and chemical properties of the compound formed from those of the true carbocyclic compound. Also, just as the five- and six-membered carbocyclic structures are the most stable and the most common, so the five- and six-membered heterocyclic compounds are the most stable and the most important. Also, we have observed that certain of these compounds are very closely related to some corresponding aliphatic compounds: Epoxyethane (Ethylene oxide)

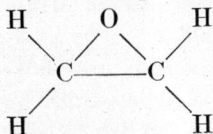

$$\begin{array}{ccc} H & O & H \\ \diagdown & \diagup \diagdown & \diagup \\ & C\!\!-\!\!-\!\!-\!\!C \\ \diagup & & \diagdown \\ H & & H \end{array}$$

(p. 80) may be considered as an internal ether;

succinic anhydride

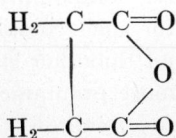

$$\begin{array}{c} H_2\!\!-\!\!C\!\!-\!\!C\!\!=\!\!O \\ | \qquad\qquad \diagdown \\ \qquad\qquad O \\ | \qquad\qquad \diagup \\ H_2\!\!-\!\!C\!\!-\!\!C\!\!=\!\!O \end{array}$$

(p. 107) reacts to produce succinic

acid derivatives, the same as the acid itself; the basic ring structure of succinic anhydride is repeated in the very important compound phthalic

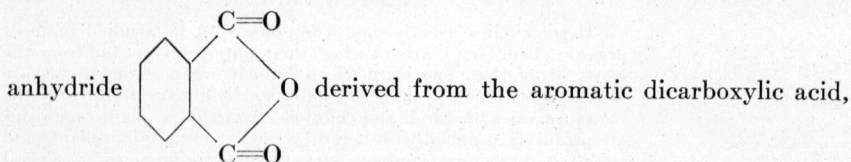

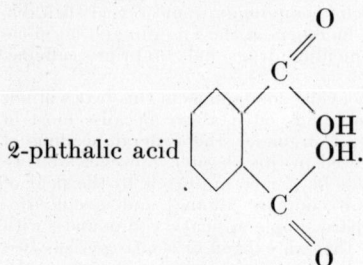

anhydride   derived from the aromatic dicarboxylic acid,

2-phthalic acid

It follows then that:

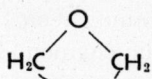

is an ether

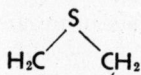

is a thioether

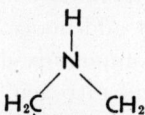

is a secondary amine

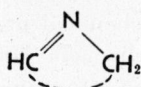

is related to tertiary amines

in which the ether, thioether, and amino properties are slightly influenced by ring structure. Chemically the five- and six-membered rings are more stable and less reactive than the aliphatic compounds of the same class. A three-membered ring, however, is easily cleaved, because of strain (p. 175), and hence is more reactive.

Properties. The more important heterocyclic compounds are those possessing unsaturated linkages corresponding to benzene. Both in physical and chemical properties, these compounds are similar to aromatic compounds. The opportunity for such complete resonance as in benzene does not exist in the ether and thioether structures. Hence the tendency for cleavage of the ring is more pronounced. For purposes of quantitative comparison the resonance energy (in kilocalories) for benzene is 39. The value for thiophene (p. 239) and pyrrole (p. 241) is 31, and for furan (p. 241) is but 23.

On the other hand pyridine, because it contains a double-bonded, —C=N—, linkage and because the nitrogen carries an unshared electron pair, permits (1) resonance in the same manner as benzene and (2) addi-

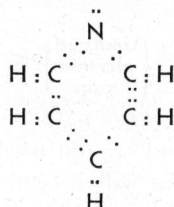

tion of positive groups on the nitrogen atom. The reactions of pyridine are therefore quite diverse and the ring is difficult to cleave. In fact, the following example shows that upon oxidation, the benzene ring cleaves before the pyridine ring. The resonance energy of pyridine is 43 kilocalories.

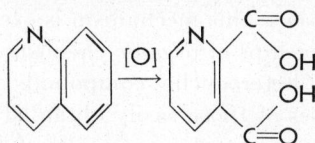

As would be expected, the carbon atoms in heterocyclic ring structures behave in practically the same manner that they do in the similar aromatic or alicyclic compounds. Those closest to the oxygen or nitrogen are always influenced in their reactivity by the foreign element. Side-chains attached to heterocyclic aromatic nuclei react like side-chains on carbocyclic aromatic compounds. Accordingly the differences to be observed in these compounds are dependent, in the main, upon the nature of the bonding to the element other than carbon and to the electron arrangement.

In combinations of carbon atoms, each bond consists of a pair of shared electrons and double bonds represent two pairs of shared electrons:

In rings containing N, O, and S, one or more sets of electron pairs are

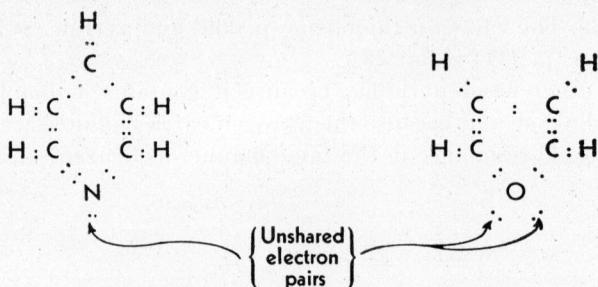

associated with these atoms and are not used in the bonding mechanism. The resonance within the ring will be influenced only slightly by the sharing of these electronic pairs with, for example, hydrogen ion. Hence there is a relative influence upon (1) cleavage of the ring by another reactant (p. 80) and (2) addition of, for example, HCl to produce an ionic compound (p. 146) and to induce water solubility. The structure of nitrogen-containing compounds is particularly favorable for the second type of reaction which, in the example cited, indicates its character as a proton acceptor, i.e. a base. This mechanism is exceedingly important in the physiological relationships of organic chemistry.

The importance of heterocyclic compounds is largely dependent upon the commercial uses of those easily obtained from natural sources, and upon those whose structural features are found in more complex materials of physiological importance. The organic chemist has learned to synthesize many of the required heterocyclic groups as steps in the synthesis of vitamins and drugs.

I. Heterocyclic Compounds of Oxygen

Preparation. In previous chapters several compounds in which oxygen exists in a ring structure have been presented.

1. 2-Chloroethanol condenses in the presence of sodium hydroxide to form epoxyethane (ethylene oxide) (p. 80).

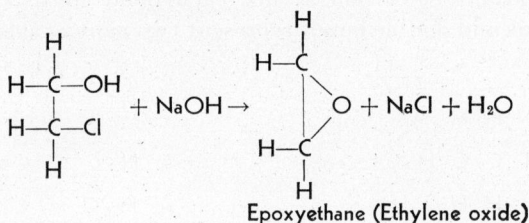

Epoxyethane (Ethylene oxide)

2. The strain in a three-membered ring is relatively great so ethylene oxide polymerizes in an acid medium to form dioxane (p. 80), a six-membered ring.

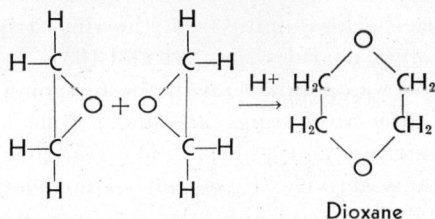

Dioxane

3. Butanedial, merely upon being heated with water, forms the important five-membered ring compound furan (p. 241). This reaction probably occurs so readily because of the reactivity of the enolic tautomer (p. 86). The formation of furfural from any pentose illustrates a similar

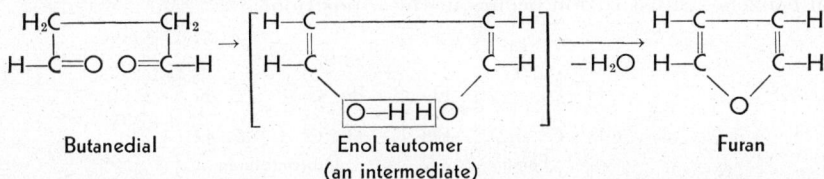

Butanedial   Enol tautomer   Furan
(an intermediate)

structural change. Furfural for commercial use is prepared in this manner

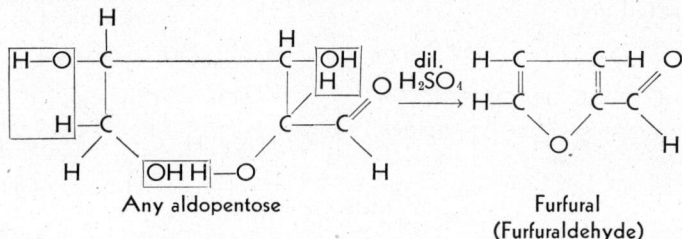

Any aldopentose   Furfural
(Furfuraldehyde)

from pentoses in oat and rice hulls. Furfuraldehyde is extensively used as a solvent, an insecticide, a fungicide, an intermediate in synthesis, and a vulcanization accelerator.

4. Pentoses and hexoses are simple monosaccharides existing as tautomeric open-chain and ring structures (pp. 156 to 159).

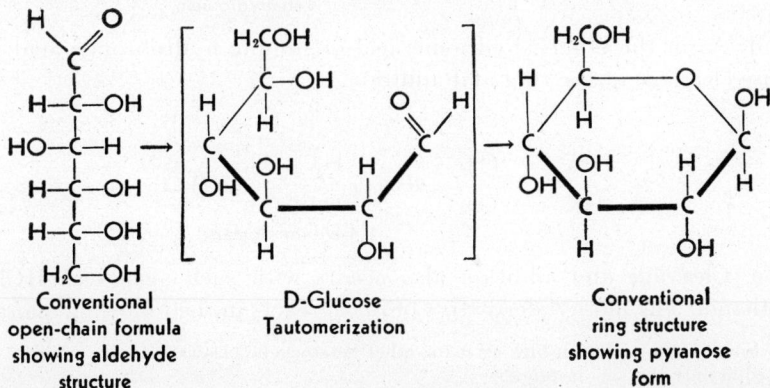

Conventional
open-chain formula
showing aldehyde
structure

D-Glucose
Tautomerization

Conventional
ring structure
showing pyranose
form

Simple monosaccharides unite from the ring structure to form disaccharides and polysaccharides (pp. 161 and 163).

Reactions. 1. As was pointed out in the foregoing section, ethylene oxide and other three-atom rings are under considerable strain and readily undergo cleavage and addition. The reactions shown on p. 237 further illustrate this activity. Five- and six-membered rings are relatively more difficult to cleave, but the reactions which do occur are similar.

2. The properties of heterocyclic compounds in general (p. 234) are readily illustrated by the reactions of furan.[1]

a. Bromination results in the formation of substitution products, as in benzene. Substitution occurs in the α-position.

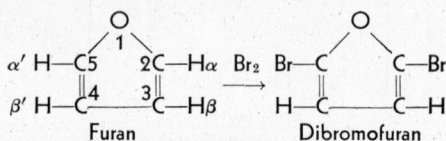

Furan   Dibromofuran

b. Oxidation causes cleavage and the formation of a dicarboxylic unsaturated acid.

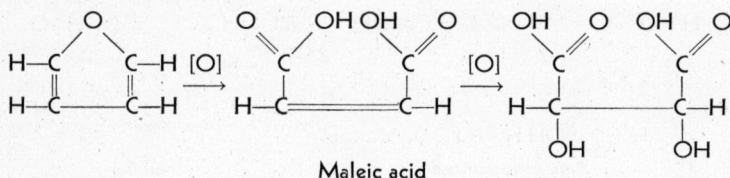

Maleic acid

c. Hydrogenation using palladous oxide as a catalyst produces a cyclic ether.

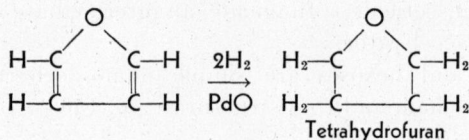

Tetrahydrofuran

d. As in the ethers, hydriodic acid or fuming hydrobromic acid will cause cleavage of the ring and addition.

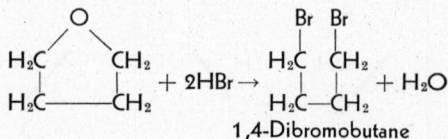

1,4-Dibromobutane

e. Cleavage and addition also occurs with such agents as HCl in methanol. The methyl derivative of furan reacts in the following manner:

[1] Since similar relationships exist for other heterocyclic classes, reactions will be presented in only this one instance.

$$H_3C-O-\underset{\underset{CH_2}{|}}{\overset{\overset{CH_3}{|}}{C}}-H \quad \underset{\underset{CH_2}{}}{\overset{\overset{O}{||}}{C}}-CH_3$$

Methylfuran + 2CH₃OH $\xrightarrow{HCl}$ 1,1-Dimethoxy-4-pentanone

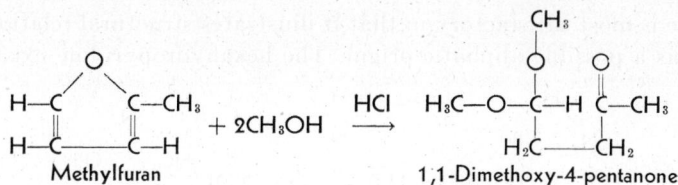

II. Heterocyclic Compounds with Sulfur

The sulfur analog of furan is thiophene. There are several methods of synthesizing this compound. One is by heating a mixture of sodium succinate and phosphorus trisulfide.

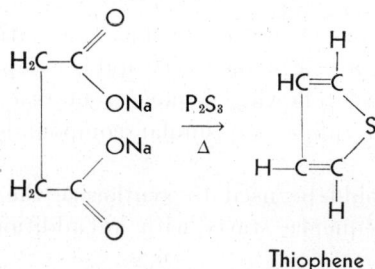

Thiophene

Since the reactions of thiophene are similar to those of furan, and since the more important sulfur compounds also contain nitrogen in the ring, further presentation will be given in a later part of this chapter.

III. Heterocyclic Compounds with Nitrogen

The heterocyclic compounds with nitrogen offer an interesting contrast with those containing oxygen and sulfur. In the first place, combinations with —O— and —S— represent a divalent linkage; those with nitrogen are trivalent or pentavalent presenting the possibility for structural combinations:

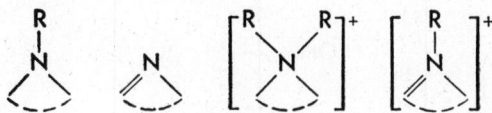

Unshared electron pairs in these heterocyclic compounds are favorable for cleavage reactions and for coordinate bonding (p. 236).

The simplest and the most important heterocyclic compound of nitrogen is pyridine, an evil-smelling aromatic substance with boiling point of 115°C. It is miscible with water in all proportions. This compound will accordingly be used to illustrate the preparation and reactions of the group.

Preparation. 1. The commercial source of pyridine and its homologues is by separation from coal tar (p. 178).

2. There are many methods for synthesizing pyridine. The following

sequence is most satisfactory in that it illustrates structural relationships as well as a possible aliphatic origin. The hexahydropyridine exists as a

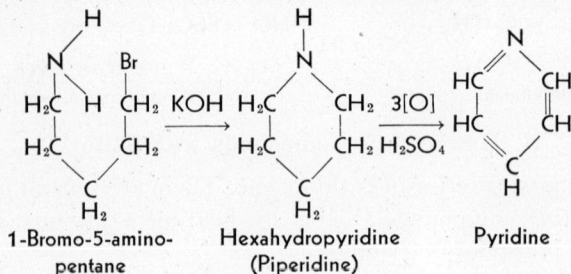

1-Bromo-5-amino-
pentane   Hexahydropyridine   Pyridine
(Piperidine)

"puckered" ring (p. 174). Oxidation produces aromatic structure capable of resonance. The result is a planar, very stable compound.

Reactions. The two following sequences of reactions illustrate the typical behavior of pyridine and similar compounds in the formation of their derivatives.

1. As pyridine would be used to synthesize the vitamin, nicotinic acid (p. 271). The sequence starts with an addition reaction and an ionic compound.

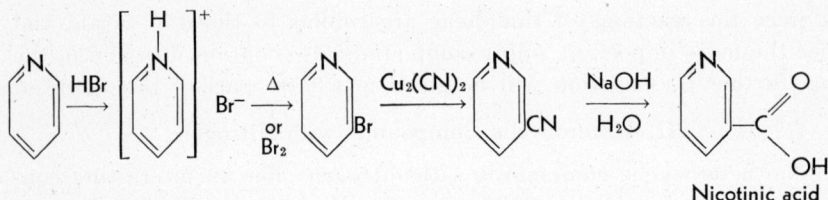

Nicotinic acid

2. The sequence which starts with the addition of a nonionic compound and yields a pyridone.

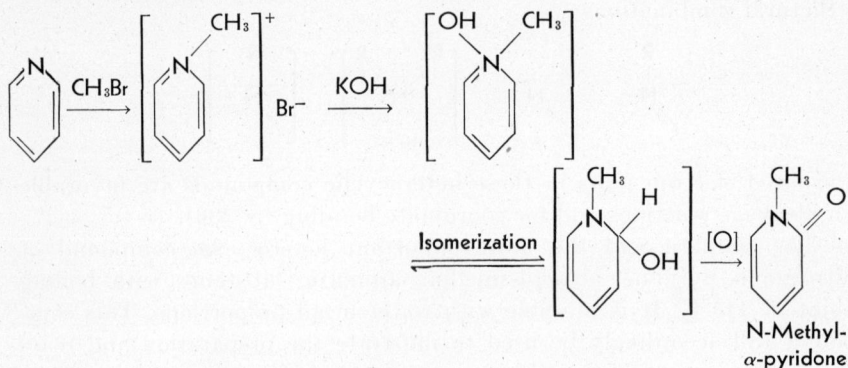

N-Methyl-
α-pyridone

IV. Important Heterocyclic Compounds

Just as there are heterocyclic compounds of five- and six-membered rings, containing one atom other than carbon, so there are structures that contain more than one atom other than carbon. Dioxane is shown on p. 237. A list, by no means complete, of the more important nuclear rings will illustrate heterocyclic structures. Numbers and Greek letters are indicated in order to show how substituted groups are designated.

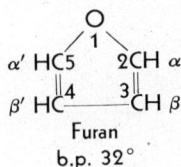

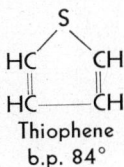

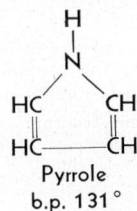

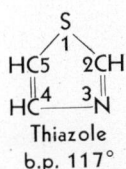

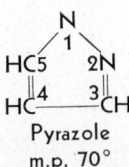

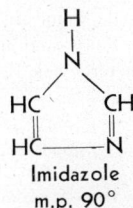

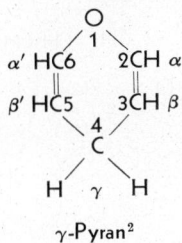

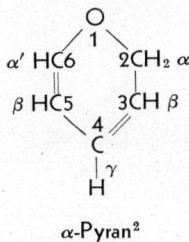

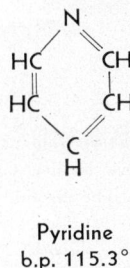

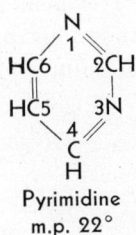

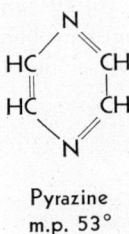

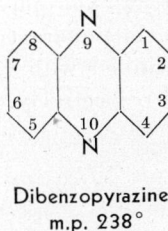

Furan
b.p. 32°

Thiophene
b.p. 84°

Pyrrole
b.p. 131°

Thiazole
b.p. 117°

Pyrazole
m.p. 70°

Imidazole
m.p. 90°

γ-Pyran²

α-Pyran²

Pyridine
b.p. 115.3°

Pyrimidine
m.p. 22°

Pyrazine
m.p. 53°

Dibenzopyrazine
m.p. 238°

The foregoing basic ring structures appear again and again in compounds of vital importance. Many form parts of complex materials as yet unanalyzed by man. It has been said that about one third of all known organic compounds and one half of the products of living organisms are heterocyclic in nature. A few examples from selected fields will serve as illustrations.

² γ- and α-pyran are unstable but their stable derivatives are known.

Penicillin g, the antibiotic, has a thiazol ring

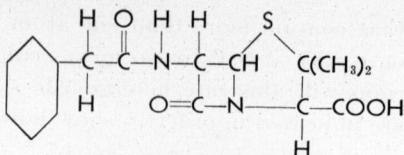

Atabrine, the antimalarial prepared in large quantities for the armed forces, has dibenzo-pyridine structure

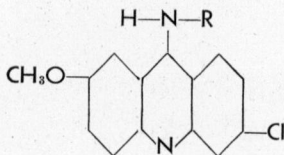

Uric acid, found in small quantities in the body fluids and urine of animals, contains two heterocyclic rings

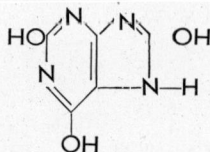

Vitamin B$_1$ contains a thiazole and a pyrimidine ring

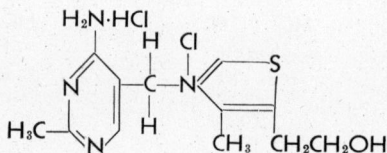

Many heterocyclic compounds are colored. The dye Indigo, C. I 1177, is such a substance. The insoluble ketoform is shown

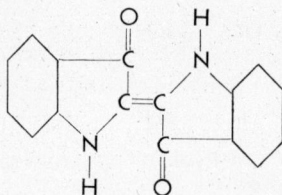

Green chlorophyll of plants (p. 10) and red Hemin I of hemoglobin of the blood are complex structures believed to contain four pyrrole rings linked with carbon and cross bonded with magnesium and iron atoms respectively.

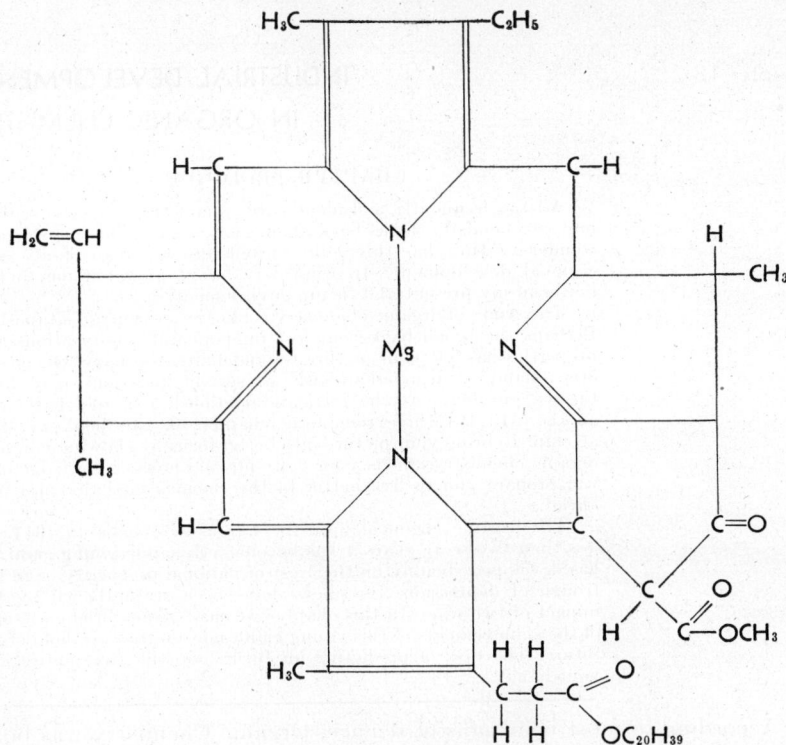

Chlorophyll (a)

Study and Review Questions

1. Show by an equation how dioxane is formed.
2. Show by an equation how trioxane is formed from methanal.
3. Explain the formation of furan from butanedial.
4. Why can *any* pentose and acid be used in making furfural?
5. Write the graphic formula of furan and pyran and compare them with the furanose and pyranose formulas of glucose.
6. Show how a heterocyclic ring such as succinic anhydride may be cleaved by water.
7. Show with graphic formulas the equation for the cleavage of furan by HI.
8. Compare the graphic formulas of furan, thiophene, and pyrrole.
9. How is thiophene synthesized?
10. Write a series of equations showing how pyridine may be synthesized?
11. In two sequences of reactions show typical behavior reactions of pyridine.
12. Name six compounds containing heterocyclic structure, identifying the chief heterocyclic ring in each.

Chapter 18

INDUSTRIAL DEVELOPMENTS
IN ORGANIC CHEMISTRY

CHAPTER PROLOGUE

All too frequently a student, after completing a course such as organic chemistry, looks back at his work during the semester and wonders: "How has this course fitted into the plan of my educational development? In what way has it prepared me to live better in my present and future environment?"

The study of organic chemistry is like the weaving of a tapestry. Perhaps the broad background of fundamental classifications and properties has left your picture too indefinite, too hazy. Or, on the other hand, your overemphasis of specific reactions may have thrown out of focus the intricate relationship of one part with another. In the three remaining chapters of this book we shall attempt to bring your picture into better focus by showing you how organic chemistry affects your daily life and how a knowledge of it will prepare you to live better in the chemical age that lies just ahead.

The organic chemical industry was a mere baby in 1914. In less than two score years it has become a dynamic young giant. It has developed steadily on the crest of national prosperity or in the trough of depression. Its sinews have been strengthened by the impact of two wars. In this chapter we shall glance briefly at some of the achievements of this young giant, now on the threshold of his future. The task of predicting his future we will leave entirely in your hands.

According to most historians of science, Organic Chemistry was born in 1828 when Woehler first synthesized urea from inorganic materials. The first productive effort of Organic Chemistry dates from the establishment of the dye industry following Perkin's discovery of mauve in 1856. During World War I the overnight conversion of Germany's dye plants into factories to feed the machines of destruction announced that the science of Organic Chemistry had acquired a new stature and importance in the life of mankind.

The Social-Economics of Organic Chemistry

During the past twenty-five years in the United States the productive efforts of industrial chemists in the organic field have revealed a new young giant among the industries, a giant capable of enriching peacetime living, and an aggressive powerful defender in the days of war. New, cheap, and in most cases vastly superior products have emerged from research laboratories to add comfort to daily living. Raw materials wasted by industry at the turn of this century are now the stock piles of profitable enterprises.

In our country, we are gradually acquiring a new independence in our manner of living. No longer are we dependent upon the silk worm of the Orient, the rubber trees of Asia, or the dyes from Germany. From our own factories emerges a kaleidoscopic array of man-created articles: colorful,

synthetic fabrics for fine garments; shockless transportation and oil resistant coatings aided by varieties of rubber not found in nature; light plastics
to serve an unlimited variety of tasks; and drugs that were undreamed of
40 years ago relieve our pain, cure our diseases, and strengthen our bodies.
Since these things are new to man, he has found his pattern of living altered. The older members of society have readily adopted these new things
because living becomes easier and more diversified; the younger generation has accepted them as a matter of course.

Industry has perhaps felt the press of the encroaching Chemical Century more strikingly than the individual. Century old industries have
toppled to be replaced by the creations of the young giant. The influence
upon world economics has been dramatic and, in a sense, ruthless—in a
few decades the synthetic dye industry "wiped out the ancient madder-
plant industry of France, the still more ancient indigo industry of India,
and the Tyrian purple coming down from immemorial time."[1] The production of oil has caused mass shifts in population and the expenditure of
fortunes which in turn brought unbelievable wealth to many—yes, and
such a scramble of nations for potential oil fields that wars have followed.
The force of chemical industry has crept into the heart of all other industry becoming the center of an economy in which new industries are
created, and decadent industries are revived. In the past decade chemical
plants have more than doubled in size. Production has averaged more
than two and one-half times the 1935–39 average, compared with an average of less than two for all industry. Plastics, practically unheard of 30
years ago, are now produced in the amazing amount of one and one-half
billion pounds a year—more in weight than any nonferrous metal. The
sales of the biggest companies in chemical industry are astronomical:
DuPont, the leader and now the largest company in the world is selling
goods for over a billion dollars a year; Union Carbide and Carbon achieved
sales of $631,600,000.00 in 1948; eighteen other chemical companies in
that same year each showed sales of over 25 million dollars. The amazing
thing in this dollar-picture is that some 40 per cent of these sales were for
products undeveloped, and in some cases not even known, only 15 years
ago.

This industrial picture has been summarized in *The Chemical Century*[2]
as follows: "To understand U.S. industry in the second half of the twentieth century, you must understand the chemical industry. It is not too
easy, for this is the area in which industry disappears into pipes, and
becomes hard to follow. Yet chemistry is the exemplary industry of the
age. Its intricate processes, working silently night and day in plants of
great geometric form, convert some of the most common materials of
earth into an array of higher products that daily grows more prodigious.

[1] "The Chemical Century," *Fortune*, March 1950, p. 71.
[2] Idem, p. 69.

It is almost wholly an industry founded, built, and run by chemists and engineers, men trained in the sciences, who have learned to live with ceaseless change, regularly plowing a large portion of income into research, and as regularly bringing forth new processes, new products, and revolutionary growth."

Market for Organic Chemicals

The market for organic chemicals is based upon the necessity of food, clothing, and shelter for man, upon his desire to enjoy good health, to move conveniently and quickly from place to place, for pleasure and business, to communicate with his fellow men in the interchange of ideas and to enjoy a richer, fuller life.

Food. The Organic Chemical industry has contributed widely and wisely to man's basic need for food. Soils are fertilized and fumigated and parasitic insects, fungi, and weeds are destroyed by organic chemicals produced cheaply within the industry. New chemicals have been produced for cooking, processing, and preserving foods. Today these foods are cleanly packaged in organic chemical materials, flavored with synthetic additives, and fortified with healthful synthetic vitamins.

Clothing. New synthetic fibers, new methods of processing older fibers, and easier and safer methods of cleaning clothing offer new markets for organic chemicals. Our shoes and clothing today are better than at any time in man's existence because better organic chemicals are available to our industries.

Shelter. The modern home is an ever-broadening market for organic chemicals. Roofs, walls, and floors are covered by organic materials. Wood is preserved and protected against fire. Surface coating and plastic gadgets, new materials for upholstery and drapery, insulation materials, floor waxes and deodorants all claim a large volume of organic chemicals.

Medicine. Another broadening market for organic chemicals is in the field of medicine. In average retail value per pound, this market is the most lucrative in the entire organic chemical industry. The use of antiseptics, germicides, and insecticides has become routine in most households. Anesthetics, antibiotics, and hormones are now familiar terms and vitamins are not only procurable in any drug store for individual use, they are also used in quantity by the food processing industry for the reenrichment of products.

Transportation and Communication. The largest market for organic chemicals, however, is in the field of transportation and communication. Transportation on land or sea or in the air is served by organic fuels, containing antiknock chemicals. Lubricating oils of ever-increasing efficiency keep the wheels of transportation rolling. Antifreeze and synthetic engine coolants permit the use of internal combustion motors at opposite extremes of temperature. Hydraulic fluids of exacting specifications serve modern transportation in diverse ways. Automobiles require large quantities of synthetic rubber, not only for tires but throughout each car. Airplanes and automobiles use plastics, synthetic upholstering materials, and surface coatings. Their use of synthetic insulating materials for electrical parts is extremely large and increasing from year to year.

Other Uses. It would be impossible to list with any degree of completeness the various markets of organic chemicals. It will be sufficient to mention only a few large consumers of organic materials: the tobacco industry for processing tobacco, the photography industry for films and photographic chemicals, the cosmetic and perfume industry, the plastic industry, the telephone and radio industries, and the explosive industry.

Raw Materials of the Organic Chemical Industry

Raw materials for the organic chemical industry come chiefly from six sources, coal, petroleum, cellulose, natural gas, carbohydrates, and fats or oils. Nearly five million tons of synthetic organic chemicals are produced annually from these raw materials.

Coal. At the mid-point of the century coal remains the leading producer of organic chemicals but it is probable that when production records for 1950 are completely available, petroleum chemicals will be in first place. Coke for metallurgical processes is still the chief product from coal. Fifty years ago other by-products of coke-making were wasted. Today benzene from coal leads the long list of primary products of the organic chemical industry. From benzene are prepared (1) styrene for a large segment of our rubber production and polystyrene plastics, (2) phenol for plastic and other uses, (3) aniline for dyes, (4) alkyl-benzene for some types of detergents, and (5) adipic acid for the production of nylon. In addition to benzene, coal distillation also produces huge quantities of toluene, xylene, naphthalene, cresol, phenol, and kindred compounds. Coke also is an important source of organic materials. When combined with sulfur as carbon bisulfide, coke serves as a raw material for producing viscose rayon, cellophane, and freon refrigerants. When coke forms calcium carbide, it serves as a raw material (1) for acetylene in the production of chloroprene, acetaldehyde, and the host of vinyl products, or (2) for calcium cyanamide in the production of melamine plastics and sodium cyanide. Heated coke with water yields water gas, a mixture of carbon monoxide and hydrogen which may be further hydrogenated by the Fischer-Tropsch method to form Kogasin oil and its many aliphatic derivatives.

Petroleum. In second place as a source of raw materials for organic chemicals is petroleum. It is at present our cheapest source of alkanes and ethylene and their important derivatives (p. 25). Ethylene especially has become an important source of such basic materials as ethyl alcohol, acetaldehyde, acetic acid, polyethylene plastics, glycols, cellosolves, tetraethyl lead, ethylene dibromide, and vinyl compounds.

Cellulose. Third place as a source of raw materials for organic chemicals belongs to cellulose. Viscose rayon and cellophane, plastics and film are the chief chemical products of cellulose.

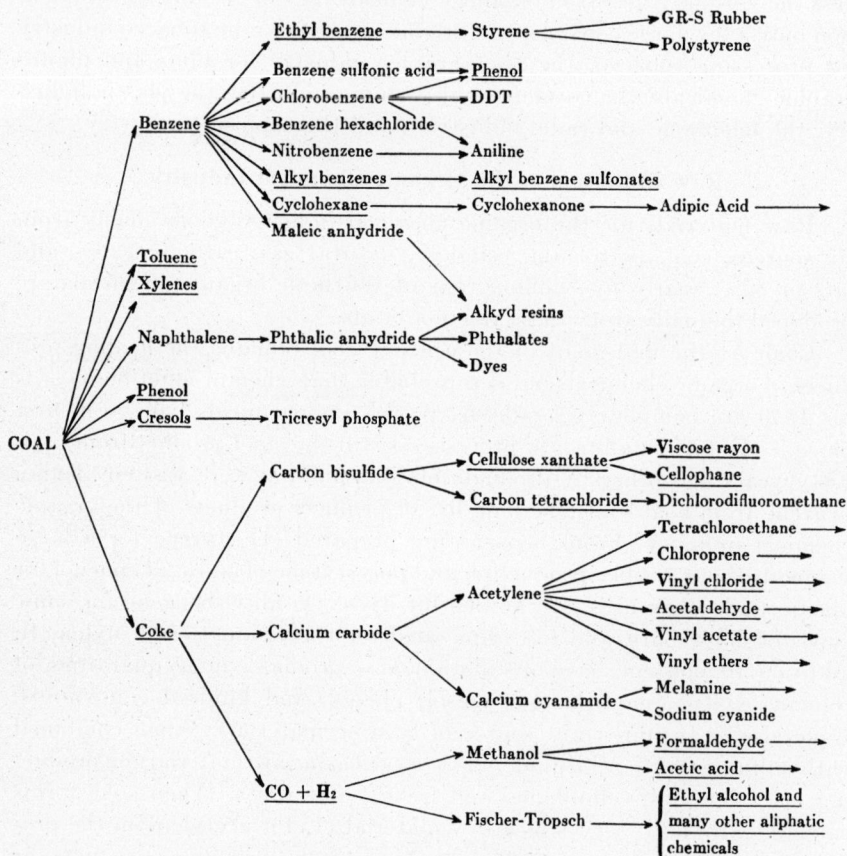

Fig. 35. Organic chemicals from coal (over 50,000,000 lbs. per year). (Courtesy of the Vortex and Dr. Raymond H. Ewell, Stanford Research Institute.)

Natural Gas. Natural gas ranks fourth as a source of organic chemicals, and its future as an increasing source of chemicals is most promising. From natural gas propylene, butylene, isobutylene, and cycloalkanes are readily obtained. A new method of preparing acetylene by the dehydrogenation of methane is expected to reduce the price of acetylene considerably, thereby restoring its competitive value as a primary product.

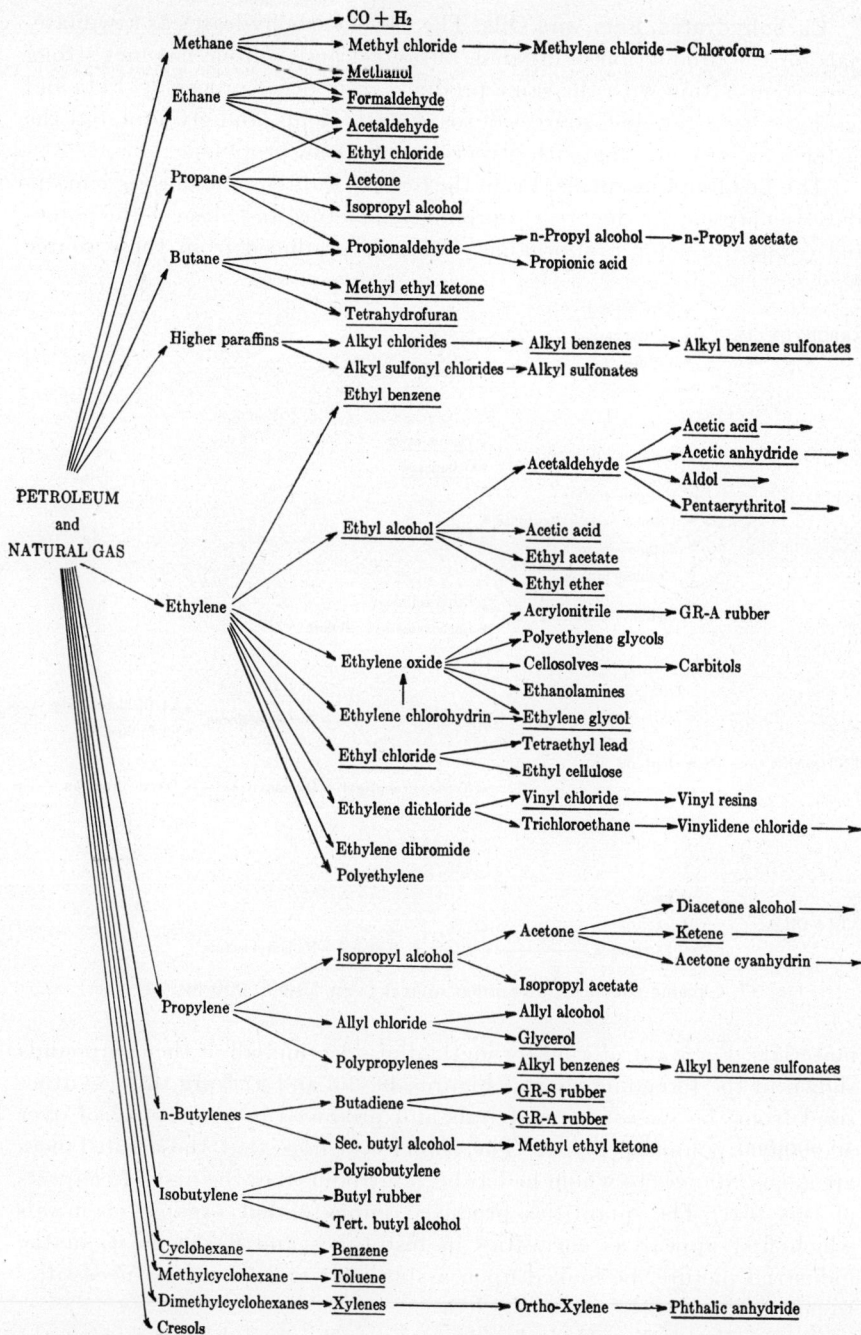

Fig. 36. Organic chemicals from petroleum and natural gas (over 50,000,000 lbs. per year).

Carbohydrates, Fats, and Oils. The use of carbohydrates as raw materials for the production of ethanol, butanol-1 and acetone has met strong price competition with the same products from other processes. Fats and oils have been our chief sources of soaps, fatty acids, and glycerol, but the latter now is competing with glycerol made from propylene.

The Leading Chemicals. From the foregoing discussion of source materials we may survey the present production picture and observe the potential competition for raw organic chemicals produced from these source

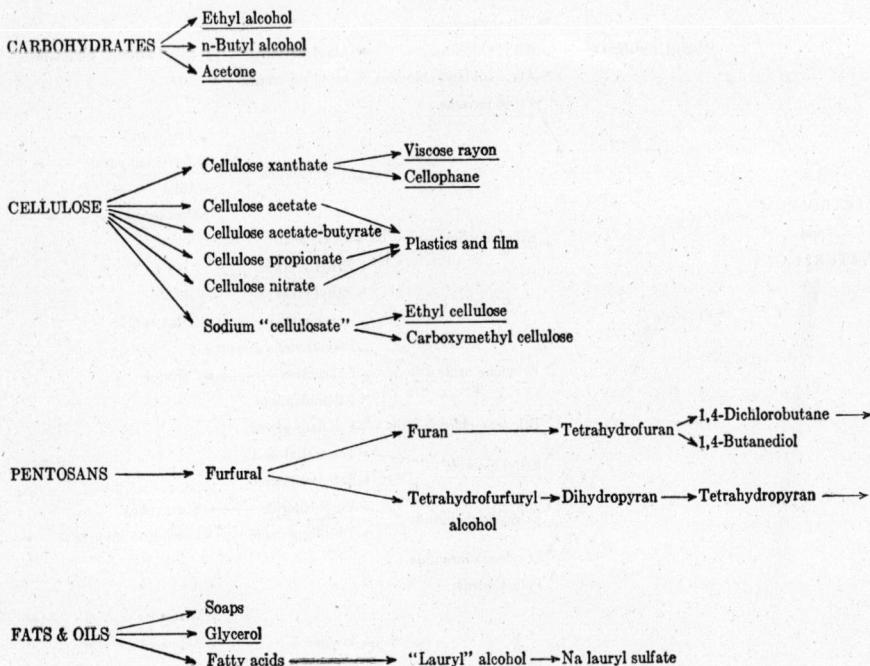

CARBOHYDRATES
- Ethyl alcohol
- n-Butyl alcohol
- Acetone

CELLULOSE
- Cellulose xanthate
  - Viscose rayon
  - Cellophane
- Cellulose acetate
- Cellulose acetate-butyrate
- Cellulose propionate
- Cellulose nitrate
  - Plastics and film
- Sodium "cellulosate"
  - Ethyl cellulose
  - Carboxymethyl cellulose

PENTOSANS ⟶ Furfural
- Furan ⟶ Tetrahydrofuran
  - 1,4-Dichlorobutane ⟶
  - 1,4-Butanediol
- Tetrahydrofurfuryl alcohol ⟶ Dihydropyran ⟶ Tetrahydropyran ⟶

FATS & OILS
- Soaps
- Glycerol
- Fatty acids ⟶ "Lauryl" alcohol ⟶ Na lauryl sulfate

Fig. 37. Organic chemicals from other sources (over 50,000,000 lbs. per year).

materials. For want of a better method of selection, all of the compounds shown in the foregoing charts (Figures 35, 36, and 37)[3] are those synthesized from the six source materials and produced in a quantity of over 50,000,000 pounds per year. The reader will note that these substances are acquaintances to whom he has been introduced in the first 17 chapters of this text. The quantities produced indicate that organic chemicals which first appear as curiosities in test tubes and flasks must, in the industrial picture, be looked upon as tank-car and train-load necessities to our modern world.

Since World War II, it has been generally agreed that each of the

[3] These charts are used by courtesy of Dr. Raymond H. Ewell, Stanford Research Institute, and the *Vortex*, Mar. 1949.

four primary products—ethylene, propylene, butylene, and acetylene—
can be utilized efficiently as a source of practically all organic chemicals.
In the competition for markets in the future all four materials will be
used but the greater markets will fall to those materials whose processes
of synthesis are carried out at a lower cost.

Economic Importance of Organic Chemical Industry. The organic
chemical industry in the United States has been since its inception a
highly competitive industry.

Naturally there has been a competition for the same market among
products chemically dissimilar. There has also been another type of competition. Not only has there been competition between different raw
materials to produce the same products by similar processes, there also
has been competition in the use of the same raw material to produce the
same product by using competitive processes, and competition to produce
the same product by using different raw materials in different processes.

The encouraging note about the organic chemical industry is its rapid,
ever-increasing volume of products and sales. As compared with other
fundamental industries in our national economy, its sustained rate of
increase is phenomenal. Even during the depression years, the dollar
volume of sales showed only slight downward deviation as compared with
the financial loss in sales of other industries.

Specific Industrial Products. Many new organic chemical products
have been marketed during the past twenty-five years. Obviously only a
few specific industries can be discussed in a short text. The four industries
selected for consideration cover the production of dyes, plastics, fibers,
and rubber. They have been chosen because each of them has initiated a
new era in organic chemistry, and each has continued to develop during
the ensuing years. Students must remember that all organic chemical
industries overlap, the product of one process enriching the others. As an
example, the synthetic fiber and the plastic industries are indebted to the
dye industry for the color appeal of their products.

Organic Chemistry of Color

The Dye Industry. We have seen that Perkin's discovery of mauve in
1856 started industrial organic chemistry. Sir William Henry Perkin, then
a lad of eighteen, was endeavoring to prepare artificial quinine. He was
oxidizing some aniline oil and one of those co-called accidents of science
occurred—Perkin's mixture turned black and tarry. Disappointment soon
disappeared, however, when a beautiful purple solution formed as he was
washing his apparatus with alcohol. Another phase of the "accident" lay
in the fact that his aniline was impure and a little toluidine was also present; had it been pure, the dye "mauve" would not have been formed. A
year later, Perkin was in the synthetic dye business. His mauve was
bringing the same price as platinum. This is the story of the start of the

aniline dye industry. Aniline colors provide infinite variety, one of their chief advantages.

By 1914, Germany was the major producer of dyes and most of the world was dependent upon her exports. Over night when World War I started, Germany's gigantic dye plants were converted to munitions factories illustrating the strategic part played by an established chemical industry in national defense. By the close of the war a new dye industry in the United States had made America practically independent of foreign dye sources. These same industries were converted to the manufacture of war materials for use during World War II. The latest production figures for American dyestuffs stands at 201,376,000 lbs. for 1948.

There are several main classes of dyes.[4] They are often classified in industry according to their method of use as acid, basic or tannin, chrome, developed, vat, mordant, direct, and food dyes. For our purpose it will be more interesting to consider some specific examples classified in relation to chemical structure. In examining the formulas which follow, the student should note: (1) the structural group providing the basis of classification, (2) the chromophore groups responsible for color (p. 231). The principal

chromophoric groups are $\diagup C{=}C\diagdown$, $={=}C{=}O$, $\diagup C{=}N{-}$, $-N{=}N{-}$,

$-N{=}N{-}O{-}$, $-NO_2$ and $\diagup C{=}S$. (3) The auxochrome groups which

serve both to intensify the color and to enable the colored compound to become fixed to the article being dyed. The principal auxochromes are $-N(CH_3)_2$, $-OH$, $-NH_2$, and to a lesser degree, $-OCH_3$, $-NH{\cdot}CH_3$, $-NH{\cdot}OH$, $-NH{\cdot}NH_2$ and the halogens.

Aniline Dyes. Many dyes are obtained from aniline. One of the two comprising Perkin's mauve is now known as pseudomauveine.

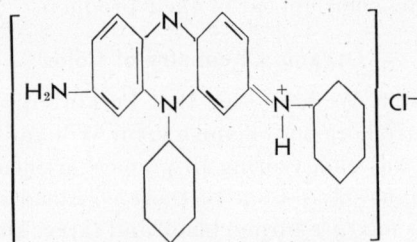

Pseudomauveine, a violet dye

Rosaniline of fuchsin is a red, aniline-derivative from the class called triphenylmethane dyes. Crystal violet is also a triphenylmethane dye.

[4] Other references to dyes and related compounds in this text are found on pp. 200, 208, and 341.

Dimethylaniline is the parent compound. The methyl groups shift the absorption characteristics so that the color is violet.

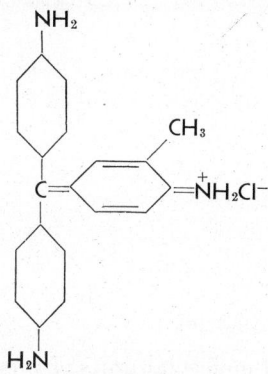

Rosaniline (red)   Crystal violet (violet)

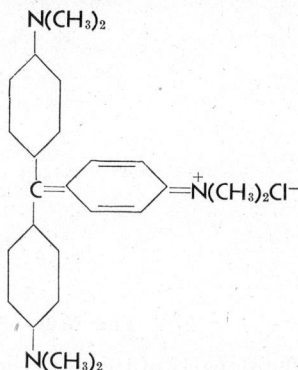

The Phthaleins. Baeyer is credited with the discovery of the first phthalein dye in 1871. They are formed by the action of phthalic anhydride on phenols. Phenolphthalein (p. 230) the common indicator, since it lacks an auxochrome, is not a dye; eosine (p. 231), its close relative with four bromine atoms, is a true dye.

Azo Dyes. Among the most important of all dyestuffs are those possessing azo structure, ($-N=N-$). The process of diazotization (formation of azo structure) was discussed on p. 211. Amines and phenols couple readily with azo compounds. Two interesting members of this class are congo red and butter yellow. Congo red (p. 212) is interesting particularly since it may be used successfully as a red dye, as a biological stain, and as a pH indicator in the laboratory.

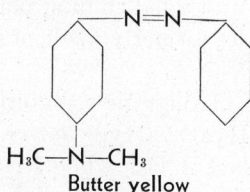

Butter yellow

Ethylenic Dyes. The ethylenic dyes are not of so great importance in industry as they are in biology. Reference to vitamin A (p. 175) derived from carotenoid pigments will show that these natural colored agents contain conjugated ethylenic double-bonded systems. Colors may range from yellow to bronze to red. Gold fish and carrots owe their color to these pigments.

Other Dyes. There are more than 20 classes of dyes, pigments, stains, and indicators based on structure. Three of these classes, -azo (44%), indigoid (16%), and anthroquinone (14%) account for three-fourths of all dyes produced.

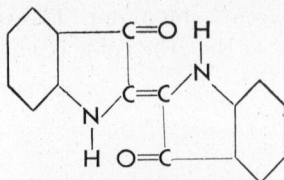

Indigo

An *indigoid* dye

with the chromophore

group

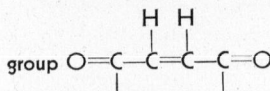

The Macro-molecule Industries

In many parts of this text the existence of molecular units of very high molecular weights has been mentioned. These molecular units are called macro-molecules. They are found in nature as cellulose (p. 164) and related material and as proteins (p. 133). During the first half of this century, man has built upon and altered nature's macro-molecules; he has also learned to fabricate his own, making them more specifically to his liking from simple materials like carbon, air, water, and natural gas. On this foundation he has built the macro-molecule industries—synthetic rubber to make us independent of the trees of Asia; nylon to displace the lowly silkworm; and plastics to match and improve upon existing natural resins.

The Plastics Industry. The public mind has been more deeply stirred by the advent of plastics on the counters of our stores than by any other commercial product. The fact that plastics production has reached an estimated rate of over one and a half billion pounds per year and is now used to make so many articles formerly made of metal and wood, is everywhere apparent.

The story of the origin of these new materials takes us back to the workshop of John Wesly Hyatt, a type setter who made billiard balls during his spare time. In fact it was while striving to produce artificial ivory for billiard balls that he mixed nitrocellulose and camphor and with heat and pressure produced celluloid. Much later L. H. Baekeland carefully studied the condensation reaction between phenol and formaldehyde producing the plastic, Bakelite. These two discoveries form the foundation for this intriguing industry. To understand the industry it is necessary to distinguish between the two products. They are representative of the two major classes of the materials known as plastics.

Classification of Plastics. The distinction between the major classes of plastics is that some, like celluloid, may be softened again by heating and then pressed into some new desired shape; others like bakelite are converted into a permanent solid mass that cannot be refabricated. The first is known as a *thermoplastic* material; the second is a *thermoset plastic*. Soon after the introduction of celluloid and bakelite to the world, it became apparent to industrialists and organic chemists that other materials would behave in a similar manner, and the search was on for new and better combinations that could be molded into useful articles under heat and pressure. The great variety of different types of plastic materials, each designed to serve some specific purposes better than competitive materials, is the fruit of that search.

In terms of quantity produced, diversification of useful articles fabricated, and low unit cost, phenol-formaldehyde resins and cellulose resins hold first place in their respective classes.

Thermoset Plastics. Reactions presented on p. 217 show that when phenol and formaldehyde react an addition product is formed. This simple addition product loses water and joins to other like molecules with bonding between adjacent units through the activated ortho- and para- positions shown in the equation at "a." When the change is carried out in the presence of an alkaline catalytic agent under controlled conditions, the

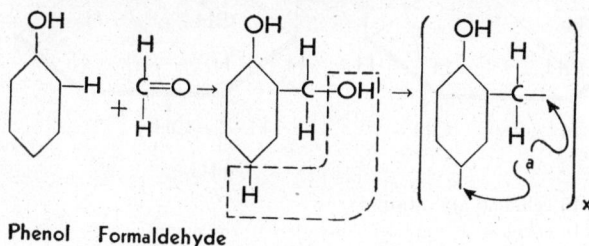

Phenol   Formaldehyde

number for "X" is small and the resin is soft, viscous, and is soluble in alcohol and acetone. By controlling the degree of polymerization, "X" may be kept at such a value that comparatively little spacial cross-linking occurs and the pre-resins thus formed may be used for protective-coating solutions. Upon further, but still incomplete polymerization, solid molding powders are formed. These, mixed with a plasticizer (a material to make the mass flow upon being heated), a pigment for color and fillers (woodflour, fiber, cloth, asbestos, mica, etc.) and subsequently heated to about 175°C and pressed at as high as 2000 lbs. per sq. in., result in multiple, three-dimensional, cross-linkages forming a permanent hard resin surrounding and interlocking the particles of filler much as cement bonds about sand and gravel to form concrete. The result is a very diversified plastic, hard and permanent, with properties, depending upon the filler used, adapting the material for its use for innumerable articles of commerce—from simple inexpensive bottle caps to heat resistant handles, from small electric outlets to huge, high-voltage housings, from radio panels to calendar-pad holders.

Phenol will also couple with other aldehydes, like the cyclic aldehyde furfural (p. 237), and with urea (p. 122). Each type of product being well adapted to a multitude of specific uses.

Cellulose Plastics. Thermoplastic materials may best be illustrated by the cellulose plastics. The earliest was Hyatt's celluloid. This material, although it was fabricated into innumerable articles had many drawbacks —it was too inflammable and too ready to absorb moisture. Why not, then, improve upon celluloid by converting the cellulose into esters with other acids than nitric acid?

The first answer was found by converting cellulose to the acetate ester, using acetic anhydride instead of the acid. Other conversions and improvements followed until it would appear that the products have forgotten their origin as a fiber of cotton. Most of the common inexpensive plastic gadgets in stores today are made of cellulose acetate.

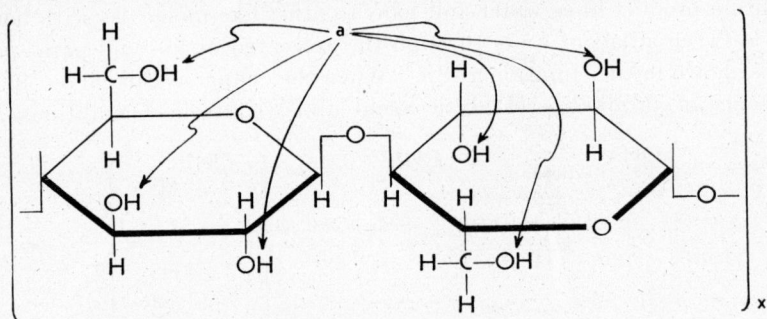

Most of the / In celluloid are nitrated.
—OH groups { In cellulose acetate are converted to the acetate ester.
at "a" \ In cellulose acetate-butyrate, are esterified by means of acetic-butyric
             anhydride.
          \ In methocel are methylated, forming an ether.

Other Thermoplastics. After cellulose acetate had proven its worth, organic chemists created many new thermoplastic materials. Different combinations are appearing on the market every year each with characteristics to make it more valuable than others for certain specific uses. The scope of this text will not permit a detailed report on any of these materials. They have invaded the paint industry providing better coatings; they have revolutionized the adhesives industry, providing better glue and better, more permanent bonding of plywood; they have improved upon glass by making it shatter-proof and have provided a product better than glass for military airplanes. The chemistry of some of the other plastics is given in Table 10. Examples are selected from the many available to show briefly the chemistry of certain well-known plastics and to show a variety of simple starting materials.

Synthetic Fibers. Closely related to the production of plastics is the production of synthetic fibers. These have wrought a revolution in the textile industry. Since prehistoric times man has woven fabrics from the

Table 10

MACROMOLECULAR PRODUCTS

1–9 form Plastics; 7, 8, 9 form Textile Fibers; 10, 11, 12 form Synthetic Rubbers

1. *Polythene*—
   (one of the simplest, but one of the most difficult to produce).

   Cracked hydrocarbons $\xrightarrow[\text{from petroleum}]{}$ $\xrightarrow[\text{200°C + cat.}]{\text{1200 at. press.}}$

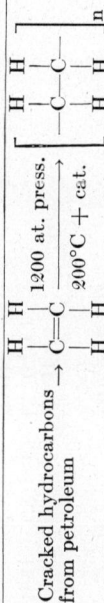

2. *Polystyrene*—
   (plastic from benzene and ethylene).

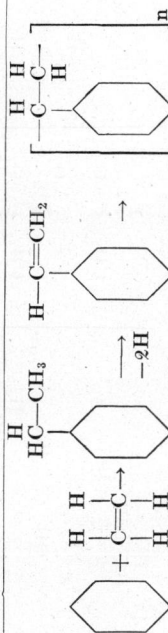

   Ethyl benzene   Styrene   for plastics and films

3. *Coumarone-indene*—
   (polymer from coal tar distillates).

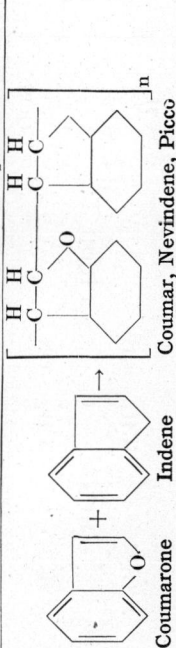

   Coumarone   Indene   Coumar, Nevindene, Picco

4. *Silicones*—
   (Silicon polymers for plastics and lubricants).

   $SiCl_4 + RMgX$

   $RSiCl_3 \xrightarrow{H_2O} RSi(OH)_3$

   $R_2SiCl_2 \xrightarrow{H_2O} R_2Si(OH)_2$

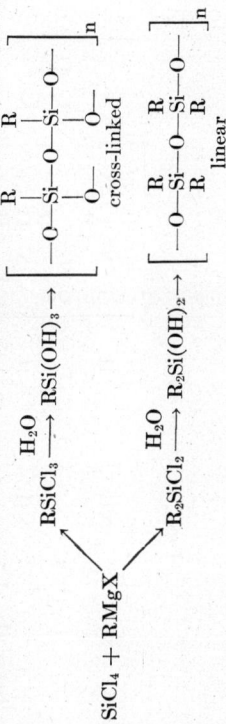

   cross-linked

   linear

Table 10—(Continued)

5. *Urea-formaldehyde—* (from coal, air and water).

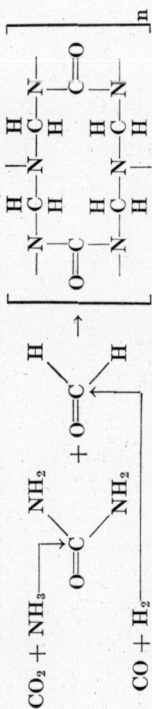

$$CO_2 + NH_3 \quad \quad CO + H_2$$

6. *Plexiglass and Lucite.*

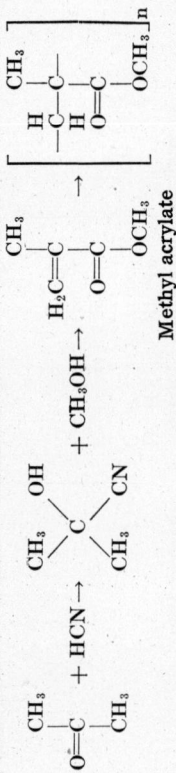

$$CH_3C(=O)CH_3 + HCN \rightarrow \quad + CH_3OH \rightarrow$$

Methyl acrylate

7. *Cellulose acetate* (from cotton for) *Rayon and Plastics.*

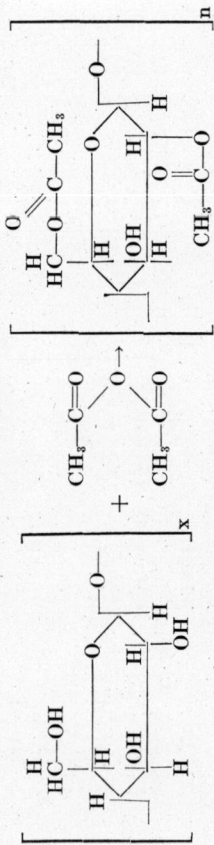

8. *Saran*
for fibers, upholstery fabrics and hemp substitute and plastics.

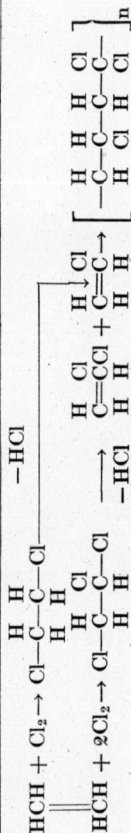

$$HCH + Cl_2 \rightarrow Cl\text{-}C\text{-}C\text{-}Cl \quad \quad -HCl$$

$$HCH + 2Cl_2 \rightarrow Cl\text{-}C\text{-}C\text{-}Cl \quad -HCl$$

9. *Nylon* (from coal, air and water)
Start with Adipic acid (p. 108).

Reduction

$$\rightarrow H_2N-\overset{O}{\underset{}{C}}-(CH_2)_4-\overset{O}{\underset{}{C}}-NH_2 + H_2N-(CH_2)_6-NH_2 \rightarrow \left[-\overset{O}{\underset{}{C}}(CH_2)_4-\overset{O}{\underset{}{C}}-\underset{H}{N}-(CH_2)_6-\underset{H}{N}- \right]_n$$

Adipamide   1,6-Diaminohexane   Nylon,—a protein-like polymer

10. *Buna S* or *GR-S* of war years.

$$H-\overset{H}{\underset{H}{C}}=\overset{H}{\underset{H}{C}}-\overset{H}{\underset{H}{C}}=\overset{H}{\underset{}{C}}H + HC=\overset{H}{C}\text{(benzene ring)} \rightarrow \left[\overset{H}{\underset{H}{C}}-\overset{H}{\underset{}{C}}=\overset{H}{\underset{}{C}}-\overset{H}{\underset{H}{C}}-\overset{H}{\underset{}{C}}-\overset{H}{\underset{}{C}}\text{(ring)} \right]_n$$

11. *Thiokol A* *Thiokol B* (Sulfur rubbers).

$$ClCH_2CH_2Cl$$

$$\underset{CH_2CH_2Cl}{\overset{CH_2CH_2Cl}{O}} + Na_2S_4 \quad\rightarrow\quad [-CH_2CH_2-S_4-]_n$$

$$\rightarrow [-CH_2CH_2-O-CH_2CH_2-S_4-]_n \quad\left. \right\} \text{ Probable S Linkage}$$

$$\left[\overset{S}{\underset{H}{C}}=\overset{S}{\underset{H}{C}}-\overset{H}{\underset{H}{C}} \right]$$

12. *Neoprene* (from C_2H_2 and HCl).

$$2H-C\equiv C-H \rightarrow H-C\equiv C-\overset{H}{\underset{}{C}}=\overset{H}{\underset{H}{C}} + HCl \rightarrow \overset{H}{\underset{}{C}}=\overset{Cl}{\underset{}{C}}-\overset{H}{\underset{}{C}}=\overset{H}{\underset{H}{C}} \rightarrow \left[\overset{H}{\underset{H}{C}}-\overset{Cl}{\underset{}{C}}=\overset{H}{\underset{}{C}}-\overset{H}{\underset{H}{C}} \right]_n$$

Vinyl acetylene   Chloroprene   Neoprene

fibers of plants—cotton, hemp, and linen—and from the fur and filaments of animals—wool and silk.

Why not then imitate the silk worm? The synthetic fiber industry started in 1855 by the spinning of artificial threads, formed from solutions of cellulose nitrate in ether and forced through tiny orifices. Thirty-five years later the cupra-ammonium process utilized Schweitzer's Reagent (an ammoniacal solution of cupric hydroxide (p. 165)) as the cellulose solvent. This process resulted in the regeneration of the cellulose as a fiber by extruding the filaments into water and then neutralizing the copper ammonia complex with dilute sulfuric acid. Then came the *viscose process* utilizing sodium hydroxide and carbon disulfide to dissolve cellulose from wood and other inexpensive sources. The product, cellulose

$$\text{xanthate } (C_6H_7O_2(OH)OC\underset{\textstyle SNa}{\overset{\textstyle S}{\diagup\!\!\diagup}}\diagdown \quad),$$

upon neutralization with acid regenerates cellulose in the form of the first true rayon. The popular packaging material, cellophane, is produced by forcing the viscose syrup (cellulose xanthate solution) through long narrow slits into the regenerating acid bath. The next step, carrying the product still farther from pure cellulose, was the *acetate process* based upon the use of cellulose acetate discussed under plastics (p. 256).

Then the questions arose, "why stay with cellulose? Why not make fibers of purely man-made proteinlike molecules? Why not make fibers more like silk and wool?" The answer to the making of fibers like silk has already come in the form of nylon. It is probable that a new wool-like fiber with revolutionary characteristics will be available to the public by the time this book is published. The name now given to this synthetic wool-like fiber is *orlon*.

Formulations for the preparation of some of these fibers will be found in the chart on p. 258. The names Vinyon, Saran, and Dynel are trade names that will become household names of the near future. When the various contributions of organic chemistry are summed up, it appears that the greatest single growth area is in the field of textiles. The influence of these new products upon the garments in which man clothes himself and in turn upon the social environment is indeed profound.

Synthetic Rubber. The foregoing sections describe developments that today's students may view as ancient history, since these industries had their inception before most of the present college students were born. There is one synthetic organic chemical development however whose start we all remember. Only a few years ago we could not get sufficient tires for our cars.

The development of synthetic processes for making rubber has affected not only our own national economy, its repercussions have seriously af-

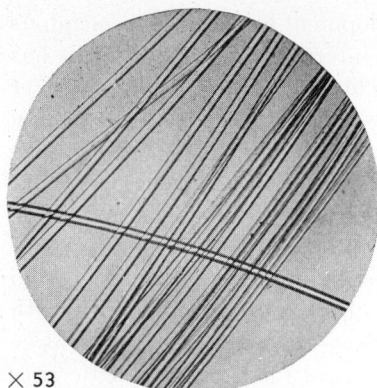

× 53

Note how translucency of nylon is demonstrated where filaments cross one another.

Nylon

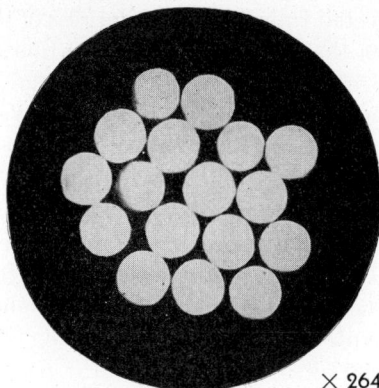

× 264

Cross section of a thread for nylon stockings. The thread is composed of 17 filaments.

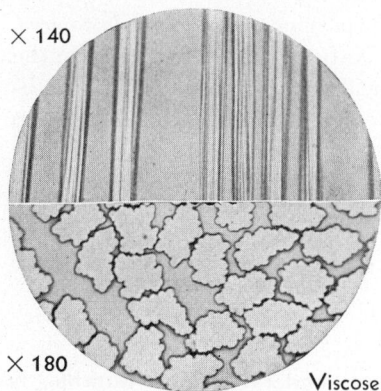

× 140

× 180

Viscose

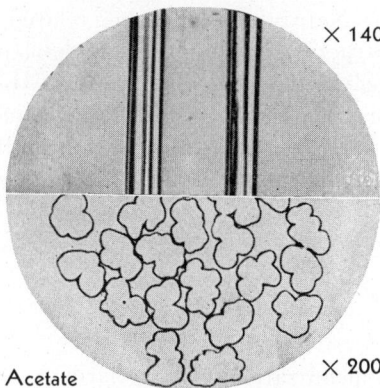

× 140

Rayon

Acetate

× 200

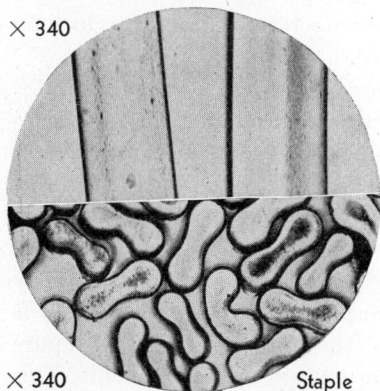

× 340

× 340   Staple

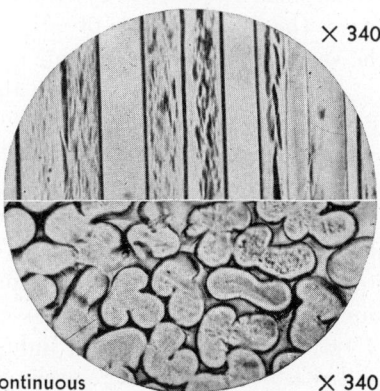

× 340

Orlon

Continuous

× 340

Fig. 38. Man-made textile fibers are unique in shape. Photomicrographs of several synthetic filaments. In the four lower photographs, the top portion shows a longitudinal view of filaments and the lower portion shows a cross-sectional view. Approximate magnification is given by each illustration. (Courtesy, E. I. du Pont de Nemours & Co.)

fected Indonesia. The impress of this development has been so tremendous on the world economic situation and our international policy that it has been found advisable to continue the purchase of natural rubber for over half our present requirements. We can now produce more than one million tons of synthetic rubber per year, enough to oversupply our national requirements. The development of "cold rubber" in 1947 has achieved tire treads superior to those from natural rubber because synthetic tires made at lower temperatures are more resistant to heat and abrasion.

Another example of the far-reaching economic effect stemming from the development of a synthetic chemical, is the use of synthetic fibers with synthetic rubber in the making of tires. One of the South's most extensive markets for cotton was the use of cotton cord for tires. Synthetic tires have forced the use of stronger, heat-resistant rayon. Uniquely enough rayon may soon have to bow to the use of even more efficient nylon for the making of cord. Who knows what the next step will be?

Natural rubber and its related synthetic polymers are called *elastomers*. Structurally they have much in common. Natural rubber results from the polymerization of isoprene, C_5H_8. Chemically it is 2-methyl-1,3-butadiene. Under heat and catalytic action the conjugate double bonds shift to the 2-position leaving the extremities of the molecule free to unite with its neighboring molecules and so polymerization occurs.

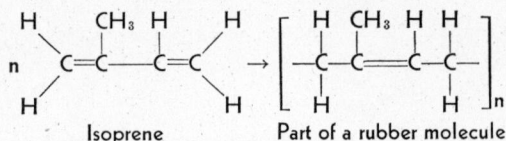

Isoprene   Part of a rubber molecule

Vulcanization, which hardens rubber and renders it more durable, is accomplished by heating with sulfur, the sulfur forming cross-links between the double bonds of adjacent, linear macro-molecules interlocking the smaller parts in a permanent structure.

The chief materials used in the making of synthetic rubber are butadiene, ethylene, styrene, and acetylene—all raw products made inexpensively from petroleum or other base chemicals. Several methods of preparing synthetic elastomers are given in Table 10.

The Future Depends on Research. Such is the story of achievement in the organic chemical industries. The curve of production rises ever more steeply. New products and improvement of old products show steady gains year after year. Progressive research programs require larger staffs of trained personnel. Some industries aid the universities in their pure research by generous scholarship grants for studies in specific fields. This pure research leads ultimately to new and improved products for industry. An added advantage exists for the young men awarded scholarships and selected for research on these industry sponsored projects—the opportunity of being more readily absorbed into industry.

The constant procession of new products from industry has amazed the American public. It is no longer a question of when will it end but rather what will be next. Innumerable materials on the organic chemical market today were undreamed of 40 years ago. There is no longer any doubt that progress and improvement will continue. Industrial firms to keep their position in industry must promote new research programs, thereby providing a proving ground for trained organic chemists with imagination and initiative—for men capable of progressing with the industry they serve. Luck will still play its role in the discoveries to come but luck will hardly favor the individual or the enterprise that is content to stand still—that is not seeking for something new and something better.

Ernest H. Volwiler, President of the American Chemical Society, speaking at a symposium at Notre Dame University in 1950, closed his remarks about the chance achievements of chemistry with words that are the key to progress in modern science and technology. Dr. Volwiler said, "Few men can live on chance alone—to take advantage of chance, hard work must give chance an opportunity to develop. There is no telling where or how we may stumble onto a new idea, but no one stumbles over anything while he is standing still."

Chapter 19

CHAPTER PROLOGUE

During the past three decades there has been an amazing development in knowledge about a variety of organic substances possessing significant physiological properties. In this short period the uses of these complex materials has become so general that their names have assumed a role in the daily conversation of the layman as well as of the specialist. Such terms as "miracle drugs," vitamins, hormones, antihistaminics, and enzymes are not new to any of us.

The importance of recent biological developments in organic chemistry can be strikingly evaluated in terms of Nobel prizes awarded in the fields of chemistry and medicine. During the past 20 years more than 40 per cent of the possible prizes in these fields have been awarded for work on compounds to be discussed in the first sections of this chapter.

Peculiarly enough, from the viewpoint of molecular structure, these compounds overlap, so that almost every class of organic compound and derivative is represented among them. Also, the majority of these compounds are polyfunctional. Hence their classification, instead of being based primarily upon molecular structure, is determined by their source and their physiological characteristics. There is a specific relationship between each group of compounds and the function that each compound within the group serves.

Our knowledge of biochemical processes has progressed at a rapid pace during the past two decades, yet we have only entered upon the threshold of the mysteries of life. May they to whom it is given to penetrate its inner sanctum be humble men and women capable of evaluating evidence without prejudice from the past or fear of the future.

Organic chemicals of biological importance as considered in this chapter are of three general classes: (1) Essential materials that control body processes, (2) materials of biological origin that man has learned to extract and use, and (3) synthetic products of biological importance. Subdivisions of these three classes are shown below.

Classes of Compounds of Biological Importance

I. *Materials that Are Known to Be Essential for Normal Life Processes.*

Vitamins   Dietary necessities obtained chiefly from plant foods.

Hormones   "Chemical messengers" secreted by the organism, chiefly in the ductless glands. These agents exert specific and striking influence preparing the body for necessary physiological processes.

Enzymes   Organic catalysts secreted in the body. Their chief function is to catalyze specific digestive and metabolic processes.

II. *Materials of Biological Origin that Man Has Learned to Extract and Use.*

Alkaloids   Heterocyclic compounds of plant origin with powerful physiological action.

Antibiotics   Chemotherapeutic agents which are products of metabolism, principally of microorganisms, bacteria, and mold.

III. *Synthetic Materials of Biological Importance.*

Drugs of all types as the "sulfa drugs," metalorganic compounds, antihistaminics, anesthetics, antiseptics, and others of specific physiological activity.

Role of the Organic Chemist. In the development of biological applications of organic chemistry, the role of the organic chemist is that of assistant to the men of medicine and biology. When accumulated records indicate the existence and importance of some complex material, the chemist plays his part through the four processes of extraction, purification, analysis, and synthesis. The final proof of the quality of the chemist's work is that the product which he synthesizes will produce the same physiological effects as the original material. The task is by no means simple. Often years are required for each of the four steps. This is particularly true of those substances of vital importance in sustaining normal physiological and mental functions and which occur in extremely small amounts in the natural mixture wherein their existence has been first suspected.

As typical of these steps, the history of Thiamin (vitamin B_1), the antineuritic, antiberiberi vitamin,[1] may be cited:

1885 The deficiency disease beriberi had long been a scourge of the Japanese navy. Takaki prevented the occurrence of this disorder by changing the dietary ration and providing increased quantities of meat, barley, and fruit.

1893-97 Eijkman, experimenting with the diet of fowls, produced experimental polyneuritis by using polished rice which had been a staple food in the Japanese navy. He then prevented the disease by dietary means.

1912 Funk administered water extracts of rice bran to rats suffering from dietary polyneuritis and thus cured the condition.

1926 Jansen and Donath isolated the vitamin in crystalline form from rice bran.

---

[1] Beriberi is a disease produced by a deficiency of thiamin in the diet. Thiamin is essential for the normal metabolism of carbohydrates in the cells, and is therefore essential for the normal functioning of the body as a whole. Beriberi has been very common in the Orient where the diet consisted largely of polished rice. Symptoms of thiamin deficiency are lameness, numbness, motor and sensory nerve disorders, and respiratory disturbances. Many people in the United States suffer from a marginal supply of thiamin.

Fig. 39. (Top) Birds too must have their vitamins. (Bottom) Thiamin crystals cured the pigeon. (Courtesy, Golden State Company, San Francisco.)

1931   Windaus and co-workers isolated the pure vitamin from yeast. They obtained a sufficient quantity and studied its chemical characteristics, finally establishing its formula.

1936   The vitamin was synthesized by Williams and Cline and also by Andersag and Westphal.

The sequence or reactions of the synthesis of thiamin as now employed follows:

Ethyl-3-ethoxypropanoate

Ethyl-2-formyl-3-ethoxypropanoate

2-Methyl-5-ethoxymethyl-
6-hydroxypyrimidine

2-Methyl-5-ethoxymethyl-
6-chloropyrimidine

2-Methyl-5-ethoxymethyl-
6-aminopyrimidine

4-Methyl-5-(2-hydroxy-
ethyl)-thiazole

2-Methyl-5-bromomethyl-
6-aminopyrimidine hydrobromide

Vitamin B₁ bromide hydrobromide

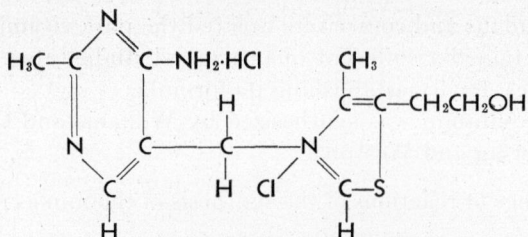

Vitamin B_1 chloride hydrochloride
2-Methyl-5-(4-methyl-5-[2-hydroxyethyl] thiazolium
chloride)-6-aminopyrimidine hydrochloride

The scope of this text will not permit a detailed discussion of the chronology and synthesis of other vitamins, of hormones or of other substances described in this chapter. Nor is the student expected to learn such detail. The presentation given at this point should be noted as an indication of the tremendous task facing organic chemists in solving the problem of synthesis of biologically important chemicals. Imaginative consideration of this example shows that the research open to organic chemists in this field is unlimited. The discovery of vitamin B_1 and its importance in the welfare of mankind is an example of the challenge and of the reward extended to those who select organic chemistry as a profession.

I. ESSENTIAL MATERIALS THAT CONTROL BODY PROCESSES

A. Vitamins[2]

Mankind has become vitamin conscious. Probably no other group of substances has ever received so rapid a meteoric rise to popular recognition and usage. The discovery that vitamins are so essentially a part of our dietary requirements, although they occur only in minute quantities, has had a profound influence upon medicine, industry, and national food habits.

The Polish chemist, Funk (p. 265), is credited with coining the name. Funk observed that a substance extracted from rice polishings gave the characteristic test for an amine. Since he also established that this material was essential to life, he noted it to be a "vital-amine" and coined the term *vitamine*. Further discovery of other accessory factors in the diet, similar from the physiological viewpoint, but containing no amine group, led to dropping the *e*, so the class is now known as vitamins. The vitamins

[2] The interested student should consult some of the standard, more comprehensive reference works on the fascinating vitamins. The following are suggested: Rosenberg, *Chemistry and Physiology of the Vitamins*, Interscience Publishers, 1945; Harrow, *Textbook of Biochemistry*, ed. 5, Saunders, 1950; Harris and Thimann, *Vitamins and Hormones*, Academic Press, 1946; Holmes, *Have You Had Your Vitamins*, Farrar and Rinehart, 1938; Borsook, *Vitamins*, Viking, 1941; Sherman, *Chemistry of Food and Nutrition*, Macmillan, 1946; Williams, *What to Do About Vitamins*, University of Oklahoma Press, 1945.

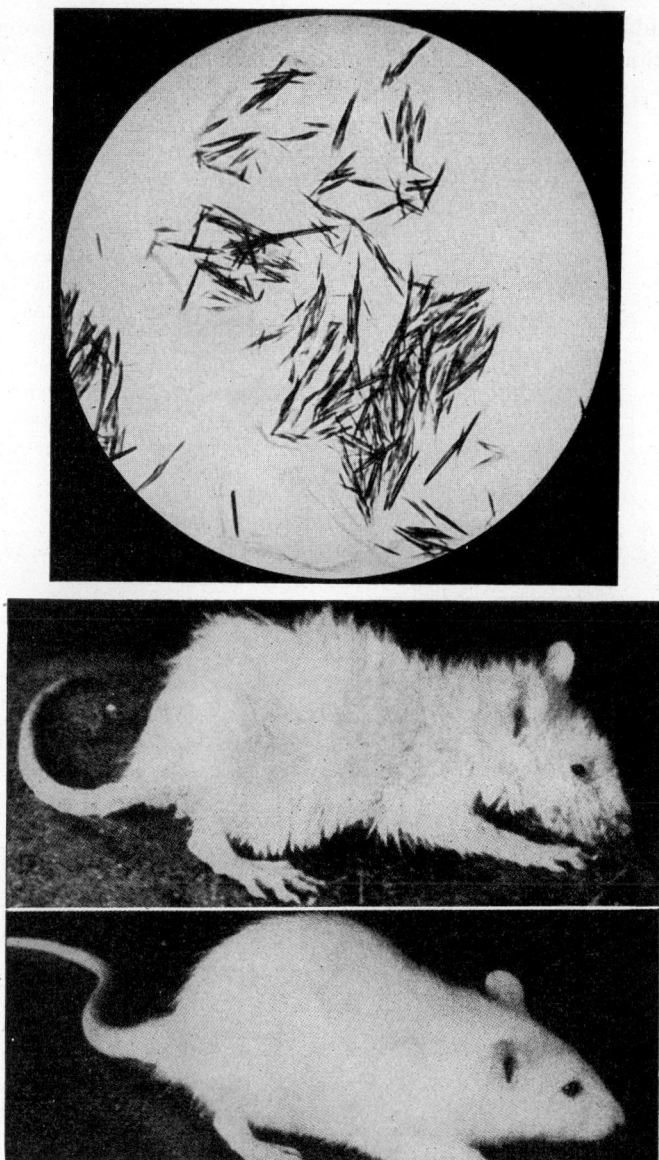

Fig. 40. (Top) Riboflavin crystals. (Bottom) Rat with severe dermatitis is cured by riboflavin. (Courtesy, Golden State Company, San Francisco.)

are all organic materials, but there is no functional organic structure common to the specific vitamins that have been identified as compounds. The early basis of classifications A, B, C, D, etc., is gradually giving way to the use of the chemical name of the pure isolated compound. Before chemical nomenclature could be employed, however, proof of structure had to be established.

Formulas. The striking difference in structure will be noted upon examination of the established formulas which follow:

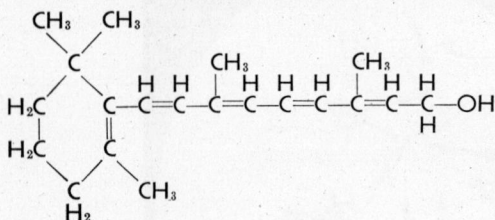

Vitamin A ($C_{20}H_{30}O$)

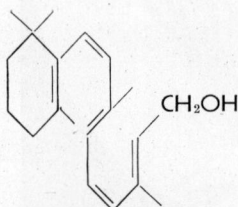

Skeletal formula
sometimes used

Synthesized in 1937.

Occurs in fish liver oils and butter.

Can be formed by the body from carotinoids, yellow or red compounds found in vegetables and fruits.

Lack of vitamin A produces an eye disease (xerophthalmia), night-blindness, lowered resistance to infection, loss of weight, and possible sterility.

Fat soluble; heat stable.

There are also vitamins A_2 and A_3.

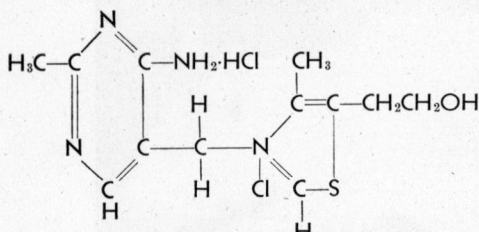

Thiamine hydrochloride[3]
(Thiamin) (B_1)
($C_{12}H_{17}ON_4SCl \cdot HCl$)

Synthesized in 1936.

Occurs in plants, particularly in the germ of cereal grains.

Lack of thiamin produces polyneuritis, beriberi, paralysis, loss of weight, and general nervous and digestive trouble.

Water soluble; stable to most cooking operations.

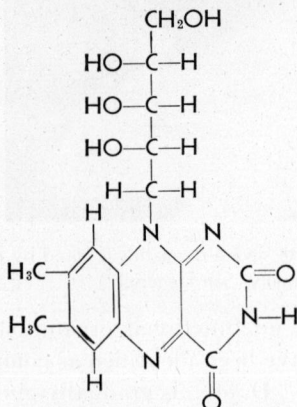

Synthesized in 1936.

Occurs in eggs, liver, meat, yeast, cheese. Added to bread to "enrich" it.

Lack of Riboflavin produces weakness, loss of weight, and loss of hair.

Slightly water soluble; fairly stable.

Riboflavin (B_2)(G) ($C_{17}H_{20}N_4O_6$)

[3] See p. 267 for synthesis equations and strict organic nomenclature.

Pyridoxin (B_6) ($C_6H_{11}O_3N$)

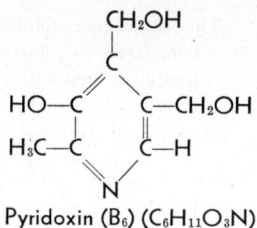

Synthesized in 1939.
Yeast and rice polishings are rich sources.
Also occurs in seeds and cereals, fish, and mammalian livers.
Small amounts in milk, eggs, lettuce.
Lack causes impaired growth.
Water soluble; optically active.

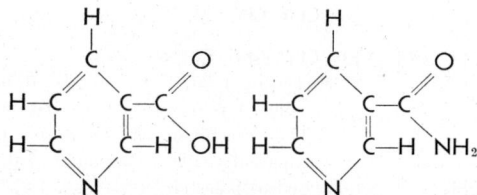

Nicotinic acid (Niacin) (PP factor) (Pellagra preventive factor) ($C_6H_5O_2N$) and its amide, Nicotinamide ($C_6H_6ON_2$)

Synthesized in 1873.
Occurs in liver, kidney, eggs, milk, yeast, and wheat germ.
Lack produces gastrointestinal and mental disturbances. Acute condition is pellagra.
Water soluble, heat stable.

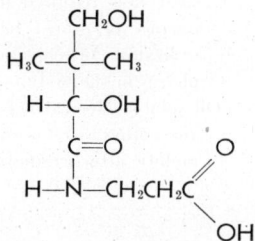

Pantothenic acid (Bx) ($C_9H_{17}O_5N$)

Synthesized in 1940.
Occurs in egg yolk, liver, milk, yeast.
Essential for nutrition and growth of rats and dogs. Prevention of dermatitis in chicks.
Usually found bond chemically to protein material.

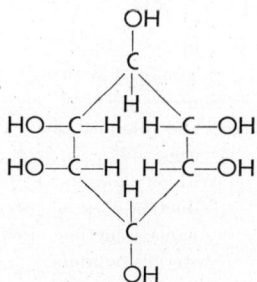

Mositol, Cyclohexanehexol ($C_6H_{12}O_6$)

Composition has long been known.
Found as normal cell constituent of practically all plant and animal tissues. Usually in combination as an ester.
It is not commercially important.

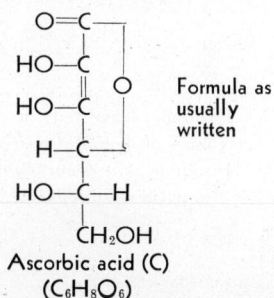

Ascorbic acid (C) ($C_6H_8O_6$)

Formula as usually written

Synthesized in 1933.
Sources: Citrus fruits, tomatoes, raw leafy vegetables, peppers.
Deficiency symptoms:
    Mild: tenderness of joints, lowered resistance to infection.
    Acute: scurvy.
Water soluble, alkali sensitive, easily oxidized.

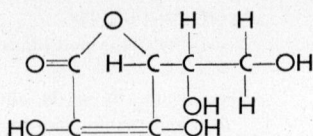

Another method of writing
the graphic formula

The structural similarity to carbohydrates is illustrated by the formula as first shown. The second method of writing the formula shows the furanose configuration.

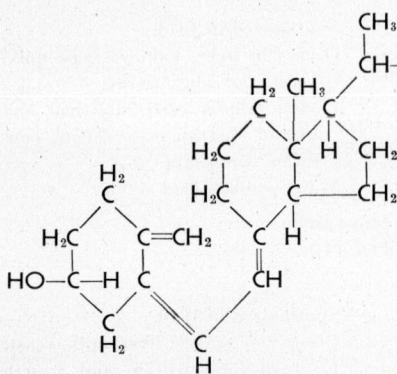

Calciferol (D₂) (C₂₈H₄₃O)

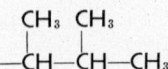

Structure related to sterols from which it is formed by irradiation. The precursor of calciferol is apparently the compound 7-dihydro-cholesterol.

Source: Fish liver oils, milk, eggs.

Acute deficiency produces rickets since it is required for the maintenance of proper balance in the utilization of calcium and phosphorus in the body.

Oil soluble; derivatives of ergosterol and other sterols become antirachitic after irradiation.

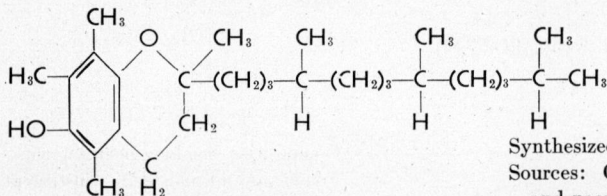

α-Tocopherol (E)
(C₂₉H₅₀O₂)

Synthesized in 1938.

Sources: Oils of cottonseed, corn and peanuts, wheat germ, lettuce, eggs, milk.

Without vitamin E normal reproduction does not occur, hence the name sometime used is the anti-sterility vitamin.

Oil soluble, stable to heat, alkali and acid.

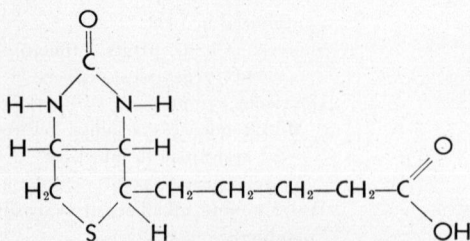

Biotin (H) (C₁₀H₁₆O₃N₂S)

Formula suggested in 1942 on a basis of degradation reactions.

Insoluble as it occurs in combination in natural sources.

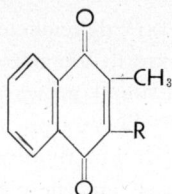

K-vitamins (antihemorrhagic vitamin)

Synthesized in 1939.
Sources: Alfalfa, spinach, kale, tomatoes.
Hemorrhage; acute in infants at birth.
Fat soluble.

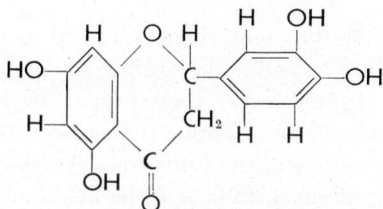

5,7,3',4'-Tetrahydroxyflavanone
Citrin (P) (P is the crude extract) (Permeability vitamin)

Lack of this vitamin results in decreased capillary resistance, followed by skin and kidney hemorrhage.

Nonidentified vitamins are given the names B_3, B_4, B_5, $B_7(I)$, B_8 (may be Adenylic acid), B_c, B_p, J, L_1, L_2, M, T, U, Folic acid and groups of related material known as vitagens.

Table 11

VITAMIN UNITS AND REQUIREMENTS

Vitamin	Quantity Equivalent to One International Unit	Recommended Daily Minimum for Normal Adult*
Vitamin A..................	0.6 microgram (0.0006 mg.) of β-carotene	5,000 I.U.; 8,000 I.U. recommended for pregnant and nursing women
Thiamin chloride.............	3.3 micrograms	700 to 900 I.U.
Riboflavin..................	None established	2 mg.
Nicotinic acid...............	None established	4 to 20 mg.
Ascorbic acid...............	50 micrograms of l-ascorbic acid	75 mg.
Vitamin D..................	0.025 microgram of crystalline vitamin D_2 (calciferol)	Adequate; 400 to 800 for children and pregnant women

* During pregnancy and lactation the recommended requirement of most vitamins is notably higher.

Vitamin Assays. An examination of the foregoing formulas and a consideration of the comparatively recent dates of synthesis indicate that for such complex substances a quantitative chemical analysis for vitamin content is impractical. These substances are recognized by the deficiency diseases which result from their absence in the diet and are subject to biological assay. Rats, guinea pigs, and chickens are most frequently used in this work because they respond to dietary deficiencies and develop symptoms similar to those suffered by mankind. A control group of the selected animals is fed a normal diet. Additional groups are fed diets normal in every respect but the content of the vitamin being studied. Thus the minimum quantity of essential vitamin required for normal growth and well-being has been established. From these results for the more important vitamins a system of International Units (I.U.) given in Table 11 has been adopted.

Biological assays are not only costly but may require several months. Accordingly, as soon as a pure vitamin is isolated in sufficient quantity, its physical-chemical properties are studied. From these properties, rapid systems of analysis have been devised. For example, it has been found that vitamin A absorbs a particular wave length of ultra-violet light (328 Å, angstrom units). When a sample of shark liver is to be examined for vitamin A content, the oil is extracted and dissolved in 2-propanol. Light of the particular wave length is passed through the dilute solution; the vitamin potency is proportional to the quantity of light absorbed. In this manner, a sample of fresh liver can be analyzed in 15 minutes.

B. Hormones

Many functions of the human body are controlled by minute amounts of complex organic substances known as hormones. These materials are secreted principally by the ductless (endocrine) glands, and are released into the blood and lymph as needed. A proper balance of hormone is essential to normal functioning.

Studies now in progress indicate that many of the disorders of man, such as those producing symptoms of old age, may be dependent upon the hormone balance within the body. For example, hormones of the pituitary gland influence hormone secretion by other glands and the related functioning thereof. Also a balance must exist between thyroid and gonad secretions; as a consequence hyperthyroidism may be controlled by administering extracts from the gonads. This is an example of chemical equilibrium within the body.

When physiological conditions occur that interrupt the flow of these secretions, serious pathological symptoms result. Like the vitamins, the existence of these substances has only been recognized for about 40 years, the quantity required is minute; their influence is, in general, catalytic. Hormones are different from vitamins in that they are normally secreted by the organism itself.

In man, the glands which secrete the most interesting and best known hormones are the pituitary glands, the thyroid, the pancreas, the adrenals, and the sex organs. The pituitary, a small gland located at the base of the brain, is occasionally spoken of as "the master endocrine gland" because it secretes various hormones that control the secretion of hormones by nearly all of the other organs.

It is an interesting commentary on the search for these "chemical messengers" that while this text was being prepared for publication the dramatic story of the work of Edward C. Kendall[3a] of the Mayo Foundation appeared in the June 19, 1950, issue of *Chemical and Engineering News*. It was Kendall who, in 1914, first isolated crystalline thyroxine, the hormone of the thyroid gland. Thirty-four years later, after 20 years of continuous research, Kendall and his co-workers succeeded in isolating cortisone recognized as "Compound E of the secretion of the adrenal

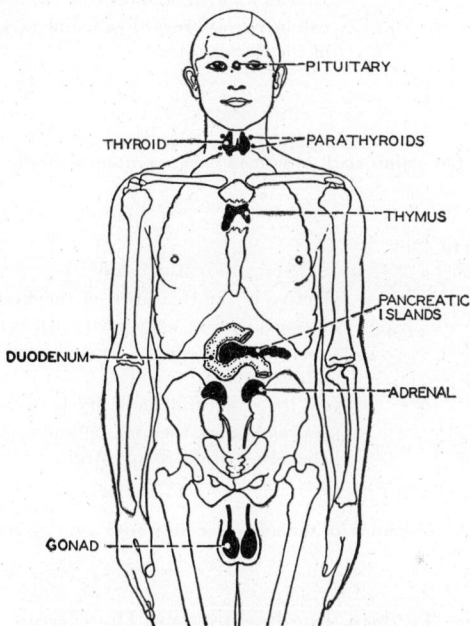

Fig. 41. Location of endocrine glands. (Courtesy, Neal and Rand: "Comparative Anatomy," Philadelphia, The Blakiston Company.)

cortex." The former compound has successfully been used to aid those with thyroid disorders (goiter, cretinism, myxedema); the latter compound is the most promising aid to those afflicted with rheumatoid arthritis, a disease which, as stated in the previously mentioned article "has been considered to be relentless and progressive, with death the only avenue of escape from pain and deformity."

Specific Hormones. The chemical nature and some facts of interest about the best known hormones is sufficient for our consideration in this text. Interested students should read supplementary material in standard texts on endocrinology and physiological chemistry.

[3a] Dr. Kendall shared the Nobel Prize for Medicine in December 1950, for the work herein outlined.

Regulates the rate of metabolism (p. 293) in the body.

Deficiency causes "cretinism," i.e. stunted growth and lack of mental development.

Excess is associated with goiter and abnormally rapid oxidation of body tissues.

Research with isomeric compounds, in which the —OH group and —I are located at different points on the benzene rings, has shown markedly less activity. Research with radioactive iodine, I^{131}, has given extensive data regarding iodine metabolism and cancer of the thyroid gland.

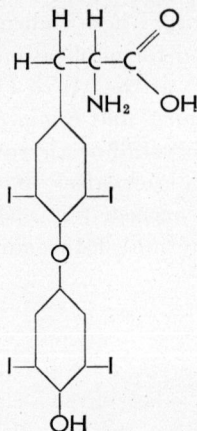

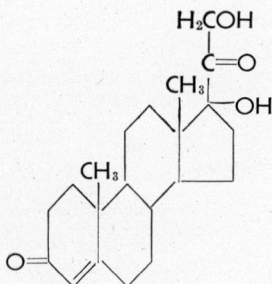

Thyroxine (an amino acid, produced from a protein from the thyroid)

Effective in the treatment of rheumatoid arthritis.

Addison's disease associated with a degeneration of the cortex, successfully treated with extracts of the adrenal glands.

Note the structural similarity between cortisone and the sex hormones (below), vitamin D (p. 272), and the alicyclic hydrocarbon, estrane.

Cortisone (hormone of the adrenal cortex with sterol structure)

Sex Hormones. To date, five hormones of the estrone (thielin) group, which are female sex hormones, have been isolated. At least four male sex hormones have also been identified. These sex hormones are associated in development of physical sex characteristics of the normal male and female and with the functional cycles in sex development such as menstruation. Recent therapeutic use of certain of these hormones has greatly simplified the difficulties of the menopause. Examples are:

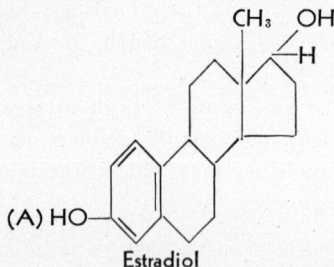

Estradiol, produced in the ovaries. The benzoate ester (at A) is used in medicine for relief of nervous symptoms during the menopause.

Estradiol

CH₃ OH (B)

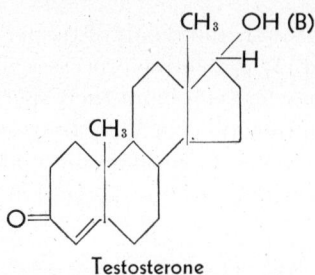

Testosterone, the most potent of the male hormones. For therapeutic use the propionate (at B) is employed, since the esterified product is more slowly absorbed and eliminated than the pure hormone.

Testosterone

Insulin. In 1923 the Nobel Prize was awarded to Banting and Macleod in recognition of their preparation of an active extract of pancreatic tissue containing the protein-hormone, insulin. The pure crystalline protein was first obtained in 1926. It is secreted by the "Islets of Langerhans," small endocrine glands embedded in the pancreas. This hormone is associated with the metabolism of carbohydrates (p. 294) and it regulates the concentration of glucose in the blood. The sugar concentration of the blood rises to dangerous levels when insulin is deficient.

Diabetes is a disease induced by some pathological condition of the pancreas which prevents it from secreting sufficient insulin to produce the normal catabolism of glucose. The rise of the glucose content of the blood is a *symptom* of the disease. Injection of insulin into a vein in the correct amount, by increasing the speed of catabolism of the glucose, prevents the lethal effects of hyperglycemia.

Insulin is not a cure for diabetes however, i.e. it does not appear to bring back a "lazy" pancreas to normal activity as thyroxine sometimes does for a defective thyroid. Hundreds of thousands of diabetics living practically normal lives today would have died were it not for insulin.

Plant Hormones. A number of plant hormones, known as *Auxins*, have important growth stimulating properties. Several have found use in recent years to serve effectively in speeding the rooting of plant cuttings and in improving seed germination. Some are used to prevent premature drop of apples and pears. A particular man-made auxin, 2,4-D(2,4-dichlorophenoxyacetic acid and its salts and esters) (p. 193) is extensively used as a weed killer.

C. Enzymes

Closely related to vitamins and hormones in terms of their catalytic biochemical action is the group of compounds known as enzymes. Although their chemical nature has not been definitely established, some have been obtained as, or closely associated with, crystalline proteins. These compounds are produced as a result of cellular activity. Their catalytic action, however, is independent of living cells, that is, the pure enzymes will cause the same chemical change outside the body as in the body. In naming enzymes the suffix -ase is used.[4] Human enzymes catalize

---

[4] Many enzymes whose specific effect was noted long ago still retain their early names, e.g. pepsin, tripsin, rennin, etc.

reactions associated with (1) digestion, (2) growth and repair of tissue, (3) oxidation of energy-producing material, and (4) protective processes such as the coagulation of the blood. Many of these enzymes and their specific function are given in Chapter 20 on Organic Reactions in Vital Processes.

The effect of zymase from yeast, which causes the production of alcohol from glucose during fermentation, is an example of enzyme action (p. 73).

II. Materials of Biological Origin that Man Has Learned to Extract and Use

A. Alkaloids

Since antiquity man has used in a crude form a group of substances of vegetable origin, known as alkaloids. The alkaloids have pronounced physiological action which accounts for their use as drugs.

Table 12

Important Alkaloids and Their Salts*

Alkaloid	Alkaloidal Salts	Crude Drug	Source
Atropine...............	Atropine sulfate .	Belladonna	Roots and leaves of *atropa belladonna* plant
Cocaine................	Cocaine hydrochloride	Coca	Leaves of *coca*, a South American shrub
Codeine................	Codeine phosphate	Opium	Dried milky exudate of capsule of white poppy, *papaver somniferum*
Emetine................	Emetine hydrochloride	Ipecac	Roots of *cephaelis ipecacuanha*, a tropical plant
Ephedrine..............	Ephedrine hydrochloride	Mahuang	Twigs of the Chinese plant, *ma huang*
Morphine..............	Morphine sulfate	Opium	See Codeine
Physostigmine..........	Physostigmine salicylate	Physostigma	Seeds of *calabar bean*, a climbing plant of Africa
Pilocarpine.............	Pilocarpine nitrate	Jaborandi	Leaves of the South American shrub, *Pilocarpus Jaborandi*
Quinine................	Quinine sulfate Quinine bisulfate Quinine dihydrochloride	Cinchona	Bark of the South American *cinchona* tree
Strychnine.............	Strychnine sulfate	Nux vomica	Seeds of *Strychnos nux vomica* tree

* Goostray: Introduction to Materia Medica. Copyright 1924, 1939 by the Macmillan Company and used with their permission.

Chemically, the alkaloids are heterocyclic compounds (p. 242) containing nitrogen. They are basic compounds (hence the name alkaloids) and are optically active (p. 134). They taste bitter. Many are poisonous; very small doses being lethal.

Since the names and action of the more important alkaloids or their corresponding crude drugs are relatively familiar to the layman, only the formulas of noteworthy members of this group of substances will be presented here. The student should note the similarity of molecular structure to the heterocyclic ring structures described in Chapter 17.

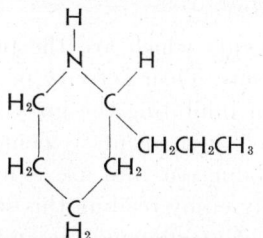

Coniine (α-Propylpiperidine)

Coniine is obtained from hemlock. Students of Greek history will recall that Socrates was forced to drink hemlock juice.

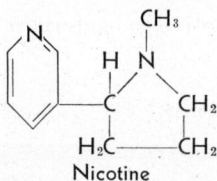

Nicotine

Oxidation with fuming nitric acid produces the vitamin, Nicotinic acid (p. 271)

This is another example of the great stability of the pyridine ring: the five-membered ring (related to pyrrole) is oxidized and the pyridine ring is not broken or nitrated by the nitric acid.

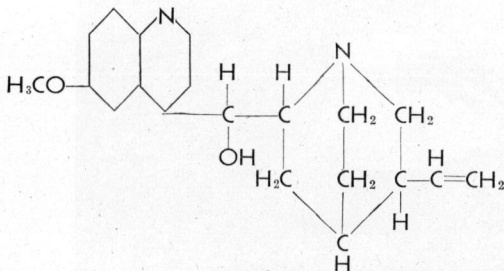

Quinine

Ex GI's are well acquainted with the use of quinine in treatment of malaria. Quinine has local anesthetic action and is antiseptic. Two synthetic antimalarials are Atabrine and Plasmochix.

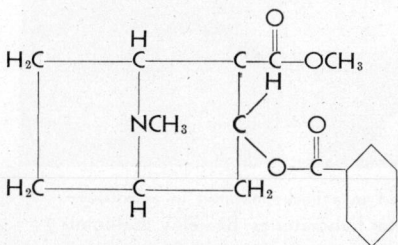

Cocaine

The local anesthetic cocaine is a habit forming drug. It is prepared from coca leaves.

Caffeine

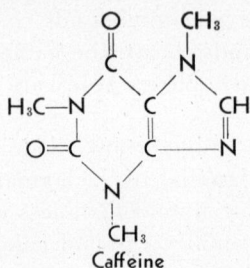

Although its physiologic reaction is slight in comparison with other compounds of this group, caffeine may be considered to be an alkaloid. Tea and coffee and the familiar "coke" contain appreciable amounts. Caffeine is a stimulant to the heart and the central nervous system. It is also a diuretic.

B. Antibiotics

Antibiotics are chemotherapeutic agents which are the products of metabolism, principally of micro-organisms. Their activity in the curing of disease is based upon the property of inhibiting the growth of other micro-organisms. The most important of these agents is penicillin. The story of its discovery, development, production, and use is indeed dramatic. Premedical students will especially enjoy reading this story.[5] It is generally well known that the diseases of pneumonia, meningitis, diphtheria, scarlet fever, gonorrhea, syphilis, and the infections of streptococci and staphylococci, usually yield to treatment with the penicillins. The remarkable thing is that an agent so destructive to bacteria possesses such low toxicity to higher animals.

Penicillium chrysogenum

(a) —born in a test tube, reared in a flask, matured in a bottle—
Fig. 42. Penicillin. (Courtesy, Cutter Laboratories, Berkeley, California.)

[5] See Ratcliff, *Yellow Magic*, Random House, 1945.

Penicillium chrysogenum

(b) —then transplanted in giant fermentation vats, there to grow and produce the crude penicillin—

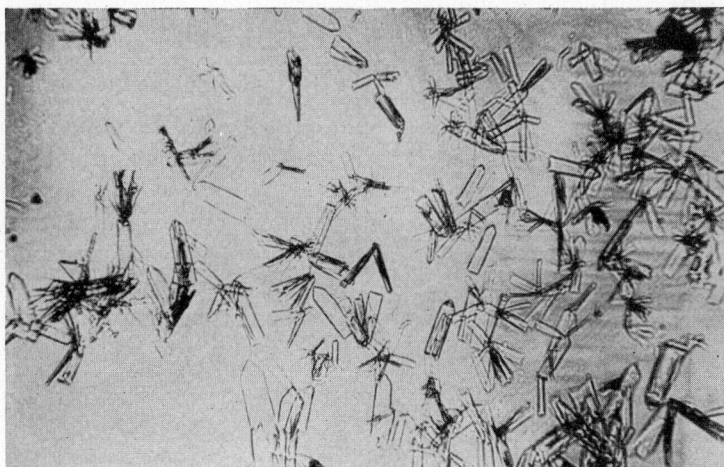

(c) —whose purified, picket-shaped crystals will serve to cure man's diseases.
Fig. 42—(Continued). (Courtesy, Cutter Laboratories, Berkeley, California.)

The Penicillins; derived originally from the mold, *Penicillium notatum,* and now from several related strains.

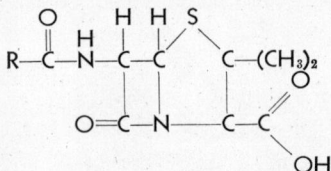

Wherein R is:

in penicillin $F = CH_3CH_2CH{=}CHCH_2{-}$
$G = C_6H_5CH_2{-}$
$K = CH_3(CH_2)_6{-}$
$X = \text{p-}HOC_6H_4CH_2{-}$

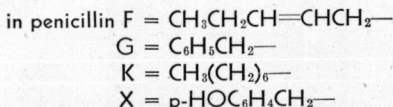

The most effective form against most of the organisms of the afore-mentioned diseases is penicillin G.

A soluble addition compound with prococine is less rapidly eliminated (in the urine) and hence is effective for up to 24 hours.

Discovered by Fleming in 1929 while studying the growth and properties of staphylococcus.

Extensive use of the agent on man did not come until World War II.

Production of penicillin rocketed by December 1943 to 470 times the production during June 1943. Production has continued to increase steadily since that time and now there is enough for all needy patients.

In 1945 the Nobel Prize in Physiology and Medicine was awarded jointly to Fleming, Florey, and Chain for their work on penicillin.

Streptomycin is perhaps the most promising of the antibiotics now under investigation. It inhibits the growth of a number of gram-negative bacteria,[6] a group on which penicillin has little effect.

III. Synthetic Materials of Biological Importance

The foregoing sections of this chapter deal with dramatic biological developments in organic chemistry related to the products of living organisms. Of course many of these substances are now chemically synthesized. This section will briefly introduce other groups of biological agents which have contributed greatly to human welfare and to biological study, but that are more commonplace and that are purely man's creations. The one notable exception is the sulfa-derivatives sometimes popularly referred to as "miracle drugs." In the main, the compounds to be discussed here are synthetic materials.

A. Synthetic Drugs

Sulfa Drugs. Although sulfanilamide has been known since the turn of the century, the first use of sulfa drugs in treating infection was reported in the early 1930's. Pharmacological studies showed that sulfa-

[6] Gram-positive bacteria can be stained by the method of Hans Gram, Danish bacteriologist; gram-negative bacteria are not stained by this method.

$$NH_2$$

$$H_2N-\underset{}{\bigcirc}-N{=}N-\bigcirc-SO_2NH_2$$

Prontosil

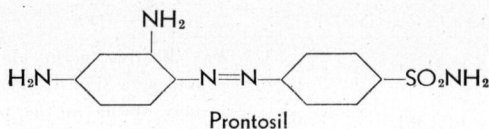

nilamide (formed from Prontosil) in the body, was responsible for the effectiveness of the dye in treating coccal infections, particularly the streptococcal. Hence the question, "Why not use sulfanilamide direct?" Synthesis of sulfanilamide is now accomplished from acetanilide as follows:

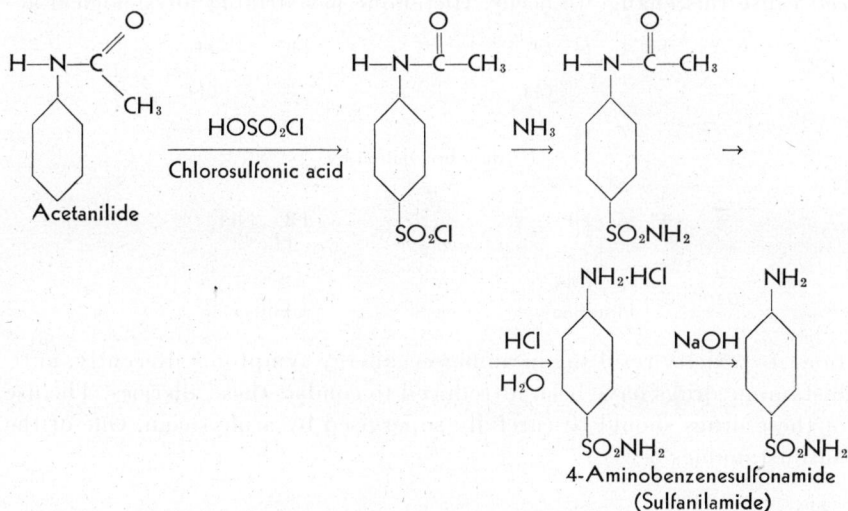

$$H-N-\overset{O}{\underset{CH_3}{C}}$$

Acetanilide

$$\xrightarrow[\text{Chlorosulfonic acid}]{HOSO_2Cl}$$

$$H-N-\overset{O}{C}-CH_3$$

$$SO_2Cl$$

$$\xrightarrow{NH_3}$$

$$H-N-\overset{O}{C}-CH_3$$

$$SO_2NH_2$$

$$\rightarrow$$

$$\xrightarrow[\text{H}_2\text{O}]{HCl}$$

$$NH_2{\cdot}HCl$$

$$SO_2NH_2$$

$$\xrightarrow{NaOH}$$

$$NH_2$$

$$SO_2NH_2$$

4-Aminobenzenesulfonamide
(Sulfanilamide)

The other sulfa drugs, sulfapyridine (p. 208), sulfathiazole, sulfadiazine, and sulfaguanidine, are synthesized in similar manner. Each has certain advantages. Among the diseases and infections controlled by these drugs are erysiplas, mastoiditis, pleurisy, pneumonia, gonorrhea, and septicemia.

Metalorganic Drugs. In the treatment of some infections and diseases certain synthetic organic compounds have proved effective. Among these are compounds of arsenic, mercury, antimony, bismuth, and silver. Others have been investigated but have not been found to be effective antiseptics except for a few specific diseases. Two of the best known examples are:

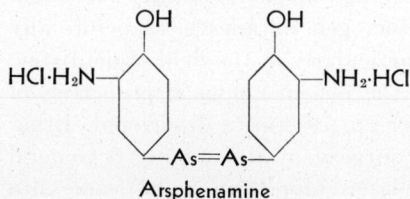

$$HCl{\cdot}H_2N-\underset{OH}{\bigcirc}-\underset{}{\bigcirc}-\underset{As{=}As}{\bigcirc}-NH_2{\cdot}HCl$$

Arsphenamine

Ehrlich's compound "606" prepared in 1909, which is sold as Salvarsan and is used in the treatment of syphilis.

CH_3

—ONa

—HgOH

NO_2

Metaphen

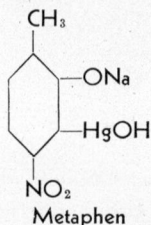

An effective germicide of low toxicity, used as a skin disinfectant. Metaphen is more stable in the presence of protein matter and blood serum than the earlier mercurial, mercurochrome.

Antihistaminics. Histamine is formed in the decarboxylation of the essential amino acid, histidine (p. 295). In the intestine, certain strains of *coli* cause this change to occur. Histamine has striking physiological ac-

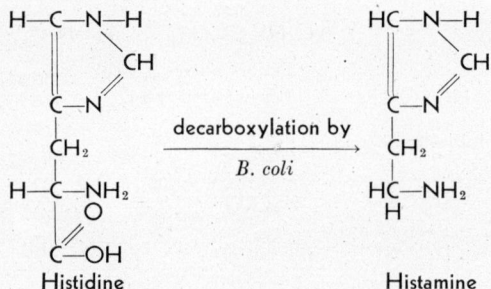

H—C—N—H
        CH
C—N
CH_2
H—C—NH_2
    O
C—OH
Histidine

decarboxylation by
———————————→
B. coli

HC—N—H
        CH
C—N
CH_2
HC—NH_2
H
Histamine

tion, its toxicity resulting in numerous allergy symptoms. Recently, antihistaminic drugs have been introduced to combat these allergies. The use of these drugs should be carefully supervised by a physician. One of the antihistaminics is:

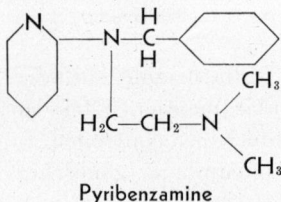

N
        H
    N—C—
        H
            CH_3
H_2C—CH_2—N
            CH_3
Pyribenzamine

Pyribenzamine is one of these drugs that has been found to relieve allergic symptoms, such as hay fever and asthma, due to histamine.

Pain and Chemistry. One of man's most profound searches has been for agents that relieve pain. It seems incredible to one living today to consider that nearly all operations performed a hundred years ago were conducted without anesthesia. Truly, the development of anesthetics and their use in surgical cases is a story to thrill the students who will seek it out. Uniquely enough, nitrous oxide (laughing gas), the only successful inorganic anesthetic, was used to produce general anesthesia before any organic compound, and is still used, particularly in the field of dentistry. Slightly over one hundred years ago, the pharmacological properties of the synthetic compound, diethyl ether (p. 79), were discovered. Ether was first used to produce anesthesia for surgery, at Massachusetts General Hospital on October 16, 1846. A little later chloroform (p. 63) came into general use. The search for more effective and safer general anesthetics has been continuous, but diethyl ether still retains first place.

Among other effective general anesthetics are two hydrocarbons, the unsaturated compound, ethene (ethylene) $CH_2:CH_2$ (p. 35) and the alicyclic hydrocarbon, cyclopropane,

$$H_2C \overset{\displaystyle CH_2}{\diagup \diagdown} CH_2 \ (p.\ 174).$$

The most recent efforts to secure better, safer anesthetics have been through the combining of the favorable effects of these hydrocarbons with those of the ethers by preparing mixed compounds. Noteworthy examples are:

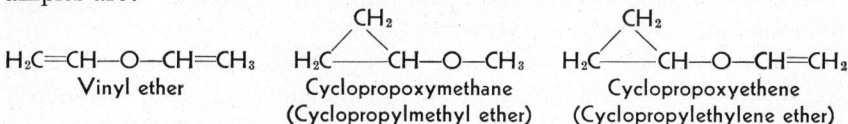

$H_2C{=}CH{-}O{-}CH{=}CH_3$
Vinyl ether

$H_2C \overset{CH_2}{\diagup \diagdown} CH{-}O{-}CH_3$
Cyclopropoxymethane
(Cyclopropylmethyl ether)

$H_2C \overset{CH_2}{\diagup \diagdown} CH{-}O{-}CH{=}CH_2$
Cyclopropoxyethene
(Cyclopropylethylene ether)

Local anesthesia can be obtained with a variety of agents. As examples, ethyl chloride (p. 58) serves for minor local surgery because its rapid vaporization freezes tissue surfaces. Cocaine (p. 279) is a variety of alkaloidal type agents anesthetizing locally by blocking the nerve responses when injected in a proper area. Habit forming narcotics such as opium and morphine (p. 278) also relieve pain and produce drugged sleeping.

Aspirin (p. 220) and acetanilid (p. 206) relieve pain generally and locally and are classed as analgesics and anodynes. Substances like chloral hydrate (p. 88) and the barbiturates (p. 106) produce nearly normal sleep and are classed as soporifics and hypnotics.

There are numerous other types of drugs with specific physiological activity. These drugs offer a field of study to the interested student. The common indicator, phenolphthalein (p. 230) is an important laxative. Caffeine (p. 280) increases the flow of urine and is classed as a diuretic. Quinine (p. 280) is a febrifuge or antipyretic, that is, it acts to reduce fevers. Common carbon tetrachloride (p. 64) is a vermicide which kills and causes the expulsion of worms from the intestine. Potassium antimonyl tartrate (tarter emetic) causes vomiting. Atropine dilates the pupil of the eye. Nitroglycerine (p. 77) and the alkyl nitrites (p. 141) lower blood pressure. There are many drugs such as quinine (p. 279) which are specific for certain specific diseases. "War gases" are drugs possessing certain types of action; for example 2-propenyl isothiocyanate (mustard oil) is a vesicant which blisters the skin. *The student must realize that the foregoing substances are all toxic agents and accordingly should not be used without the supervision of a physician.*

Disinfectants and Related Substances. The use of disinfectants has become general and important to modern civilization. The field of related substances includes antiseptics for the sterilization of wounds, and bactericides and germicides for the killing of microorganisms. One of the earliest disinfectants to be employed successfully was phenol (p. 214). Accordingly the germ killing (germicidal) power of substances is rated in relationship to the power of phenol to kill the same type of organism. In collateral reading the student may find the notation: "the phenol coefficient of mixed cresols is 2.5." This means that under the same conditions and for the same type of organism, the concentration of cresols needed to kill the organism in the same length of time as phenol is only 1/2.5 that of the phenol.[7] Cresols have largely replaced phenol in the drug trade because they are more effective and less toxic and corrosive to man and livestock. Lysol is essentially a soapy solution of cresols. Other related phenol derivatives serving as effective disinfectants are thymol (p. 220), hexylresorcinol and salicylates (p. 221).

We have already considered the antibiotics (p. 280), the sulfa drugs (p. 282), and certain metalorganic compounds (p. 283). Mercurochrome, crystal violet (p. 252), and other dyes possess antiseptic properties. Formaldehyde (p. 92), formerly used in large quantities as a fumigant, destroys organisms due to the hardening (coagulating) effect upon protein. Urotropine (p. 94) is an effective disinfectant of the urinary tract since it liberates formaldehyde in an acid solution. Sodium acid phosphate is usually administered with urotropine in order to insure an acid urine. Iodoform (p. 63) and the chloroamines (p. 196) are effective disinfectants.

Recently, a number of quaternary ammonium derivatives (p. 147) have become widely accepted as effective materials for the sterilization of dishes in restaurants and bars and for use in the dairy industry. These materials are so new that their use has required special action by the legislatures of the states or by the boards of health in such states as designated by law. A typical such product is Hyamine 1622.

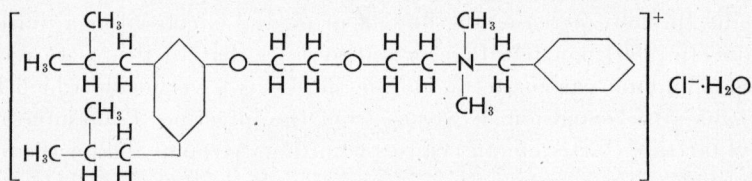

Di-isobutyl-phenoxy-ethoxy-ethyl-dimethyl-benzyl-ammonium-chloride monohydrate

[7] *Bacillus typhosus* and *Staphylococcus aureus* are the organisms most frequently used In some tests the time factor is held constant and the concentrations of the phenol and the disinfectant are compared.

Chapter 20

ORGANIC REACTIONS IN VITAL PROCESSES

CHAPTER PROLOGUE

Beyond any doubt the past century has witnessed an extraordinary change in man's method of living. New materials have been produced and older materials have been improved for our added comfort and convenience. Modern discoveries in biological chemistry have led to an improved understanding of life in its various forms. Man has altered his method of looking at things in the world about him, but through it all the central figure in the world today, man himself, has changed very little, if any, in his basic needs since our prehistoric ancestor first beheld carbon in its elemental form. During the ages, and especially during the past century, man has accumulated a fund of accurate, organized knowledge about vital processes in the human body and that knowledge has helped immeasurably in producing a healthier people. During the past decade we have not only acquired knowledge about the physiological importance of food in maintaining physical well-being but we have gained a new insight into the psychological importance of food to mental and moral well-being. With this complicated information food is rapidly becoming an international weapon of peace, bringing a new challenge to the organic chemists of our day, the challenge to provide more and better foods.

May they who have contributed so much to a healthier, happier, and more efficient race also meet this challenge successfully and give us a more peaceful world.

In the preceding chapters we have become acquainted with many classes of organic compounds. Some of these make up the list of substances essential to the human organism. Fats, carbohydrates, and proteins are the principal organic components of the food we eat. Vitamins are accessory factors necessary in the diet. Hormones and enzymes are substances secreted by glands and organs within the body; their function is to control body processes. Since physiological chemistry is essentially the study of the reactions which these organic compounds undergo or control within the body, this brief presentation giving the fundamental nature of the changes involved will serve as a foundation for the complex structure of physiological chemistry. Upon reading this chapter the student will gain an understanding of his own body chemistry.

I. The Functions of Food

We eat in order that we may live; in living we grow and work. What we are and what we become are in part dependent upon the food we eat.

The obvious functions of food are to provide (a) tissue-building material, (b) energy for muscular activity, and (c) energy evolved as heat for maintenance of body temperature. In addition our bodies require relatively small quantities of materials, both organic and inorganic, to regulate body processes. Carbohydrates (Chapter 11) and fats (Chapter 8) serve principally as energy foods. Proteins formed from amino acids (Chapter 10) are the chief tissue builders. Vitamins (Chapter 19) are provided in the normal diet, and serve to regulate certain body processes whereas hormones and enzymes are secreted by the organism itself and serve as agents that keep the body processes in balance and catalyze the chemical changes within the body. The inorganic materials in food consist largely of water and salts. Water provides many physiological services; but herein we shall consider only that of reacting chemically with carbohydrates, fats, and proteins during digestion. Minerals regulate body functions and build specialized tissue such as bone. The oxygen which we breathe is used to oxidize carbohydrates and fats, thus providing energy.

The organic components of our food undergo many complex chemical changes during the processes of digestion and metabolism. The chemical nature of the food we eat is not exactly like that of the tissue which is formed. Our interest lies in the mechanism and the nature of the characteristic reactions which foods undergo during the processes of digestion, anabolism, and catabolism. Essentially, these changes are of the same type as the characteristic reactions observed for each class of compound as previously studied. These reactions are possible because of the existence of certain functional groups in the molecules concerned. Digestion is primarily a process of hydrolysis; anabolism, the building of tissue, is essentially the reverse of digestion and during anabolism large molecules are again synthesized; catabolism, the breakdown of tissue, consists largely of complex oxidation and decomposition reactions during which energy is released and the tissue broken down into smaller units again, becoming waste products that must be eliminated. In order to expedite these changes, the body organs secrete specialized organic catalysts known as enzymes (p. 277).

II. Digestion Is Hydrolysis

The process of digestion as previously mentioned, consists essentially of the hydrolysis of complex molecules to form simpler, soluble molecules that are small enough to pass through the intestinal wall. The process of hydrolysis is so closely associated with individual enzymes, that a study of the chemistry of digestion must consider the specific enzymes that catalyze each reaction.

As examples we may consider the change occurring in a specific material for each of the main classes of food material.

Carbohydrate

$$C_{12}H_{22}O_{11} + H_2O \xrightarrow[\text{(Intestinal juice)}]{\text{Sucrase}} C_6H_{12}O_6 + C_6H_{12}O_6$$

Sucrose   Glucose   Fructose

Fat

$$(C_{17}H_{35}COO)_3C_3H_5 + 3H_2O \xrightarrow[\text{(Pancreatic juice)}]{\text{Lipase}} 3C_{17}H_{35}COOH + C_3H_5(OH)_3$$

Tristearin   Stearic acid   Glycerol

Lean beef

$$\text{Complex protein + water} \xrightarrow[\text{(Gastric juice)}]{\text{Pepsin}} \text{Proteoses (etc.)}$$

$$\text{Proteoses + Water} \xrightarrow[\text{(Pancreatic juice)}]{\text{Trypsin}} \text{Polypeptides and finally amino acids}$$

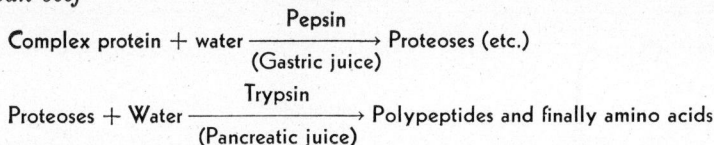

The fact that these catalyzed reactions are examples of typical hydrolysis should be noted.

Digestion in the Mouth. Thorough mastication of food is the prelude to the chemical changes that are to follow. In the process, the carbohydrate, fat, and protein are mixed to a paste with saliva. The saliva consists largely of ptyalin, a carbohydrate enzyme (amylase). Solution of water-soluble food begins and the digestion of starch is started. Since the food stays in the mouth for too short a time to allow much digestion to occur, catalytic action of the ptyalin continues in the stomach. Ptyalin acts best in a slightly acid medium (pH 6.6). Conversion of starch starts in the mouth and this action in the main food mass continues in the cardiac (see Fig. 43) portion of the stomach until the much higher acidity of the gastric juice penetrates the mass and inhibits further hydrolysis by ptyalin. The digestion of starch thus starts in the mouth, continues for a time in the stomach, but is not completed until the material reaches the intestine.

There are no fat- or protein-splitting enzymes present in the saliva. It does contain a small amount of maltase by which some maltose is converted into glucose. In addition to the water content of saliva (approximately 99.5 per cent) and the aforementioned enzymes there are a small amount of mucin and traces of the organic compounds, cholesterol, urea, and uric acid, as well as traces of ammonia, sodium, calcium, magnesium, and potassium as bicarbonates, chlorides, and phosphates. Since these components vary in amount, the pH of normal saliva may range between 6.5 and 7.2. Approximately 1500 cc. is the quantity secreted every day.

Digestion in the Stomach. Upon being swallowed the protein, fat, sugar, and partially digested starches pass into the stomach where they are gradually mixed with the strongly acid gastric juice. The food remains in the stomach for from 2 to 7 hours, thus allowing time for many important changes to occur. The cardiac portion of the stomach serves as a reservoir holding the main mass of food in which salivary digestion continues. In the pyloric portion food is thoroughly mixed with gastric juice, starting protein digestion and making the mass liquid enough to pass through the pyloric valve. The high acidity of the gastric juice provides bactericidal action.

The amount of gastric juice secreted, normally from 2 to 3 liters per day, is roughly proportional to the amount of food eaten. The gastric juice consists principally of water, 99 per cent, hydrochloric acid, 0.15 to

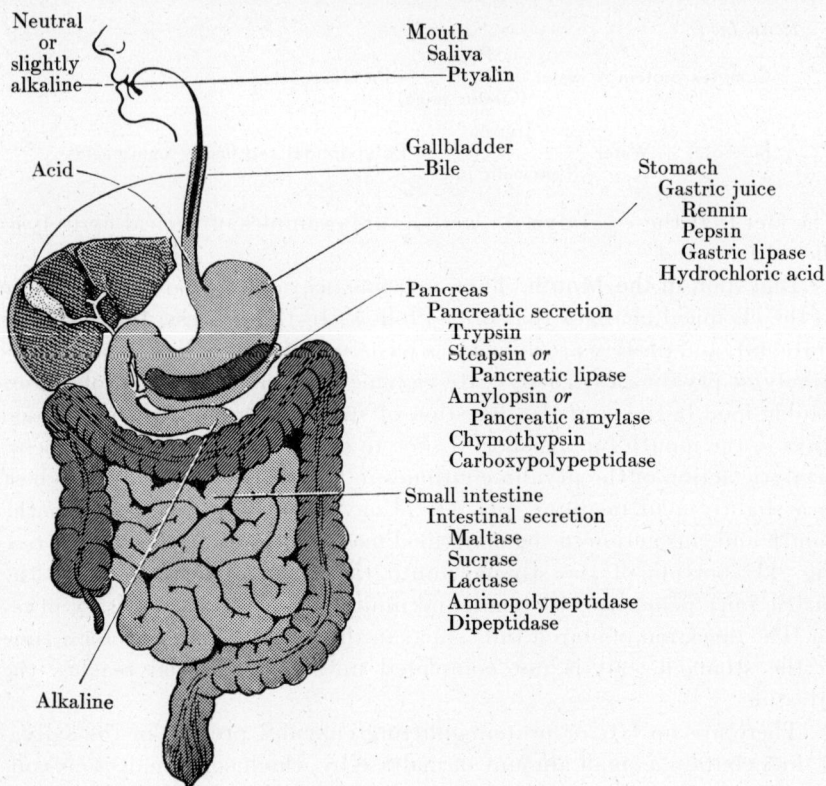

Fig. 43. The organs of digestion.

0.4 per cent (pH 1.0 to 3.5), and two important proteases. The first of these, *pepsin*, is excreted from the walls of the stomach in an inactive form, pepsinogen, which is activated by the hydrochloric acid. The pepsin begins the digestion of protein molecules, splitting them to simpler products known as proteoses and peptones. The second protein enzyme is *rennin* which is especially abundant in infants. With the aid of the calcium in milk, rennin precipitates the milk protein, *casein*, forming curds.

Although there is no carbohydrate enzyme present in gastric juice, the ptyalin of the saliva, as has already been noted, continues to act for some time. This furthers the digestion of starch to form maltose. A fat-splitting enzyme, lipase, is present but it has slight, if any, action in the acid medium.

Digestion in the Intestine. The semiliquid, partially digested food (chyme) next passes through the pyloric valve into the intestine. In the first portion of the intestine, the duodenum, the food is acted upon by intestinal secretion, pancreatic juice, and bile. The bile from the gallbladder and the secretion of the pancreas are alkaline in reaction (pH 7.8

to 8.6 and 7.5 to 8.0, respectively), and as these secretions neutralize the acid chyme, the activity of the pepsin is stopped.

The pancreatic juice contains enzymes which split protein, fat, and carbohydrate. The flow of pancreatic juice is stimulated by *secretin*, a hormone of the intestinal mucosa. The enzyme *trypsin* hydrolyzes proteins to the polypeptides, proteoses, and peptones. The protein hydrolysis is then continued by carboxypolypeptidase, aminopolypeptidase, and dipeptidase. The pancreatic secretion provides the first of these and its action is upon polypeptides with free carboxyl groups. The aminopolypeptidase from the intestinal juice acts upon those polypeptides having free amino groups. These hydrolysis steps produce dipeptides that are acted upon as they are formed by dipeptidase from the intestinal juice. This results in the formation of the end products of protein digestion, namely amino acids.

Amylopsin (a pancreatic amylase) acts much like ptyalin and hydrolyzes starches. When the starch has been reduced to the disaccharide, maltose, it is further split to form glucose by maltase of the intestinal juice. Sucrase and lactase act upon the disaccharides sucrose and lactose respectively, forming their respective monosaccharides. As was·observed in Chapter 11 on Carbohydrates (p. 161), sucrose forms glucose and fructose, while lactose forms glucose and galactose.

The enzyme steapsin, a pancreatic lipase, hydrolyzes fats to fatty acid and glycerol. This process is greatly aided by the bile which contains salts of complex organic acids. The function of these bile salts is to emulsify the fats. The bile does not provide digestive enzymes. In addition to water and salts, it contains excretory products of the liver—the bile pigments (bilirubin and biliverdin, which are red and green respectively), cholesterol, lecithin, urea, and other waste products. The pigments originate from spent red blood cells.

Thus, in the intestine, the process of digestion is completed and the three classes of food become the simple, soluble, building blocks from which new tissue may be formed:

Carbohydrates	become	*Simple sugars* (monosaccharides)
Fats	become	*Fatty acids*[1] *and Glycerol*
Proteins	become	*Amino acids*

As such these products are ready to be absorbed into the blood stream and lymphatic system.

Absorption. Tiny fingerlike projections known as *villi* rise from the inner surface of the intestine. These villi contain blood capillaries and lymphatic vessels. The digested food thus comes in contact with a tremendous surface area through which absorption occurs. Here the chemical law of mass action is in effect. As digestion end-products are absorbed,

[1] The fatty acids are rendered more soluble by combination with the alkali in the intestine forming soaps. Bile, pancreatic juice, and intestinal juice have an average pH of 8.

Table 13

THE HYDROLYSIS OF CARBOHYDRATES, FATS, AND PROTEINS OCCURRING IN DIGESTION

	pH Range	Carbohydrates	Fats	Proteins
Mouth	5.75–7.05	Starches + water $\xrightarrow{\text{ptyalin}}$ Soluble starch + maltose Soluble starch + water $\xrightarrow{\text{ptyalin}}$ Erythrodextrins + maltose Erythrodextrins + water $\xrightarrow{\text{ptyalin}}$ Achroödextrins + maltose	No change	No change
Stomach	1.6 or higher	Action of ptyalin continues until the pH becomes too low	Emulsified fats + water $\xrightarrow{\text{gastric lipase}}$ Fatty acids + glycerol	Proteins + water $\xrightarrow{\text{pepsin}}$ Proteoses + peptones Casein + water $\xrightarrow{\text{rennin}}$ Paracasein Paracasein + calcium $\longrightarrow$ Calcium paracaseinate
Small intestine	6.8–8.0	Dextrins + water $\xrightarrow{\text{amylopsin}}$ Maltose Maltose + water $\xrightarrow{\text{maltase}}$ Glucose + glucose Sucrose + water $\xrightarrow{\text{sucrase}}$ Glucose + fructose Lactose + water $\xrightarrow{\text{lactase}}$ Glucose + galactose	Fats + water $\xrightarrow[\text{(bile)}]{\text{steapsin}}$ Glycerol + fatty acids	Proteins + water $\xrightarrow{\text{trypsin}}$ Proteoses, peptones, and peptides Proteoses, peptones, and peptides $\xrightarrow{\text{erepsin}}$ Peptides + water Peptides + water $\longrightarrow$ Amino acids
Absorbed as:		Glucose, fructose, galactose	Fatty acids and glycerol	Amino acids

Goostray-Schwenck: A Textbook of Chemistry, p. 307. Copyright, 1950, by The Macmillan Company, and used with their permission.

their concentration decreases, and the remaining complex molecules break down more readily and are in turn absorbed.

Amino acids are absorbed into the blood vessels of the villi. Some dipeptides and simple polypeptides may also be absorbed. It is believed that certain protein allergies may thus be the result of absorption of incompletely digested proteins.

The monosaccharides are absorbed into the blood stream and carried to the liver where the excess beyond the part required to maintain the glucose equilibrium are converted into glycogen (animal starch) and stored. The glycerol and the saponified fatty acids are absorbed into the lymph system where the fats are resynthesized. Most of this resynthesized fat is carried through the thoracic duct, entering the blood stream at the jugular vein.

Inorganic salts and water are readily absorbed by the blood capillaries. It is probable that most vitamins are absorbed without undergoing any chemical change.

Detoxification and Elimination. Such food material as is not digested and absorbed, must naturally be eliminated. Accordingly cellulosic carbohydrate (which is not digestible by human enzymes) is eliminated as such. Some cellulose may be partially acted upon by the bacteria which abound in the large intestine. By bacterial action gaseous products (methane, hydrogen, and carbon dioxide) are formed and so are simple organic acids. Ordinarily the acids are absorbed and gases are eliminated. Putrefactive decomposition of protein by bacteria produces amino acids which may be further broken down by the loss of the amino group and the carboxyl group. Some of these converted proteins have marked toxic action. Under normal conditions these toxic products will be oxidized, reduced, hydrolyzed, or conjugated with body chemicals producing nontoxic substances which are then excreted.

III. Metabolism

Carbohydrate Metabolism. The carbohydrate absorbed through the intestinal wall appears in the blood stream as glucose. Its ultimate fate in the body is complete oxidation to form carbon dioxide and water. In being oxidized, each gram of carbohydrate liberates 4.1 kilogram calories of energy. The transformations which glucose undergoes before this oxi-

$$C_6H_{12}O_6 + 6O_2 \rightarrow 6CO_2 + 6H_2O + \text{energy}$$

dation is completed within the body are varied and complex.

The absorbed monosaccharides are carried to the liver where *glycogenesis* occurs, that is, they are synthesized to form the polysaccharide, glycogen. The concentration of glucose in the blood stream is maintained at about 0.1 per cent and as rapidly as the cells take glucose from the blood stream, glycogen hydrolyzes producing glucose in sufficient amounts to maintain the equilibrium. Excess glycogen may also be stored in the muscles. Excess sugar may also be converted to fat.

The delicate equilibrium of glucose–glycogen is thought to be under control of the hormones, adrenalin, cortin, insulin, and thyroxin. The control of blood sugar in diabetics, by the administration of insulin (p. 277) and the supervision of diet, is more or less familiar to the layman.

Oxidation of glucose is the chief source of the energy necessary for the vital processes of the individual. Much research has been applied in the effort to discover the exact mechanism of this oxidation. Before the full story is known much more study will be required. It is definitely recognized, however, that during muscular activity, glycogen is used up and water, carbon dioxide, and lactic acid are formed. It is also known that phosphate plays an important role in the formation of intermediate compounds. A simplified sequence of the reactions recognized at the present time is:

1. Glucose (in the blood) $\rightarrow$ Glycogen (in the muscle)
2. Glycogen + organic phosphate $\rightarrow$ Hexose diphosphate
3. Hexose diphosphate (intermediate compounds) $\rightarrow$ Phosphopyruvic acid
4. Phosphopyruvic acid $\rightarrow$ Pyruvic acid
5. Pyruvic acid $\rightarrow$ Lactic acid
6. Lactic acid
   - 80% $\rightarrow$ Glucose $\rightarrow$ Glycogen
   - 20% $\rightarrow$ (through intermediates) $CO_2 + H_2O$

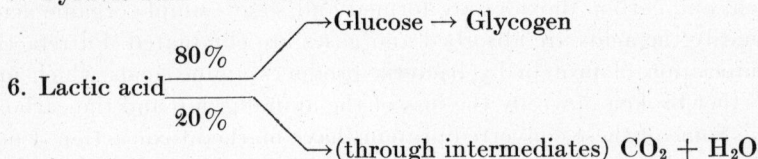

During lactation (formation of milk) some glucose is converted to galactose. Galactose and glucose then combine to form milk sugar (lactose) (p. 162).

Fat Metabolism. Fats are removed from the blood stream and are stored in the tissues as a reserve energy supply. The fatty tissue deposited around the vital organs also serves to cushion and support the organs. The catabolism of fats starts with the splitting of the molecule by hydrolysis to fatty acid and glycerol. The fatty acids are oxidized to form *beta*-hydroxy-acids, then to keto-acids. The next stage of oxidation produces acetic acid and an acid with two less carbon atoms. Thus:

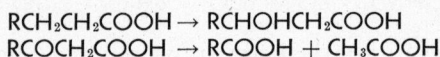

$$RCH_2CH_2COOH \rightarrow RCHOHCH_2COOH$$
$$RCOCH_2COOH \rightarrow RCOOH + CH_3COOH$$

Oxidation of the acetic acid finally produces the end products, carbon dioxide and water.

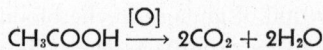

$$CH_3COOH \xrightarrow{[O]} 2CO_2 + 2H_2O$$

The glycerol portion of the fat may either be oxidized completely or, after being oxidized to glyceric aldehyde, two molecules unite to form carbohydrate. Thus:

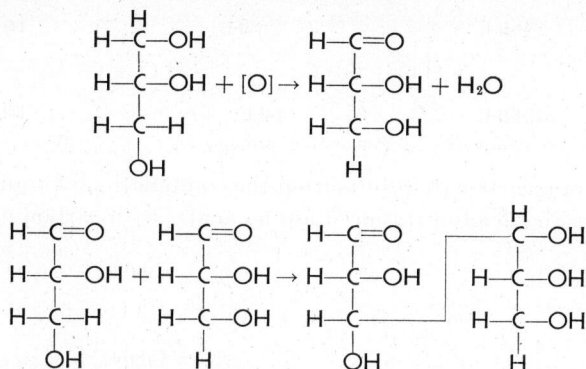

$$
\begin{array}{c}
\text{H} \\
\text{H—C—OH} \\
\text{H—C—OH} \\
\text{H—C—H} \\
\text{OH}
\end{array}
+ [\text{O}] \rightarrow
\begin{array}{c}
\text{H—C=O} \\
\text{H—C—OH} \\
\text{H—C—OH} \\
\text{H}
\end{array}
+ \text{H}_2\text{O}
$$

$$
\begin{array}{c}
\text{H—C=O} \\
\text{H—C—OH} \\
\text{H—C—H} \\
\text{OH}
\end{array}
+
\begin{array}{c}
\text{H—C=O} \\
\text{H—C—OH} \\
\text{H—C—OH} \\
\text{H}
\end{array}
\rightarrow
\begin{array}{c}
\text{H—C=O} \\
\text{H—C—OH} \\
\text{H—C} \\
\text{OH}
\end{array}
\quad
\begin{array}{c}
\text{H} \\
\text{C—OH} \\
\text{H—C—OH} \\
\text{H—C—OH} \\
\text{H}
\end{array}
$$

The reaction is a typical aldol condensation. There are other less common oxidation changes that also occur.

The changes by which fat is synthesized from carbohydrate consist of a series of oxidation and condensation reactions catalyzed by appropriate enzymes. Similarly certain proteins are converted into fat by deaminization, reduction and condensation reactions involving amino acids. The reserve fat is called upon for energy needs when the carbohydrate stores are depleted.

Protein Metabolism. At the present time it is known that the body needs 22 amino acids for its normal well-being. Ten of these amino acids are rated as indispensable: arginine, lysine, histidine, tryptophane, leucine, phenylalanine, threonine, isoleucine, methionine, and valine.[2] Of the indispensable amino acids, only arginine is synthesized by the human body though not at a rate sufficient to meet all the needs of the body. The body can synthesize the following amino acids (called dispensable): alanine, glycine, serine, norleucine, aspartic acid, glutamic acid, hydroxyglutamic acid, cystine, proline, tyrosine, hydroxyproline, and citrulline. The source of indispensable proteins must therefore be protein foods. The normal adult requirement of protein to provide the needed amino acids is about 55 grams per day. The function of several of the various amino acids is specific. Accordingly the story of the metabolism of proteins is exceedingly complex and in many cases obscure. Certain principles of protein metabolism are however quite well understood.

It is known, for example, that there is no appreciable reserve supply of protein stored in the body. This is established by the fact that in the normal adult the daily intake of protein as food and the total nitrogen eliminated in the urine, feces, and perspiration are essentially the same. Urea represents about 80 per cent of the protein nitrogen excreted by the

[2] For graphic formulas of the amino acids, consult a larger text on Organic or Physiological Chemistry.

body. It may be formed directly from certain amino acids, or it may result from ammonium carbonate by the loss of water.

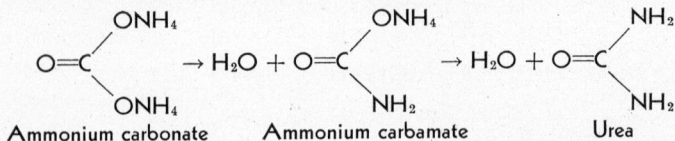

$$O=C \overset{ONH_4}{\underset{ONH_4}{\diagup}} \quad \rightarrow H_2O + O=C \overset{ONH_4}{\underset{NH_2}{\diagup}} \quad \rightarrow H_2O + O=C \overset{NH_2}{\underset{NH_2}{\diagup}}$$

Ammonium carbonate   Ammonium carbamate   Urea

The ammonium carbonate results from the combination of ammonia (produced during the deaminization of amino acids) with carbon dioxide and water.

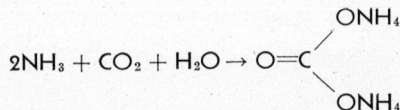

$$2NH_3 + CO_2 + H_2O \rightarrow O=C \overset{ONH_4}{\underset{ONH_4}{\diagup}}$$

As previously mentioned, carbohydrate and fat may be synthesized from certain amino acids and thus the protein may serve as energy-producing material. This condition is thought to occur on an appreciable scale only when the normal supply of carbohydrate and fat has been depleted. This occurs in wasting diseases and starvation. In the literature of physiological chemistry there is a wealth of material available on the function and fate of the individual amino acids. Our knowledge is, however, far from complete.

IV. Research Progress

In the foregoing sections it has been repeatedly pointed out that our knowledge of the exact organic reactions occurring in vital processes is by no means complete. In recent years a number of factors have stimulated research in this field.

By 1949 the use of radioactive isotopes in research studies has become a standard technique. These isotopes have proved of great value in determining the intermediate transformations and ultimate fate of ingested foodstuffs. Radioactive atoms, such as ^3H, ^{14}C, ^{24}Na, ^{32}P, ^{35}S, ^{42}K, and ^{59}Fe obtained from atomic energy plants or produced with the aid of the cyclotron, are used as "tracers." They are combined in certain foods and fed to experimental animals. The course of the digestion and metabolism of the specific food material can then be traced with a Geiger counter or by a "self photography" of tissue sections. Radioactive iodine has been used to locate portions of malignant thyroid which have spread throughout the body after the incomplete removal of cancerous thyroid. In like manner, stable ^{15}N (nonradioactive) has been incorporated in the amino acid, leucine, and fed to rats. Tests using the mass spectrograph have established that this nitrogen, in a few days' time, has been transferred to many different proteins. It now seems probable that the substances involved in biochemical processes may be drawn upon as needed for the synthesis of many compounds and that functional groups may be interchanged to meet the specific requirements of the body.

Certain unknown steps in metabolic processes are being clarified. In 1947 the husband-wife team, Dr. Carl Cori and Dr. Gerty Cori, professors of biochemistry at the Washington University School of Medicine, were awarded the Nobel Prize in medicine for their discovery of an obscure step involving phosphate in the enzymatic synthesis of glycogen. The field for such research work is unlimited.

In the past decade, a great amount of nutrition research has been recorded. Data from world-wide sources regarding starvation cases occurring as a result of World War II, have provided much knowledge. We now know that it is not alone sufficient to supply an individual with the proper number of calories of food per day. There are certain specific substances that must be included in the diet. To provide the citizens of our world with an adequate food supply is an outstanding need.

In fact, the opinion has been expressed that no other year in modern times has seen such widespread concern over the food supply of the world as did 1947. While most of the people on earth were inadequately supplied—many existing on starvation diets—we in the United States averaged a per capita consumption estimated at 3,400 calories per day. Even with our high intake of food, many of us were not properly fed.

We also know that a man on a starvation diet loses the will to resist and that his moral concepts become warped. Thus the moral fiber of the people of the world is being influenced by the food they receive. This knowledge regarding the importance of food to the mental and moral well-being of the individual, acquired through the study of the organic reactions in vital processes, is of world-wide importance. The influence upon the individual and accordingly upon the nation, may well prove to be an important factor in international events. Through intelligent application of this knowledge, a healthier, happier, and more peaceful race may evolve.

Indeed, thus to contribute to international welfare may be a vital, challenging objective for many serious students of organic chemistry.

CHAPTER PROLOGUE

History is the record of man's method of survival. It is the
story of his success or failure in improving and perpetuating his
way of life. Obstacles to this improvement or perpetuation have
caused the conflicts of history—personal, national, and international. The record of these conflicts is the history of war. Its
pages are darkened by the treachery of the traitor, quickened by
the courage of the bold, inspired by the ingenuity of genius. The
history of war tells progressively about new fear-inspiring weapons,
powerful for a time until man could formulate a defense against
them. Fortunately, the art of defense has always kept pace with the
art of offense. The swinging club, the stone-tipped ax, the hurled
stone, the two-edged sword, the horse-borne lance, and the flying
arrow begot the metal shield, the coat of mail, and the moated
castle.

Then came organic chemical warfare.

Gunpowder eventually broke the massed phalanx of Alexander
and leveled the fortress-castle, driving armies to the shelter of
trenches and man-made caves. Then other chemicals—gases—drove
the armies out of the trenches into mobile units whose broadened
horizons encompassed both tropic heat and arctic cold. Mobile
armored units ground opposition in any form beneath their churn-
ing treads, by-passing the rock-hewn forts whose protecting guns
had smugly pointed in but one direction. Air-borne explosives by
precision bombing nullified the productive effort of the enemy and
weakened his will to resist.

Then came atomic warfare.

The future of atomic warfare is not within the scope of organic
chemistry and we shall not arrogate the privilege of prophecy. At
the end of World War II there came the ominous warning attributed
to Dr. Albert Einstein: "The weapons of World War IV will be
clubs."

The next few years may hold the inscrutable answer to this
prediction.

History of Organic Chemicals in Warfare

The use of organic chemicals in warfare started at the Battle of Crécy
in 1346, unless of course one wishes to consider the cellulose in prehistoric
man's earliest weapon, a club. At Crécy the English used gunpowder with
devastating effect against the lancers and crossbowmen of the enemy.
Marco Polo had brought gunpowder to Arabia from China but it took
about 500 years before its power was harnessed in Europe to the needs of
war. Another 500 years then elapsed before a stronger explosive was discovered. Gunpowder is now rated as a low explosive because its conversion
to gases is relatively slow.

Schoenbein's discovery of guncotton in 1845 and Sobrero's discovery
of trinitroglycerol a year later produced the first organic materials classified as high explosives. Between their discovery and use in war another
quarter-century elapsed. During the interval Nobel made the first dynamite by absorbing trinitroglycerol (p. 77) in diatomaceous earth (kiesel-
guhr). In 1871 Sprengel, using mercury fulminate, detonated picric acid

(p. 201) a hundred years after its discovery by Glauber. Picric acid had been regarded as a harmless yellow dye for silk and wool although its explosive quality had been recognized as early as 1805.

The use of smokeless powder in warfare dates back to the late 1880s when Vieille in France and Nobel in Sweden first produced *poudre B* and *ballistite*, respectively. The English adopted cordite as their smokeless powder at this time. In the United States Army the use of smokeless powder did not become general until shortly after the end of the Spanish-American War.

World War I witnessed many new uses of organic chemicals. The German dye industry was converted into the German munitions industry. TNT became a household term for trinitrotoluene, a dependable explosive, safe to handle. During this war armies were mechanized and aerial warfare was started, increasing the demand for hydrocarbons as fuels and lubricants. Enormous tonnages of explosives required new methods of producing acetone and acetaldehyde. Gas warfare required the production of chloroacetophenone, chloropicrin, lewisite, and mustard gas in large quantities.

Organic materials were used so extensively in World War II that any list in a book of this size will be very incomplete. For prosecuting World War II, in addition to explosives and gases used in World War I, the organic chemical industry manufactured synthetic rubber, plastics, nylon, high octane fuels, improved lubricants, coolants, disinfectants, fungicides, insecticides, packaging materials, water purifiers, propellants (rockets), synthetic pharmaceuticals, and a host of other materials.

The organic chemistry of these materials will be considered in the remaining part of this Appendix.

Explosives

Gunpowder, often called *black powder*, was made for centuries by mixing powdered charcoal, potassium nitrate, and sulfur. The mixture was then converted to a paste with water and allowed to dry. The resulting cake was then broken into coarse or fine grains in accord with the gun in which it was to be used. When gunpowder is ignited, potassium nitrate supplies the oxygen for converting carbon and sulfur at white heat into their respective oxides. The pressure of these hot gases forces the projectile from the gun.

The most important kind of smokeless powder is *cellulose nitrate powder*. It is made by the reaction of nitric and sulfuric acids on cotton. Cellulose nitrate when mixed with acetone or alcohol-ether becomes a doughlike mass which can be cut in varying shapes and sizes. The dried grains of cellulose nitrate are smokeless powder. They do not burn like gunpowder but explode almost instantly for the oxygen, carbon, and hydrogen are all in the single molecule.

Cordite is made by dissolving cellulose nitrate, trinitroglycerol, and petrolatum in acetone. The resulting mass is rolled in sheets, cut to sizes and shapes desired, and dried.

Dynamite (p. 77), picric acid (p. 201), and trinitrotoluene (p. 201) have been mentioned in earlier chapters. Picric acid and T.N.T. are very stable compounds requiring the use of detonators for their explosion. Trinitroglycerol (p. 77), on the contrary, is readily explosive and must be carefully handled. In dynamite the trinitroglycerol particles are separated by such inactive materials as diatomaceous earth, sawdust, and flour.

Gases

On April 22, 1915, the first real gas attack in history ushered in the use of a chemical agent which broke a stalemate on the Western Front. Trench warfare made impossible the dislodgment of any enemy by the direct firing of weapons; chemical warfare with gas was the logical and inevitable outcome. On that morning in April conditions were exactly right. A gentle breeze was blowing from the German lines toward the Canadian trenches near Ypres. As the zero hour approached, 5000 cylinders of chlorine were opened. The heavy, greenish yellow gas flowed from the tanks, was blown across no-man's-land, and settled into the allied trenches. The first attack caused 15,000 casualties and opened a gap two miles wide in the Canadian ranks. It was an extremely effective attack; the value of the new weapon seemed to have been established. Protection against a repetition of the attack had to be provided.

To be sure, the first protection was crude, but, simple though it was, it proved to be noticeably effective. Pails of a solution of common hypo were kept at convenient places in the trenches. Upon the sounding of the gas alarm, the soldier quickly dipped a piece of waste or cloth into the solution and tied it over his mouth and nose. Obviously so simple a mask could not long be sufficient and so the regulation mask of World War I was devised and was supplied to the men at the front as rapidly as possible. The use of organic chemicals as "war gases" soon followed the first attack with chlorine. Throughout the war these agents proved to be exceedingly effective weapons of war.

With the advent of World War II, all nations were supplied with large reserves of these organic chemical agents. The civilian populations were fearful of impending attacks using these insidious materials. But the attacks did not come.[1] During the two decades of relief from war, the mode of warfare had become mobile and thus had changed so completely as to render military gas attacks relatively ineffective. Also, civilians were instructed in protective methods so that attacks upon centers of population would likewise be ineffective. To be prepared and to know what to

---

[1] There are but few isolated records of the use of gaseous warfare shortly before and during World War II. These were employed against unprepared and unequipped forces.

do in case of a gas attack was the best safeguard and the best assurance that such an attack would not come. The sense of security that came from understanding chemical agents, their limitations, and the protective methods to combat them was a significant factor in building morale.

Herein, then, are presented the more important organic chemical agents known as war gases, together with brief information about their physiological effects, their characteristic properties, and the recommended treatment in case of exposure.

So-called war gases are best classified according to their primary physiological effects. Most important are the *vesicants* that blister the skin; the *lung irritants* that destroy mucosa or that hydrolize and produce toxic substances absorbed by the body; the *lacrimators* that induce the flow of tears; and the *sternutators* that cause sneezing. It must be remembered that all of these agents combine a certain degree of toxicity and some overlapping of physiological effects. In Table 14 the more important agents are classified and their characteristics described.

Vesicants. Of the vesicants, mustard gas and lewisite are the most important. They are prepared as addition products to unsaturated hydrocarbons. Mustard gas, for example, results from the addition of sulfur monochloride to ethylene

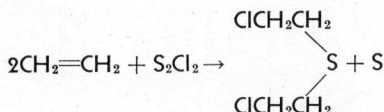

$$2CH_2{=}CH_2 + S_2Cl_2 \rightarrow \begin{array}{c} ClCH_2CH_2 \\ {\diagdown} \\ S + S \\ {\diagup} \\ ClCH_2CH_2 \end{array}$$

In chemical structure it is therefore the chloro substituted diethyl thioether. The German method of preparing mustard gas is typical of addition and substitution reactions previously studied.

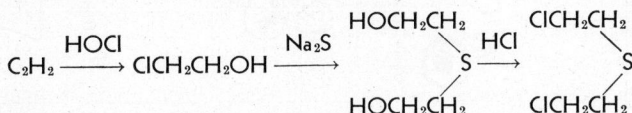

$$C_2H_2 \xrightarrow{\text{HOCl}} ClCH_2CH_2OH \xrightarrow{\text{Na}_2S} \begin{array}{c} HOCH_2CH_2 \\ {\diagdown} \\ S \\ {\diagup} \\ HOCH_2CH_2 \end{array} \xrightarrow{\text{HCl}} \begin{array}{c} ClCH_2CH_2 \\ {\diagdown} \\ S \\ {\diagup} \\ ClCH_2CH_2 \end{array}$$

The latter process produces the purer product.

Lewisite, a more deadly agent, is prepared by the addition of arsenic trichloride to acetylene. The effectiveness of vesicants is dependent upon

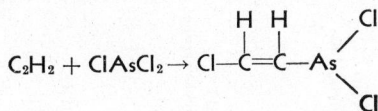

$$C_2H_2 + ClAsCl_2 \rightarrow \begin{array}{c} H \ \ H \ \ \ \ \ Cl \\ | \ \ \ | \ \ \ {\diagup} \\ Cl{-}C{=}C{-}As \\ {\diagdown} \\ Cl \end{array}$$

the rapidity of their absorption into the skin and the subsequent formation of eruptive blisters. In addition to this physiological action, the arsenic of the lewisite is absorbed and causes systemic arsenic poisoning. Mustard gas, used extensively after its initial introduction at Ypres on July 12, 1917, is less lethal.

Table 14

CHEMICAL WARFARE AGENTS KNOWN AS "WAR GASES"

CLASS	NAMES AND SYMBOLS	ODOR	PHYSIOLOGICAL EFFECT	FIRST AID [After removal from gassed area]	GENERAL INFORMATION
VESICANTS	MUSTARD Dichlorodiethyl Sulfide $S(CH_2CH_2)_2Cl_2$	Garlic, Horseradish, Mustard	Delayed effect. Burns skin or membrane. Inflammation respiratory tract leading to pneumonia. Eye irritation, conjunctivitis.	Undress; remove liquid mustard with protective ointment, bleach paste, or kerosene; bathe; wash eyes and nose with soda solution.	Liquid at room temperature; B.P. 217.5°C; M.P. 14.4°C; Sp. Gr. 1.27; Vapor density compared to air is 5.5.
VESICANTS	LEWISITE Chlorovinyl Dichloroarsine $CHClCH \cdot AsCl_2$	Geraniums	Burning or irritation of eyes, nasal passages, respiratory tract, skin. Arsenical poison.	Undress; remove liquid lewisite with hydrogen peroxide, lye in glycerine, or kerosene; bathe; wash eyes and nose with soda. Rest—Doctor.	B.P. 190°; M.P. 0.1°; Sp. Gr. 1.89; Vapor density 7.1. 0.0008 mg. per liter of air is irritating; 0.12 mg. per liter is fatal in 10 minutes.
IRRITANTS	CHLORINE Cl_2	Highly Pungent	Lung irritant.	Remove from gassed area. Keep quiet and warm. Coffee as stimulant.	Inorganic war gas, of interest because it was the first gas used effectively in warfare.
IRRITANTS	CHLOROPICRIN Nitrochloroform CCl_3NO_2	Flypaper, Anise	Causes severe coughing, crying, vomiting.	Wash eyes, keep quiet and warm. Do not use bandages.	B.P. 112°. Decomposes producing phosgene and nitrosyl chloride.

Category	Name	Smell	Symptoms	Treatment	Properties
LUNG IRRITANTS	**DIPHOSGENE** Trichloromethyl Chloroformate $ClCOOCCl_3$	Ensilage, Acrid	Causes coughing, breathing hurts, eyes water, toxic.	Keep quiet and warm. Give coffee as a stimulant.	Nonvolatile and therefore very persistent and adapted for use in shells.
	PHOSGENE Carbonyl Chloride $COCl_2$	Musty Hay, Green Corn	Irritation of lungs, occasional vomiting, tears in eyes, doped feeling. Occasionally symptoms delayed. Later, collapse, heart failure.	Keep quiet and warm, bed rest. Coffee as a stimulant. Loosen clothing. No alcohol or cigarettes.	B.P. 8°C and therefore not persistent. Hydrolizes readily and is thus destroyed.
LACRIMATORS	**CHLOROACETOPHENONE** $C_6H_5CO \cdot CH_2Cl$	Apple Blossoms	Makes eyes smart. Shut tightly. Tears flow. Temporary.	Wash eyes with cold water or boric acid solution. Do not bandage. Face wind. For skin, sodium sulphite solution.	White solid. M.P. 58–59°. Used in modern tear gas bombs.
	BROMBENZYLCYANIDE $C_6H_5CH \cdot BrCN$	Sour Fruit	Eyes smart, shut, tears flow. Effect lasts some time. Headache.	Wash eyes with boric acid. Do not bandage.	Solid. Extremely potent tear agent. Made by replacement of the side-chain bromine in p-bromobenzyl bromide with KCN.
STERNUTATOR	**ADAMSITE** Diphenylamine Chloroarsine $(C_6H_4)_2 \cdot NHAsCl$	Coal Smoke	Causes sneezing, sick depressed feeling, headache.	Keep quiet and warm. Loosen clothing. Reassure. Spray nose with neosynephrine or sniff bleaching powder. Aspirin for headache.	Addition product of arsenic chloride and diphenylamine.

Lung Irritants. Of the lung irritants phosgene and chloropicrin are most important. The former was first used in 1915 and is reported to have caused 80 per cent of the gas casualties of World War I. Relatively high concentrations of phosgene cause the lungs to stop functioning, resulting in shock that causes circulatory failure and death. The disadvantage of phosgene is that it hydrolyzes readily and is therefore easily destroyed. It is less persistent than most of the other agents. For commercial preparation chlorine is reacted with carbon monoxide at 200°C in the presence of bone-char as a catalyst. Reference has previously been made to other methods of its formation (pp. 63 and 64). Phosgene finds commercial application in the preparation of Michler's ketone (p. 228) and as a dye intermediate.

Chloropicrin also has industrial uses, primarily as a parasiticide, insecticide, and disinfectant for cereals and grains. As a lethal agent one part in 20,000 of air will prove fatal. Upon decomposing chloropicrin produces phosgene and nitrosyl chloride. The agent is manufactured by the action of bleaching powder or sodium hypochlorite on picric acid.

Tear Gases. Modern tear gas bombs used by police employ mainly α-chloroacetophenone as the active agent. It is produced commercially by the chlorination of acetophenone.

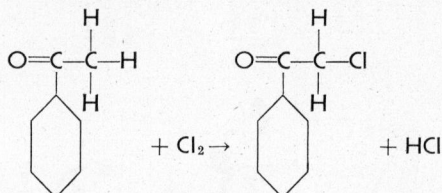

$$
\underset{\text{(phenyl)}}{O{=}C{-}\overset{\displaystyle H}{\underset{\displaystyle H}{C}}{-}H} \;+\; Cl_2 \;\rightarrow\; \underset{\text{(phenyl)}}{O{=}C{-}\overset{\displaystyle H}{\underset{\displaystyle H}{C}}{-}Cl} \;+\; HCl
$$

A wide variety of halogenated organic compounds have characteristics as lacrimators and sternutators. Those listed in Table 14 are sufficiently effective to warrant their use in warfare.

The Secret Gas Myth. Fear of gases seems innate in man, but such fears are not warranted. There are recurring rumors about secret gases that are far more lethal than those herein discussed. The known facts of chemistry indicate the improbability of the existence of such gases.

Man has studied about 500,000 chemicals; of these 9,000 have been tested and 3,000 have been found toxic enough to use for war purposes. Physical properties and availability of materials have reduced the potential number of agents to 30. These were tried in World War I and of the 12 having possible use only 6 were notably successful. We have excellent protection against these. It is inconceivable that a new gaseous agent, completely unlike these 500,000, can be produced.

Outside of the field of organic chemistry, however, the world faces the threat of two other varieties of warfare whose ultimate potentialities are not fully comprehended. These are, of course, atomic warfare and bacteriological warfare. Although it is too early to predict the extent of their use and the harm that they will accomplish, it is reasonable to surmise that in some way man will find a basic defense against them even as he has evolved a defense against other weapons. In short, the human race will survive.

Organic Chemicals in a War Economy

It is a mistake to assume that every activity of warfare is destructive and that the use of organic chemicals is limited to the direct devices of destroying life and property. True, high explosives and noxious gases find little use in peacetime civilian economy. However, many of the ingenious creations of organic chemistry produced under the stress of war emergency have continued their employment on an ever-increasing scale to the betterment of living.

In Chapter 18, among other notable examples, we saw how the United States became independent of the German dye industry following World War I; how new methods of making synthetic rubber have made us independent of the rubber plantations of the Orient, leaving world economic requirements as the basis for using natural rubber; how the increased production of nylon for making parachutes, glider towropes, and other war necessities has practically supplanted the product of the silkworm; how the need for aromatic compounds required in a war economy has led to their synthesis from natural gas and petroleum; how the need for lubricants and coolants and internal combustion fuels in extremes of climate has led to the designing of many products that will continue to serve better our daily civilian requirements.

Similar developments in other organic chemical industries were greatly stimulated by the synthesis of materials required in the total war effort and by the compounding of those materials into preparations for processes that have civilian applications. The following examples will serve as illustrations.

Surface Active Agents. The production of so-called "wetting agents" (pp. 30 and 147) was greatly stimulated by the requirements of war, first, as an adjunct to processes in war plants, and second, to provide material to take the place of fats for soap-making, the availability of fats for soap having been greatly reduced by their requirement for war purposes.

Surface active agents are effective in that they reduce the surface tension of solutions, making closer contact between reactants possible. One of the most striking illustrations of the effectiveness of this operation is shown in Fig. 44. The feathers of the duck on the left were dusted with a surface active agent; water enters the crevices between the feathers so that buoyancy is destroyed and the duck is able to keep afloat only by vigorous swimming. In synthetic detergents, many varieties of which are now household commodities, surface active agents take the place of

Fig. 44. Surface agents at work. (Courtesy, American Cyanamid Company.)

soap and improve upon the cleansing action and rinsability, particularly in hard water. In the dye industry these agents cleanse the surface more completely and bring the dye into closer contact with the material, thus insuring better and more uniform coloring. In the metal industries, mixtures containing surface active agents better prepare the surface for "pickling" operations. Their use extends widely through such diversified industries as the manufacture of paper, the tanning of leather, the preparation of paints, and chemical synthesis.

Solvents. Diversified solvents have found innumerable uses in the war effort and these uses successfully extend to peacetime industries. Thus the need for special solvents for paints and lacquers in the airplane industry has been met. Solvents have been created such as the "cellosolves" (p. 80) that combine the solvent action of ethers and alcohols. Halogenated hydrocarbons such as ethylene chloride (p. 64) were used in tremendous tonnages for the cleansing mixtures of airplane parts in repair departments. These processes and the solvents which they require are now employed in the civilian aviation and automotive industries.

Specific plastic materials (pp. 254 to 256) solved innumerable war problems better than any other material available. The familiar housing for lights and noses of military airplanes are made of methylmethacrylate (p. 258) known better as plexiglas or lucite. Peacetime applications of this material are great indeed. The General Electric Company published, in 1948, a bulletin—"Twenty-four Case Histories; How Plastics Solved War Problems." All of these "Solved War Problems" are directly or indirectly applicable to peacetime requirements.

Perhaps the most dramatic contributions of organic chemistry to the war effort, in their timeliness and in their permanent impress upon civilian welfare, were those accomplished in the biological-organic field. These applications extend to increased knowledge and availability of materials in the fields of disinfectants, insecticides, and medicine. Several examples are presented in Chapter 20. As an illustration, penicillin, now available to all, is produced in large quantities because of the strong demand for such a biochemical agent during the war. The lives saved by penicillin and the present availability of the material are results of the dramatic and untiring research stimulated by the infections of war.

Although some wars have served as catalysts for peacetime production, the chief by-products of war are always destruction, desolation, and debt. In time of conflict the struggle for survival entails unprecedented expenditures on concentrated scientific research and on quickened production, but the degree of national enrichment is puny in terms of total war costs. War does not pay in international circles any more than feuding pays among families. The logical ultimate result of feuds is family extinction. The warring nations of the past are broken monuments to their own folly. Just as families have learned to resolve their difficulties in the interests of the State, so will Nations ultimately resolve their problems in the interests of our World.

$$94.0$$
$$49.6$$
$$44.4$$

$$150$$
$$94$$
$$56$$